Organic Inhibitors of Corrosion of Metals

Organic Inhibitors of Corrosion of Metals

Yurii I. Kuznetsov

Institute of Physical Chemistry
Russian Academy of Sciences
Moscow, Russia

A. D. Mercer, Scientific Translator

J. G. N. Thomas, Scientific Editor

Springer Science+Business Media, LLC

Library of Congress Cataloging in Publication Data

Kuznetsov, Yurii I.
Organic inhibitors of corrosion of metals / Yurii I. Kuznetsov; A. D. Mercer, scientific translator; J.G.N. Thomas, scientific editor.
p. cm.
Includes bibliographical references and index.

DOI 10.1007/978-1-4899-1956-4
1. Corrosion and anti-corrosives. 2. Organic compounds. I. Mercer, A. D. II. Title.
TA462.K885 1996 96-3811
620.1′ 623—dc20 CIP

Originally published by Plenum Press, New York in 1996
MyCopy version of the original edition 1996

10 9 8 7 6 5 4 3 2 1

Preface

Industrial development and the intensification of production place considerable demands on the reliability and economical use of equipment in technical processes. In this connection, one of the most important scientific and technical tasks is the protection of metals from corrosion, the direct losses from which in the developed countries are estimated to be 3–5% of the gross national product. The need to curtail corrosion losses and the development and introduction of highly effective methods of increasing the corrosion resistance of metallic components and semi-products are constantly emphasized in considering the main directions of scientific development.

The use of inhibitors is one of the most universal and economical routes for the anti-corrosion protection of metals. In a number of cases there will be no alternative and their use, either independently or in combination with other protective measures (paints, cathodic protection, etc.), will raise efficiency of systems constructed from metals and possibly lead to the feasibility of new technological processes or to the use of cheaper and more accessible materials of construction as well as reducing the chances of accidents and the leakage of pollutants into the environment.

Inhibitors of the corrosion of metals are chemical compounds and formulations of these which, when present in small quantities in an aggressive medium, inhibit corrosion by bringing about changes in the surface condition of a metal. This process can be associated either with adsorption of the inhibitor or the formation of difficultly-soluble films which, however, are significantly thinner than protective coatings. In either case the action of the inhibitor is associated with an effect on the kinetics of the heterogeneous reactions—usually electrochemical in nature—responsible for the corrosion of a metal. If one takes into account that corrosion is caused by the chemical (electrochemical) interaction of a metal with its environment then the reduction of the rate of this interaction by modification of the composition of the environment, i.e., by inhibition, represents a fundamental method of combating this interaction. It is not surprising that inhibitors are used in all spheres of human activity; in the atmosphere, in building, machine construction, and so on. In the majority of cases the corrosive media, particularly those occurring naturally, have pH values that are close to neutral.

An important aspect of protection by inhibitors in neutral media is the close connection with one of the central problems of corrosion science, i.e., metal passivity. The simple traditional inhibitors in these media are either inorganic compounds having oxidizing properties (chromates, nitrites) or those capable of forming difficultly-soluble salt films on the metal surface. It is generally considered that organic compounds which inhibit the acid corrosion of metals as a result of adsorption are insufficiently effective in neutral solutions, where the main passivating role is played by the oxygen-containing compounds of the solvent (water). However, the capacity of many organic compounds to prevent the local depassivation of metals, i.e., primarily to provide protection from pitting, has been underestimated.

Despite the importance of studying the quantitative laws of inhibition, the selection of inhibitors for neutral media has up to the present been made empirically and studies have been limited to qualitative observations and conclusions. Of the inhibitors that can be used, organic compounds in which the chemical composition can be widely varied but containing the same main reaction centers provide a distinct possibility for establishing a theory of action of inhibitors. Such a theory would describe the effects on the protective action of the nature of activators and inhibitors, the solvent and the metal, the temperature, inhibitor concentration, and so on.

Ecological problems, which occur particularly in water supply systems, can be powerful stimuli for the development of new multifunctional reagents, e.g., phosphorus-containing complexes, that are thermally stable and of low toxicity and are capable of preventing scale formation as well as corrosion. In this book, there is, apparently, the first attempt to examine the scientific basis of the protection of metals by inhibitors based on such complexes and to describe experiences with their application.

My purpose in this monograph is to summarize not only the results of scientific studies of organic corrosion inhibitors that have been conducted under my direction in the Institute of Physical Chemistry of the Russian (former USSR) Academy of Sciences but also the ever expanding experience of their use in practice. To this end, apart from information on the mechanism of action of existing and new inhibitors suggested by the author and his coworkers, the book contains two chapters devoted to the most important tasks for inhibitors in practice. One of these two chapters provides a short review of recent achievements in the protection of various types of cooling system, and the other, describes the combating of atmospheric corrosion of metallic parts and components during their assembly, inter-operation storage, transport, and conservation. While realizing that each of these topics could be the subject of a separate monograph, I would not claim to have the necessary exhaustive information on these subjects; rather, I have attempted to identify the most promising directions for the use of these inhibitors by providing a discussion of typical examples.

The book is mainly based on the results of experimental investigations conducted by myself and my coworkers and research students at the Institute of Physical Chemistry of the Russian Academy of Sciences. I would like to thank all those who have taken part in the experimental aspects of the work, in particular, N. N. Andreev, N. P. Andreeva, V. A. Isaev, O. A. Luk'yanchikova, S. V. Oleink, and L. P. Podgornova. The author is also deeply grateful to S. S. Veselyi and T. V. Fedotova, who kindly assisted in the preparation of the manuscript for publication.

The publication of the book would hardly have been possible without the goodwill and support of Mr. A. D. Mercer and Dr. J. G. N. Thomas, who not only prepared the English language version but also suggested a number of corrections and improvements to the text. I consider it a privilege to have had such professional assistance and express my sincere thanks to them for their participation.

I hope that a wide circle of readers who encounter the corrosion of metals in their activities will find interesting information on protection by inhibitors in this book. Any comments on its contents will be gratefully accepted by the author.

Yu. I. Kuznetsov

Moscow

Contents

Abbreviations and Symbols

AN	acceptor number of a solvent; also acetonitrile
Alk	alkyl, hydrocarbon radical
a_1	activity of substance or ion 1
B	nucleophilic constant of a solvent
DN	donor number of a solvent
E	electrochemical potential
E_{st}	steady-state potential (corrosion potential)
$E_{q=0}$	zero charge potential of a metal
E_{rev}	equilibrium potential for formation of a complex
E_d	adsorption–desorption potential of organic compound
E_p	passivation potential of a metal or alloy
E_{pit}	potential of pit formation
E_B	binding energy of electrons
E_f	potential for formation of an oxide film
E_F	Flade potential
E_{O_2}	oxygen evolution potential
ΔE	protective effect measured from the shift in E_{pit}
ΔE_R	the protective effect on E_{pit} due to the replacement of hydrogen by a substituent R
F	Faraday constant, 96500 coulombs
f	hydrophobicity constant
ΔG	change in free energy
$\Delta G^{\neq}$	activation (free) energy of a reaction
ΔG_{hydr}	free energy of hydration of an anion
ΔG_{corr}	activation (free) energy of a corrosion process
ΔG_A	free energy of adsorption
ΔH	change in enthalpy of a system
Hal^-	halide ion (F^-, Cl^-, Br^-, I^-)

h	Planck's constant
J	ionization potential of a substance; also Langelier index
J_p	resonance potential of ionization
i_O	exchange current density of an electrochemical reaction
$i_{p.s}$	anodic current density in the passive state
$i_{a(c)}$	anodic (cathodic) current density
i_d	limiting current density for oxygen diffusion
K	equilibrium constant or reaction rate
MR	molar refractivity
n	number of electrons, molecules, or ions taking part in a reaction
n_C	number of carbon atoms in a molecule or ion
n_k	complex refractive index of light
P_a	anodic polarizability
P_c	cathodic polarizability
Q	quantity of electricity
R	substituent in organic molecule or ion; also universal gas constant
S	symbol for a solvent
ΔS	change in entropy
S parameter	averaged value of the conditional activity of anions in various solvents
T	temperature on the Kelvin scale
t	time
$:Y^-$	generalized symbol for a nucleophilic reagent
Z	degree of protection of a metal from corrosion (%)

Transliterated Russian Symbols

ABDM	alkylbenzyldimethylammonium chloride
AN	acetonitrile
ARC	arylcarboxylate (substituted sodium benzoate)
BTA	benzotriazole
BI	benzimidazole
CMC	critical micelle concentration
G	adsorption of a substance
GF	glycine-N,N-di(methylenephosphonic) acid
DAPP	dimethylamino-propyl(hydroxy)diphosphonic acid
DBDS	dibenzylthiazolyldisulfide
DEAK	diethylaminoproprionic acid
DEG	diethylene glycol
DFT	isodifluorant, N-(m-difluormethylthiophenyl) sodium anthranilate

DMFA	dimethylformamide
DMSO	dimethylsulfoxide
DTPA	diethylenetriamine-N,N,N′,N″,N″-pentacetic acid
EDTA	ethylenediaminetetraacetic acid
EDTP	ethylenediaminetetraphosphonic acid
EG	ethylene glycol (ethanediol)
ES	ethyl alcohol (ethanol)
GP	glycine-N,N-di(methylene phosphonic) acid
HEDP	hydroxyethane 1,1′-diphosphonic acid
HLB	hydrophilic–lipophilic balance
HMDTA	hexamethylenediamine tetracetic acid
IDAMP	imino-N,N-diacetic-N-methylenephosphonic acid
K_s	stability constant of a complex
LS	lignosulfonate
MBAP	methylbenzylamino-methylene-diphosphonic acid
MBI	2-mercaptobenzimidazole
MBT	2-mercaptobenzothiazole
MEF	sodium mefenimate (N-2,3 xylyl anthranilate)
MEK	methyl ethyl ketone
NTA	nitrilo-tris-acetic acid
NTP	nitrilo-tris-phosphonic acid
OLC	olefinic carboxylic acids
PABI	2-pentylaminobenzimidazole
PAN	sodium salt of phenylanthranilic acid
PDK	limiting acceptable concentration (in Russia) of a substance in water or air
PO	pitting formation
PUN	sodium salt of phenylundecanoic acid
SAN	N-substituted sodium anthranilate
SP	solubility product
SVZ	temporary protective
TEA	triethylamine
XPS	X-ray photoelectron spectroscopy

Greek Symbols

α and β	transfer numbers of cathodic and anodic reactions, respectively
γ	coefficient of corrosion retardation by an inhibitor
Δ	ellipsometric parameter
ε	dielectric constant
η	overpotential for any electrochemical reaction or process
θ	degree of surface coverage

κ	coefficient characterizing the probability of the occurrence of a reaction upon reaching a transition state of a system
χ_{Me}	effective electronegativity of a metal
Λ	shift in the energy level of an organic molecule (ion) arising on its approach to an adsorbent
Λ_O	that part of Λ, practically independent of the potential of ionization
π	Hansch constant of hydrophobicity of a substituent
ρ	reaction parameter
σ	Hammett constant
σ_I and σ^*	induction constants of a substituent
σ^n	Wepster's constant for an electrochemical substituent
σ_o^*	Taft constant for a substituent introduced into a position ortho to the reaction centre
τ	the induction period for pitting formation
$\phi_{Me(x)}$	electronic work function from a metal to a phase X
ψ	restoration angle for polarizaed light
ψ'	potential at the point of disposition of the centre of charge of a reacting particle into a transition state
$_{Me}\psi_{H_2O}$	Volta potential at the metal–water boundary

1

Electrochemical Aspects of the Inhibition of Corrosion of Metals

1.1. THE ROLE OF POTENTIAL AND THE CHARGE OF A METAL

Most forms of corrosion are the result of electrochemical processes; therefore, the effectiveness of corrosion inhibitors is largely determined by the electrochemical potential of the metal that is to be protected. This potential, together with the pH and the electrochemical reactivity of any component added to the corrosive environment, will determine whether oxidation–reduction reactions will occur on the metal surface. If difficultly-soluble compounds form on the metal as a result of such reactions, then the metal can become passivated and the additive is potentially a corrosion inhibitor.[1,2] If, on the other hand, there is no electrochemical transformation of the additive, then the role of the potential will be to affect the free energy of adsorption of the components of the solution. The surface concentration of the adsorbate will therefore depend on the potential and, as described by the Bronsted–Polanyi–Semenov principle, on the activation energy of adsorption. In this connection, one of the fundamental concepts of electrochemistry introduced by Frumkin in 1927, the potential of zero charge of a metal $E_{q=0}$, is of considerable importance.

In an idealized case of nonspecific reversible adsorption, such as one taking place under the action of electrostatic forces only and without desolvation of the adsorbed species, the adsorbability of the species on "clean" metal at a given potential will depend primarily on its value in relation to $E_{q=0}$. Even in this situation, the nature of the solvent must be taken into account. Antropov proposed that $E_{q=0}$ determined in aqueous solutions of surface-inactive electrolytes should be the null point of the metal, and that potentials should be reckoned from this value.[3] The value of a potential in the scale introduced by Antropov serves as an approximate measure of the charge of the metal in relation to the solution and allows, at least in principle, an assessment to be made of the electrostatic interaction of the adsorbate with the metal.

However, it is difficult, particularly when solid metals are the object of corrosion studies, to find a solution which does not contain surface-active substances. The problem is sometimes simplified by the observation of Damaskin[4], who found that for weak specific adsorption of ions the shift in $E_{q=0}$ will be below the precision of measurement. According to Antropov, for real systems it is more useful to use a scale which reckons the potential of a metal not from the null point, but from the potential of the uncharged surface measured in a given background solution.

The use of such a scale for selection of an effective adsorption inhibitor of acid corrosion has been examined in detail,[5,6] with particular attention given to the possibility of transferring the results from adsorption measurements made with a mercury electrode to other metals (the "modelled electrode" method). In this approach, the existence of a type I adsorption is proposed, which is primarily determined by the structure and properties of the actual chemical compound being adsorbed and only the *charge* of the metal, and a type II adsorption which depends essentially on the *nature* of the metal. In the first case—for the same values of potentials of different metals (using the described scale)—there will be approximately the same adsorption. Since the measurement of this adsorption on a mercury electrode is quite reliable, it then becomes possible to use such data to estimate the degree of coverage of the surface by the adsorbate, θ, or the ψ_1 potential (the drop in potential in the diffuse double layer) even on corroding metals. Thus, many quaternary ammonium salts are adsorbed from sulfuric acid solutions mainly by this mechanism, with the retardation of corrosion as the result of the appearance of an additional drop in potential, $\Delta\psi_1$, due to the adsorption. This is in agreement with the facts that cations interact with a metal surface only electrostatically and that their adsorption is determined by interaction with the solvent and with mirror-image forces. Thus, for a given charge the adsorption is independent of the nature of the metal.[7] The absence of a similar analogue in the specific adsorption of type II can have several causes, including the heterogeneous nature of the surface of solid metals, variations in the adsorbability of the components of the solution, and the nonequilibrium nature of the adsorption on a corroding metal resulting from the dissolution and renewal of the surface. Evidently, in type II adsorption, specific interactions can arise not only between the metal and the inhibitor but also between the metal and the solvent. For example, when water is the solvent, the orientation of water molecules differs for different metals.[8] As a result, there is an additional difference of potentials that requires the hydrophilic nature of the surface to be taken into account.

As a criterion of the hydrophilic nature, the difference between $E_{q=0}$ and the adsorption–desorption potential, E_d, of an organic compound (at constant concentration) may be considered. This difference can be determined from the differential capacity obtained in a solution of a surface-inactive electrolyte. However, even in the best case, such a criterion will be only qualitative since E_d is not determined solely by the free energy that is gained by replacing a molecule of water by a molecule of organic compound. Trasatti suggested the use of the value of effective electronegativity χ_{Me} for the quantitative assessment of the hydrophilicity of metals, proposing a linear dependence between the degree of orientation of the water molecules and χ_{Me}.[8] Despite the comments that have been made[7] on the deficiencies of such an approach, it was on the whole not unreasonable since electronegativity could be successfully used to characterize the bonding between chemical elements, including that occurring during complex formation, which plays a significant role in the interaction of a metal with the oxygen atoms of water.

The influence of potential on adsorption is complicated in many media, particularly those which are neutral, by oxidation of the electrode surface, deposition of the products arising from the interaction of ions of the corroding metal with components of the solution, and the formation of films of variable chemical composition. In such cases, the prediction of an inhibiting action of an additive on the basis of $E_{q=0}$ and the Antropov scale becomes extremely difficult. For example, in studying the effect of sodium benzoate on the corrosion of iron in neutral solution, it was found that the metal dissolution was inhibited only at potentials more positive than a certain critical value. The opinion had been expressed[9] that the inhibition was directly connected with $E_{q=0}$ and with the degree of surface coverage by the adsorbate. However, later investigations, including our own,[10,11] showed that the pH of the solution, the concentration of oxidizers in the solution, and the presence of an oxide film on the metal surface were more important factors.

Even with solutions of weak oxidizers such as nitrite, or phosphates—which form surface layers of salts on many metals—there is also a lack of any parallelism between their adsorption on mercury and their effectiveness in protecting iron, according to Antropov *et al.*[12] Anions which show no surface activity on mercury, such as OH^-, F^-, SO_4^{2-}, exhibit this property on iron and other solid metals, e.g., by sometimes inhibiting certain types of corrosion. The protection from corrosion of the iron group of metals by anions usually tends to be associated with a clearly-expressed specific adsorption and only to a lesser degree with the surface charge of the metal.

Despite other limitations in practice of the use of $E_{q=0}$ and the Antropov scale, their successful application in the quantitative description and prediction of the protective properties of chemical compounds in many cases may be indicated by the important role of surface charge in the inhibition of corrosion. The charge on an electrode, even in a qualitative form, should be taken into account even in systems which are characterized by chemisorption of the components. Thus, the nucleation of pitting takes place on chromium and chromium–nickel steels in 0.1N H_2SO_4 + 0.25 M NaCl at relatively positive potentials, and it was therefore a reasonable assumption by Grigor'ev and Ekilik[13] that the electrode surface would be positively charged. Not only cation-active compounds but also compounds of the molecular type are capable of being chemisorbed on the surfaces of these metals without ennobling the pitting potential (E_{pit}). An increase in the anionic character of the additive in the series $NH_2C_6H_4SO_3H \rightarrow NH_2HSO_3 \rightarrow Na_2SO_3$ intensifies the adsorption and leads to a marked ennoblement of E_{pit} in solutions containing the sulphite anion. These data indicate the important role of electrostatic forces as a component of the more complex interaction of the adsorbate with the electrode surface that takes place during pitting and its inhibition.

It is well known that one and the same component of an environment can, depending on the potential of a metal, stimulate or inhibit its dissolution. Such behavior of an additive is not necessarily related to the attainment of a certain degree of coverage of the surface by the adsorbate. Apart from the possibility, as already noted, of a change in the composition of a surface as a result of its oxidation with increase in the potential, there can also be significant changes in the adsorption bond energy.[14] In a limiting case, this can reach the value of the bond energy of the metal cation in chemical compounds, for example, in a salt or complex. If the compound is easily soluble in the corrosion medium, the adsorbate will usually lose its inhibiting action and may even accelerate corrosion. On the other hand, the formation of difficultly-soluble products arising from the interaction between the metal and the solution components can facilitate protection from corrosion. In this case, the value of the equilibrium potential of electrodes of the second kind or of the solubility product (SP) of the appropriate compound of the metal will be of special significance. The solubility products of those oxides and hydroxides which are formed upon contact of the metal with water will be of primary importance. However, soluble compounds can also take part in reactions with the metal and so compete with its direct oxidation. For the assessment of such systems from the thermodynamic position it is convenient to use the Pourbaix diagram[1] in which the dependencies of the equilibrium potentials on pH are represented for all the possible reactions in the system.

The E–pH diagram for aqueous solutions of phosphorus-containing species in their standard states and for Cu–H_2O systems at 25°C may be considered as an example (Fig. 1.1).[1,15] In Fig. 1.1a, each line reflects the equilibrium potential of a defined reaction separating the regions of thermodynamic stability of Cu, Cu_2O, and $Cu(OH)_2$. Parallel lines are given for various ionic activities, e.g. Cu^+ or Cu^{2+} from 1 to 10^{-6}g ion L^{-1}. Clearly, in the lower part of the diagram, at sufficiently negative potentials, copper does not corrode in pure water. However, as the potential is ennobled, dissolution can be expected and, depending on pH, it can be in the form of Cu^+ and Cu^{2+} in acid media, and $HCuO_2^-$ or CuO_2^{2-} in alkaline media. Protective oxides or hydroxides can form over a wide range of pH at potentials that are sufficiently positive.

Consideration of the standard states in the Cu–P_2O_5–H_2O system indicates that, at pH > 4.5, phosphates in solution will occur in the form of $H_2PO_4^-$, HPO_4^{2-}, and PO_4^{3-}. The reactions of these with copper are as follows:

$$Cu + 2H_2PO_4^- = Cu(H_2PO_4)_2 + 2e \tag{1.1}$$

$$\Delta G^\circ = -24685 \text{ J mol}^{-1};\ E^\circ = -\Delta G^\circ/193000 = -0.128 \text{ V}$$

$$Cu + HPO_4^{2-} = CuHPO_4 + 2e \tag{1.2}$$

$$\Delta G^\circ = -36905 \text{ J mol}^{-1};\ E^\circ = -0.191 \text{ V}$$

$$3Cu + 2PO_4^{3-} = Cu_3\,(PO_4)_2 + 6e \tag{1.3}$$

$$\Delta G^\circ = -126273 \text{ J mol}^{-1};\ E^\circ = -0.218 \text{ V}$$

Analogous calculations for the other possible reactions with corrections for the dependence of the potentials on pH give, for example:

$$Cu + 2OH^- = Cu(OH)_2 + 2e \tag{1.4}$$
$$\Delta G^\circ = -42300 \text{ J mol}^{-1};\ E_{pH9.5} = 0.076 \text{ V}$$

$$Cu + 2OH^- = CuO + H_2O + 2e \tag{1.5}$$
$$\Delta G^\circ = -49790 \text{ J mol}^{-1};\ E_{pH9.5} = 0.039 \text{ V}$$

$$2Cu + 2OH^- = Cu_2O + H_2O + 2e \tag{1.6}$$
$$\Delta G^\circ = -68952 \text{ J mol}^{-1};\ E_{pH9.5} = -0.062 \text{ V}$$

Thus it is seen that, from the thermodynamic point of view, the formation of phosphates is more probable since this is possible even at negative

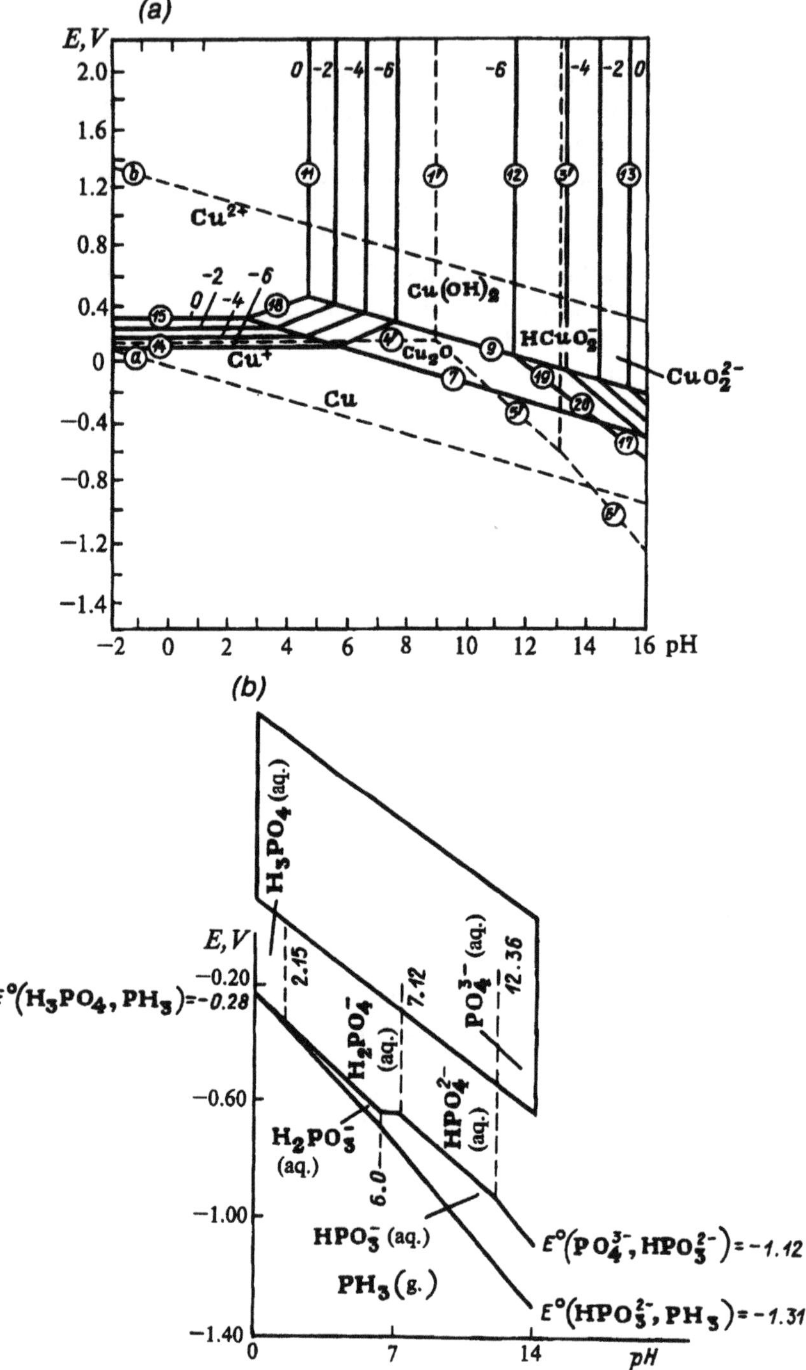

Figure 1.1. E–pH diagrams for the Cu–H_2O system, hydrated form, 25°C (a) and for aqueous solutions of phosphorus species in standard conditions (b).

potentials, particularly in the case of $Cu_3(PO_4)_2$. Actually, according to X-ray and thermographic studies[16] of the $CuO–P_2O_5–H_2O$ system, the $Cu(H_2PO_4)_2$ that is formed in aqueous solutions crystallizes with time into $Cu_3(PO_4)_2 \cdot 3H_2O$. It would therefore be expected that the soluble phosphates that form in the initial stages of the corrosion of copper could transform into the difficultly-soluble and more stable form. Copper hydroxide has a quite low solubility product: $SP_{Cu(OH)_2} = 5.6 \times 10^{-20}$ and, according to our approximate estimates, only $Cu_3(PO_4)_2$ has a lower solubility. Electron diffraction analysis of a copper surface after passivation in a phosphate solution at pH 9.5[17] has shown that surface films of $Cu(OH)_2$ and CuO do, in fact, contain inclusions of $Cu_3(PO_4)_2$.

The introduction into phosphate solutions of organic compounds capable of forming stable complex compounds with copper ions complicates still further the thermodynamic appraisal of the system. We may consider the case of a known copper corrosion inhibitor, benzimidazole (BI). Unfortunately, there are no literature data for ΔG° of formation of this compound, but data are available for the stability of its complexes which, from the value of the appropriate stability constant K_s, allow calculation of the standard potential for complex formation:

$$E^\circ_{form} = E^\circ_{Me/Me^{n+}} - 2.303\ RT/nF \log K_s \tag{1.7}$$

This system is interesting because it is a quite rare example of a metal in a low oxidation state (Cu^+) forming stable complexes. According to the work of Hawkins *et al.*,[18], $\log K_s = 4.47$ and 9.73 for the Cu^+BI and $Cu^+(BI)_2$ complexes, respectively, and 3.56 and 6.34 for the analogous Cu^{2+} compounds. Calculations show that $Cu^+(BI)_2$ can form at low potentials ($E^\circ_{form} = -0.053$ V). This can occur at potentials more negative than the steady-state potential, E_{St}, of copper in a phosphate solution, ($E_{St} = 0.15$ V at pH 9.5). From the thermodynamic point of view, if the indicated complex is of low solubility, then copper will be protected from corrosion as a result of the formation on its surface of a complex—in this case, $Cu^+(BI)_2$. This proposition has been confirmed by the results of X-ray studies of copper specimens held in 0.5M solutions of $Na_2HPO_4 \cdot 2H_2O$ containing BI.[19]

Thus, the role of the potential of a metal is not concerned exclusively with its effect on the surface charge since it can also affect the thermodynamic requirements for the formation of new surface layers of products and the interaction of the metal with its environment [to form oxides, salts, or complexes], as well as oxidation–reduction transformations of the additive and other factors.

However, any such assessment of the role of the potential, although largely valid for ideal systems, will in practice only indicate a possibility for protection, while not necessarily guaranteeing its realization. Experimental confirmations of these effects should nevertheless not be considered as chance events but rather as a result of the free energies in the real system being in some proportion to those in the ideal system. Unfortunately, in the general case the irreversibility of the corrosion process means that the explanation of the role of potential must take into account the kinetics of the partial electrochemical reactions. Since the nature of these reactions lies at the basis of corrosion, changes in their rates in one direction or another will determine the effectiveness of protection of the metal by an inhibitor.

1.2. INHIBITION OF CORROSION AS A CONSEQUENCE OF CHANGES IN THE KINETICS OF ELECTROCHEMICAL REACTIONS

The established classification of inhibitors into anodic, cathodic, and mixed types[20,21] already indicates a close association of their mechanisms with their effects on electrochemical processes. However, the character of such reactions depends significantly on the nature of the metal and the composition of its environment, particularly the pH. Therefore, an even more precise classification that would take into account the ability of an inhibitor to bring the metal into the passive state as a result of an acceleration of the cathodic process[22,23] would not be universally applicable. This conclusion is arrived at by examining the corrosion diagram in Fig. 1.2. Assume that dissolution of the metal takes place from the active state by the reaction

$$\mathrm{Me} = \mathrm{Me}^{n+} + ne \tag{1.8}$$

at an equilibrium potential E_a° and with exchange current i_{oa}, and that the electrolyte contains no depassivating substances. In this case, the anodic polarization curve will follow ABCDEF. If the cathodic reaction (for example, the reduction of oxygen) takes place under significant diffusion control (curve KBB′) and limits the corrosion process, then its rate, i_c, and value of the steady-state compromise potential E_{St} will be found on the diagram by projection onto the respective axis so that the fundamental equation of electrochemical corrosion $i_c = i_a$ is fulfilled. The introduction into the electrolyte of an anion, An^-, forming with the metal ions soluble complexes or salts

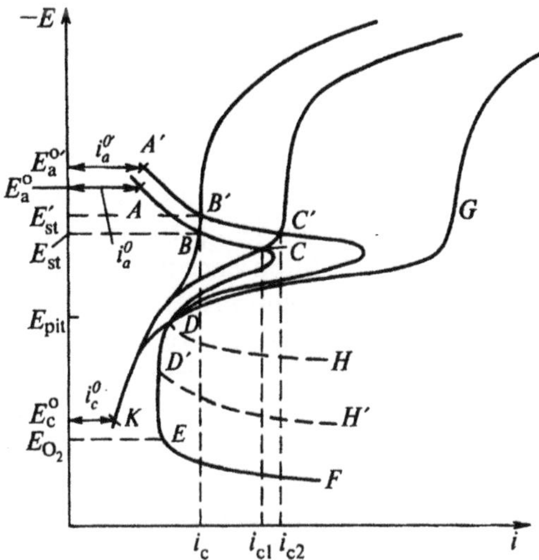

Figure 1.2. The corrosion diagram for the metal–solution system.

$$\text{Me} + m\text{An}^- = [\text{MeAn}_m]^{(n-m)+} + ne \tag{1.9}$$

will alter the equilibrium potential of the anodic reaction to $E_a^{o\prime}$ and facilitate the dissolution of the metal in the active region of potentials [A′B′C′]. However, if at the same time there is no change in the cathodic reaction rate, then there will be practically no increase in the corrosion of the metal of the new value of E'_{St}. In the case of a diffusion–kinetic control of oxygen reduction (curve KCC′), which can easily be achieved by movement of the solution, the introduction of the additive will stimulate corrosion ($i_{c_2} > i_{c_1}$). Let us assume that a further increase in the rate of the cathodic reaction (curve KDG) does not change the anodic polarization curve of the metal in both solutions.* The metal will then be in the passive state and the rate of its dissolution may become extremely low. In the system under consideration, An^- does not fulfill an inhibiting function and can justifiably be described as an anodic stimulator of corrosion. However, it is not difficult to see that even without changing the pH or the nature of the cathodic

*In the general case this is not so, although this fact does not indicate a departure from the principle of independence of the occurrence of electrode reactions, since this principle, strictly speaking, relates to reaction-rate constants. The values of the currents of these reactions can be mutually connected through the activity (concentrations) of ions and neutral molecules that participate in, or are generated during, the course of the reactions.

process, the same addition will be a corrosion inhibitor when ions, particularly Cl^-, are present that may lead to a breakdown of the passive state upon reaching a certain potential E_{pit} (section DH on the curve relating to the background solution).

Thus, the added An^- can displace E_{pit} to more positive values; it can hinder depassivation (D'H') and even completely prevent it—at least up to the potential of oxygen evolution [E_{O_2}]. It should be emphasized that the An^- ion is not a passivating addition since the critical passivation current of the metal is increased in these solutions. However, there is justification for calling it a stabilizer of the passive state—or an inhibitor of local depassivation of metals. Such an addition to the classification of inhibitors, already well established in the literature, implies that anodic inhibitors should be defined as such only when they prevent active dissolution of a metal. Moreover, any classification of inhibitors will necessarily become less formal when the composition of the solution, particularly its pH, is taken into account.

This situation also relates to cathodic inhibitors; their specificity of action is well known. Many compounds, while effectively inhibiting hydrogen depolarization, have practically no effect on the rate of oxygen reduction. This difference can also, to a large extent, be associated with the distinction between inhibitors for acid media and those for neutral media. Despite this kind of complication, which occasionally requires a different approach to effective inhibitor selection, there are undoubtedly certain principles which, although appearing to be of specific application, particularly to acid media, nevertheless are of importance in understanding the mechanism of action of corrosion inhibitors as a whole. This is demonstrated by the question of the effect of the adsorption of various compounds and ions on the kinetics of hydrogen evolution which, in general, has been the most widely studied electrode process.

The overpotential, η, the difference between the equilibrium potential of a reaction E_H and the potential of a polarized electrode, in acid media with slow discharge and taking account of the diffuse double layer,[24] is described by the equation

$$\eta_H = E_H - E = a_k + \frac{1-\alpha}{\alpha}\psi_1 - \frac{1-\alpha}{\alpha}\frac{RT}{F}\ln[H_3O^+] + \frac{RT}{\alpha F}\ln i_c \quad (1.10)$$

where a_k is a constant dependent on the nature of the cathode (electrode) and temperature of the solution, α is the transfer coefficient, ψ_1 is the potential at the point where hydrogen ions are being discharged or the potential at which the centre of charge of the reacting ion is found in the transition state of the reaction.[7] From (1.10) it follows that the adsorption of cations, which shifts the ψ_1 potential to positive values, should lead to an increase in η_H. In the absence of specific adsorption, the addition to the solution of ions formed from the dissociation of salts will increase η_H because of the increase in the ψ_1 potential. If the salt contains an anion which is specifically adsorbed on the metal, then, depending on the nature of the metal, an opposite effect can be observed. For example, the adsorption of I^- on mercury lowers, but on iron or cobalt raises, η_H.[25] In the first case, this is explained by the shift of ψ_1-potential in a negative direction and in the second case, by adsorption of the anion. The latter is the result of the formation of a covalent bond between the anion and the surface, and therefore iodide can be appropriately considered as a component part of the metal side of the double layer which, according to Frumkin, [7] will lead not only to a displacement of $E_{q=0}$ into the positive direction but also to a hydrophobization of the surface. As a result, the introduction of I^- increases η_H and also sharply increases the adsorption of organic cations, as well as facilitating to some extent the adsorption of neutral molecules. This phenomenon also appears with other anions [Cl^-, Br^-, CNS^-, HS^-] and has been discussed in detail.[6,23,26] The action of certain organic molecules is manifested by an increase in η_H over a more narrow region of potentials (close to $E_{q=0}$) and, as with the adsorption of large cations, it is associated not only with a change in activation energy for discharge but also with a screening of the electrode surface.[5] An example of the separation of these contributions is given with the adsorption of tetrabutylammonium cations on mercury, as described by Krishtalik.[27] He found that in the non-barrier discharge of H_3O^+, when ψ_1 effects are absent, the organic cation effectively loses its inhibiting properties. This allows the screening of the surface by the tetrabutylammonium to be disregarded, which agrees with the conclusions of Antropov on the nature of adsorption of similar cations on the surface of metals.[5] However, in cases where the adsorption is caused not only by Coulombic forces, the blocking effect can play the deciding role. In particular, the degree of retardation by butynediol of the cathodic evolution of hydrogen on iron in sulphuric acid solutions increases linearly with increase in surface coverage θ and is considered to be the result of a blocking action of the inhibitor.[28]

It is particularly important to take into account the effect of specific adsorption in the inhibition of the cathodic reaction by anions. The changes in η caused by this are made up of, according to Kolotyrkin,[29] an electrostatic influence on the discharge of the ions and changes in the adsorption as a consequence of the interaction of adsorbed anions and hydrogen:

$$\Delta\eta_H = (\alpha_1/\alpha)\,\Delta\,\psi_1 + \beta f\theta/\alpha \qquad (1.11)$$

where α and α_1 are transfer coefficients, β is the coefficient of asymmetry of the potential barrier for adsorption and f is the Temkin isotherm parameter, which takes into account the energetic heterogeneity of the surface. Such a model explains the difference in action of halides with different ratios of the terms in Eq. (1.11) which occurs particularly on solid metals.

The effects of surface-active ions or molecules are further complicated when the process of hydrogen evolution on the surface of corroding steels is considered. Even those additives not tending to specific adsorption such as triethylamine (TEA) can, depending on the microstructure of steel, facilitate or inhibit the cathodic reactions taking place on the surface.[30] Thus, the evolution of hydrogen on carbonyl (ferritic) steel in 1N H_2SO_4 takes place with a noticeable overvoltage, but in the presence of a pearlitic component the rate of the cathodic reaction can increase by an order or more (Fig. 1.3). The thermodynamically possible reduction of TEA cations accelerates the cathodic process on the ferritic steel for which, as for pure iron, there is a characteristically high activation energy for the direct reduction of hydrogen ions. On the other hand, the adsorption of cations, by increasing the positive value of the ψ_1 potential, leads to an increase in η_H. With increase in the content of the pearlite component in the steel the heterogeneity of the surface and adsorption of cations increase. This in its turn intensifies the inhibiting action of the TEA on the cathodic reaction although its rate on the ferritic–pearlitic steel still remains above that on the ferritic steel.

A formal theory of the action of adsorption inhibitors[5] has given an assessment of the various contributions of their inhibiting effects in the acid corrosion of metals for cases of their dissolution from the active state.[5] Based on equations of the electrochemical kinetics of anodic and cathodic processes and taking into account the fact that in the absence of external polarization $i_{cor} = i_c = i_a$ and, using the ψ_1 potentials of their displacement from the equilibrium values $[E_{eq}]$, instead of the exchange-current rate constants (i_o), Antropov obtained an equation for the corrosion rates taking place with inhibited hydrogen ion discharge.

$$i_{corr} = \left[i_{O,M} \right]^{\frac{\alpha_H}{\alpha_H + n\beta_M}} \left[i_{O,H} \right]^{\frac{n\beta_M}{\alpha_H + n\beta_M}} \exp\left[\frac{F}{RT} \frac{\alpha_H(n - z_M) - n\beta_M z_H}{\alpha_H + n\beta_M} \psi_1 \right] \exp\left[\frac{F}{RT} \cdot \frac{\alpha_H \beta_M n}{\alpha_H + n\beta_M} \left[{}_{H}E_{eq} - {}_{M}E_{eq} \right] \right] \tag{1.12}$$

In this equation, α_H and β_M are the transfer coefficients for the cathodic and anodic (metal dissolution) reactions, respectively, n is the number of

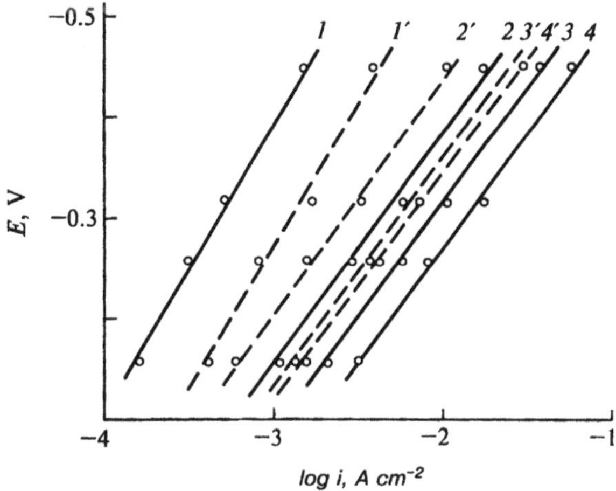

Figure 1.3. Cathodic polarization curves for steel in 1N H_2SO_4 without (1–4) and with additions of 0.1M triethylamine (1′–4′): 1,1′-ferritic steel; 2,2′-Armco steel; 3,3′-steel 20 [C~0.2%]; 4,4′ steel 45 [C~0.45%].

electrons participating in the reactions z_H and z_M are the charges of hydrogen and metal ions respectively, ${}_{H}E_{eq}$ and ${}_{M}E_{eq}$ are the equilibrium potentials for the cathodic (hydrogen evolution) and anodic (metal dissolution) reactions, respectively.

When taking into account the nonuniform distribution of an inhibitor on solid metal surfaces, two limiting cases of the adsorbed state can be distinguished. When the adsorption energy is commensurate with the repulsive forces between the inhibitor ions or molecules, these will be randomly distributed and form a skeletal surface structure. Strictly speaking, the particles of adsorbate will not be localized since there will be surface diffusion and exchange with the solution. The inhibiting effect will therefore be due to the adsorption potential. In the other case, which can occur with higher energies of adsorption, the formation of relatively dense accumulations (clusters) of inhibitor ions or molecules becomes possible. These areas of the surface then effectively no longer participate in the process of corrosion. Then, $i_{corr} = i'_{corr}$ (1-θ), where i'_{corr} is the corrosion current density on the surface that is not occupied by inhibitor.

If the electrochemical reactions causing the corrosion of a metal are under kinetic control only, as is often the case for processes with hydrogen depolarization, then with low and medium coverages ($\theta << 1$), Eq. 1.12 will become

$$i_{corr}^{inh} = \left[i'_{O,M}(1-\theta) \right]^{\frac{\alpha_H}{\alpha_H + n\beta_M}} \left[i'_{O,H}(1-\theta) \right]^{\frac{n\beta_M}{\alpha_H + n\beta_M}} \tag{1.13}$$

$$\exp\left[\frac{F}{RT} \frac{\alpha_H(n - z_M) - nz_H\beta_M}{\alpha_H + n\beta_M} \psi_1^{inh} \right] \exp\left[\frac{F}{RT} \frac{n\alpha_H\beta_M}{\alpha_H + n\beta_M} ({}_HE_{Eq} - {}_ME_{Eq}) \right]$$

This approach allows for changes in the ψ_1 potential, the exchange currents in the presence of the inhibitor, and the reduction in the free surface area of the metal, but assumes constant transport number, quantity of electrons and charge of the species participating in both reactions. The ionic strengths, ${}_HE_{eq}$ and ${}_ME_{eq}$ are unchanged as well. In this case the following equation is valid for the coefficient of inhibition, γ = the inhibited/uninhibited rate:

$$\gamma = \gamma_1\gamma_2\gamma_3\gamma_4 = \left[\frac{i_{O,M}}{i'_{O,M}} \right]^{\frac{\alpha_H}{\alpha_H + n\beta_M}} \left[\frac{i_{O,H}}{i'_{O,H}} \right]^{\frac{n\beta_M}{\alpha_H + n\beta_M}} \left[1 - \theta \right]^{-1} 10^{K\Delta\psi_1} \tag{1.14}$$

where $K = \frac{F}{RT} \frac{\alpha_H(n - z_M) - nz_H\beta_M}{\alpha_H + n\beta_M}$, γ_1 and γ_2 represent the kinetic, γ_3, the blocking or mechanical, and γ_4, the double layer or adsorption, coefficients of inhibition.

Considering the dissolution of iron in H_2SO_4 with participation of hydroxyl ions only (second order reaction with respect to C_{OH^-}) and with $\alpha_H = \beta_M = 0.5$, Antropov, analyzing the equation for this situation:

$$\gamma = \left[\frac{i_{O,M}}{i'_{O,M}} \right]^{1/4} \left[\frac{i_{O,H}}{i'_{O,H}} \right]^{3/4} \left[1 - \theta \right]^{-1} 10^{3.3\Delta\psi_1} \tag{1.15}$$

came to the conclusion that the kinetic effect, which is allowed for by the first two factors, is significant only for very small θ, and with increasing adsorption the effectiveness of the protection will be determined by the blocking and the ψ_1 potential change

$$\log \gamma = \text{const.}\ \Delta\psi_1 - \log(1 - \theta) \tag{1.16}$$

To find the contribution of these components in the protective effect, use is made of either measurements of the maxima on the electrocapillarity curves or of the differential capacity. As a criterion of the blocking or the energetic mechanism of inhibitor action, the linearity in $\gamma - \theta$ and

$\log(\gamma-1)-\theta$, respectively, has been suggested; this requires the further assumption of a proportionality between $\Delta\psi_1$ and θ.

Such an approach is undoubtedly useful for understanding the role of the double-layer structure in the inhibition of metal corrosion, although it should (particularly in the quantitative aspect) only be taken as a first approximation. Apart from the limitations already referred to in the literature, there is also a more complex dependence of electron current density on θ[31] with an effect on the electrostatic fields of the donor–acceptor complex formed on chemisorption of the inhibitor.[32] The possibility of a change in the mechanism of hydrogen evolution with the introduction of inhibitor and the direct participation of the latter in the cathodic reaction complicates the application of any formal theory for quantitatively predicting the effectiveness of the protection of metals from acid corrosion. In the majority of cases effective inhibition of the cathodic reaction—and consequently the acid corrosion of metals controlled by this—is a more complex process which is often related to the action of the inhibitor by two formal mechanisms simultaneously.[33]

The influence of surface active molecules or ions on the anodic reaction is of special interest since, in the opinion of Lorenz,[34] it is usually more marked than on the cathodic kinetics. In many cases, and particularly for metals of the iron group, it is impossible to ignore the participation in the anodic reaction not only of molecules of water or hydroxyl ions but also of other components of the solution. Although the involvement of anions in the anodic dissolution of metals had been discussed as early as in 1928, the mechanism of the phenomenon has continued to be actively discussed up to the present time. The work of Kolotyrkin and his co-workers has played a large part in the understanding of the problem. They showed that the dissolution of metals takes place via a stage of formation of surface complexes made possible by the covalent bonding of the anion with the metal surface. In this treatment, the stimulating effect of an anion on the anodic dissolution of a metal appears only with the attainment of a certain critical potential at which the strength of the bond between the metal atom and the adsorbed anion equals that of the covalent bond in the corresponding individual compound. Originally it was thought that only those ions that were weakly bonded to the surface (at this potential) could fulfill the role of an inhibitor and that these would become activators at more positive potentials. In many cases this is actually observed, with the inhibition replaced by an activating action on anodic polarization. However, even with the formation of a strong complex of the anion with the metal, the inhibition of the anodic reaction is also possible as a result of the displacement of water from the surface by the adsorbate. According to Kolotyrkin,[35] if the bond of the adsorbate with the solution is weaker than that with the metal, then the adsorbate will remain on the surface and inhibit the dissolution of the electrode.

In order to describe inhibiting effects during the active dissolution of metals, concepts of formal electrochemical kinetics may often be introduced which take into account the various reaction-rate constants and competitive adsorption of anions. Thus, if the retardation of the anodic dissolution of iron in sulfuric acid solutions by iodide is connected with a contraction of the active surface of the electrode and with the occurrence of parallel reactions involving OH^- and I^-, then at relatively positive potentials the extent of surface coverage by the inhibitor θ_{inh} will be described by the equation [36]:

$$\theta_{inh} = K(a_{I^-}/a_{OH^-})\exp(-2\alpha\, EF/RT) \tag{1.17}$$

On reaching a certain potential a noticeable desorption of I^- will occur. This potential will decrease with increase in pH as has been confirmed experimentally.

However, in the opinion of Bech–Nielsen(37) these results can be explained by the formation of $Fe^{\circ}(An^-)_2$ on active centres of the iron surface with the concentration of this species being independent of the content of An^- anions in the solution. Subsequently, the complex reacts electrochemically with water by two parallel mechanisms with rates i_1 and i_2 which depend differently on potential. Therefore, on the polarization curve—which is determined by the total current density $i = i_1(1 - \theta) + i_2\theta$—there will be two Tafel regions. Since the activity of the $Fe^{\circ}(An^-)_2$ complex can be taken as constant, then i_1 and i_2 will be inversely proportional to the concentration of the inhibiting anion, which explains the negative order for the respective reactions.

The physical meaning of a negative order of reaction with respect to the inhibitor is that the latter must have been released from the surface by an active component of the solution. This approach is widely used for explaining the inhibition of the dissolution of iron in acids. A negative order with respect to the inhibitor is also observed during dissolution in neutral or weakly alkaline media not only for iron but also for other metals(38) (Fig. 1.4). For example, the dissolution of a rotating copper disc in a weakly alkaline phosphate solution can be inhibited by small additions of benzimidazole, BI. In this case the order of reaction with respect to the inhibitor is $n = -2$. Taking into account that, in the presence of BI, the second order with respect to HPO_4^{2-} is retained, it is not difficult in the formal approach to select a kinetic scheme for the anodic process which would agree with the experimental data. Thus, it is sufficient to propose that the concentration of the inhibitor complex that is formed with copper is constant and that $\alpha = 0.5$, for the dissolution of copper to be described by the following mechanism:

$$\text{(a)} \quad [\mathrm{CuOH}]^{-}_{\mathrm{ads}} + 2\mathrm{BI} \;\mathrm{A}\; [\mathrm{Cu(BI)_2}]_{\mathrm{ads}} + \mathrm{H_2O} + \mathrm{e} \tag{1.18}$$

$$\text{(b)} \quad [\mathrm{Cu(BI)_2}]_{\mathrm{ads}} + \mathrm{HPO_4^{2-}} \;\mathrm{A}\; [\mathrm{CuPO_4}]^{2-}_{\mathrm{ads}} + 2\mathrm{BI}$$

$$\text{(c)} \quad [\mathrm{CuPO_4}]^{2-}_{\mathrm{ads}} + \mathrm{HPO_4^{2-}} \;\mathrm{A}\; \mathrm{CuHPO_4} + \mathrm{PO_4^{3-}} + \mathrm{e}$$

$$i_a = K a^2_{\mathrm{HPO_4^{2-}}} \cdot a_{\mathrm{BI}}^{-2} \exp(\alpha FE/RT)$$

The formation of complexes of univalent copper by reaction (a) is also at least a two-stage process. It is suggested that the inhibitor is capable of forming various types of bond in the surface complex such as ionic (on account of the dissociation of the NH-group) and donor–acceptor. The process of stage (c) is rate-determining and the Tafel slope of the polarization curve b_a has been calculated on the basis of the proposed mechanism as 0.116V, which is in good agreement with the experimental value. This approach well illustrates the fact that the mutual displacement of the components of a solution from an electrode surface plays an important role in the inhibition of the corrosion of metals. Unfortunately, its computation from aspects of the formal approaches and hypotheses is not possible because of the considerable difficulties in determining the surface concen-

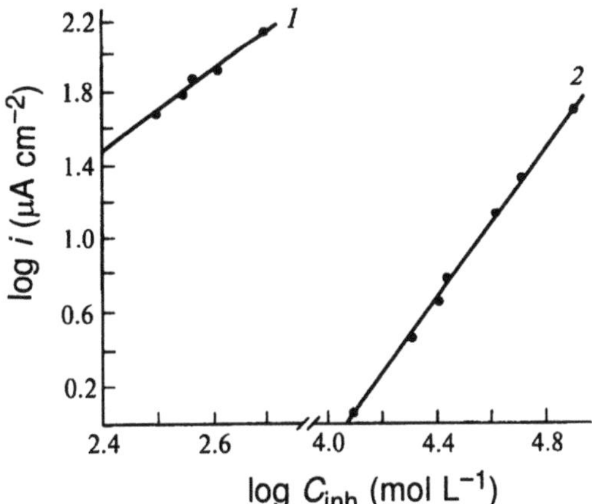

Figure 1.4. Dependence of log i on log C_{inh} for (1) zinc, at $E = -0.8$V, and (2) copper, at $E = 0.14$V, rotating disc electrodes ($f = 408$ rad sec-1) in 0.5M phosphate solution at pH 9.5 containing benzimidazole as inhibitor.

tration of the adsorbing particles. For instance, in (Eq. 1.18) there is a description of the effect of BI on the dissolution of copper only in the region of low concentrations, i.e. $C_{inh} < 7.5 \times 10^{-5}$M. It is significant that, as is often the case, a linear dependence is assumed for the surface concentration of the components of the solution and the volume concentration, following Henry's isotherm. Obviously, this can be true only to a first approximation. Moreover, as pointed out by Antropov et al.,[5] an increase in volume concentration will be accompanied by an increase in the packing density of the adsorbed inhibitor; clusters form and the blocking effect will begin to predominate. Recognizing that, in general, considerations of adsorption phenomena are significantly more complex, many investigators have attempted to introduce corrections using other isotherms. For example, the adsorption of the intermediate product of the dissolution of iron in sulfuric acid solution has been described by an adsorption isotherm for a non-uniform surface and differences in the rates of the forward and reverse reactions have been calculated.[39] In this case it was found that the order of reaction with respect to $[OH^-]$ for small and large degrees of coverage was significantly different. This approach, of which some aspects were considered in detail,[40,41] has also been used by other workers. Thus, Reshetnikov[28] explained fractional orders of reaction with respect to anions of acid solutions and inhibitors (tetraethylammonium perchlorate, triethanolamine, and butynediol) by proposing that adsorption of intermediate products is described by the Temkin isotherm and that these are distributed at various centers on the surface although the heterogeneity factors are close. In this case, the orders of reaction with respect to the components of the solution and the Tafel slope of the polarization curve are mainly determined by the ratio of the degrees of coverage by the various products, which by suitable selection can provide good agreement between the calculated coefficients and experimental data. Without analyzing this approach in detail, it must be pointed out that with the present level of experimentation, the question of the rigorous determination of θ and therefore the type of adsorption isotherm for a corroding iron electrode remains unanswered. While it is unclear for which cases agreement with the Temkin isotherm should be expected, it is known that adsorption on heterogeneous surfaces, including those of iron, is often described by the Langmuir isotherm.[26] The latter is explained by the compensating action of the exponential terms of the Temkin and Frumkin isotherms resulting from the decrease in free energy of adsorption with increasing θ and with intensification of the attraction between the particles of the adsorbate.

As in the theory of catalysis, in the theory of corrosion inhibition the individual properties of the active centre sometimes turn out to be more important than the collective electron properties of the solid body lattice.[42]

A first requirement will therefore be to understand the chemistry of the processes taking place on the metal surface. Thus, in our work based on the concept of complex formation, the dissolution of metals and inhibition are examined as a consequence of the nucleophilic substitution of ligands in the surface complex. First of all, such an approach allows a quantitative analysis to be made of the role of the chemical nature of the components of the solution and how this role affects the controlling stage of the complex heterogeneous process which is corrosion. It also, as we shall try to show in the following chapters of this book, can be applied to a more general examination of the effect of an inhibitor on the kinetics of electrode reactions, including passivation of metals.

1.3. PASSIVATION OF IRON IN NEUTRAL SOLUTION

The transfer of a metal into the passive state as a result of an appropriate change in the composition of the solution is usually accompanied by an abrupt ennoblement of the potential of the metal and is characterized by the absence of any Tafel dependence of the dissolution rate on potential (the section DE on the polarization curve in Fig. 1.2). Thus, the mechanism of such protection must differ from that involved in any reduction of the corrosion rate taking place in the active dissolution region. When passivation occurs at a potential at which the formation of oxide (hydroxide) or other difficultly-soluble layers on the electrode surface is thermodynamically possible, the rate of metal dissolution is found to be lowered by 1–2 (often more) orders of magnitude. The oxides of the majority of metals (iron, nickel, chromium, copper, aluminum, zinc) have their lowest solubility in neutral and weakly-alkaline media and therefore passivating inhibitors are widely used in these situations. Thus one can understand the approach taken by Lomonosov and Faraday in associating oxidation processes with the development of passivity. Significant contributions to this film theory of passivation have since appeared in the works of Kistyakovskii, Izgaryshev, Akimov, Evans, Tomashov, Sukhotin, Bonhoffer and his school, and Cohen and Sato. The structure and composition of surface layers formed on iron in neutral solutions have been particularly widely studied with borate-buffer systems.

A better understanding of the results obtained under these conditions has been gained by studying the effect of the components of borate buffers on the anodic behavior of metals. We have investigated this question in deaerated solutions with various borate contents, $C_{H_3BO_3} = 0.25–1.0$ M and $C_{Na_2B_4O_7} = 0.0097–0.1960$M[43] at constant pH.

Assuming that the initial stages of passivation are controlled by electrochemical reactions and that the rate of dissolution of iron in the passive area is low compared with that on the active surface, one can write an equation for the measured anodic current density i_{meas}

$$\log i_{\text{meas}} = \log k_a + \frac{\beta_a^* FE}{2{\cdot}3\, RT} - \frac{k_p\, RT}{2{\cdot}3\beta_p^*\, FV} \cdot \exp\left[\frac{\beta_p^*\, FE}{RT}\right] \tag{1.19}$$

where β_a^* and β_p^* are the apparent transfer coefficients of the anodic reactions, subscripts a and p relate respectively to the reactions of iron ionization and the electrochemical stage of the passivation, k_a and k_p are the apparent reaction-rate constants, T is temperature, and V is the rate of the linear potential sweep.

The value of the last term in (Eq. 1.19), now identified as A, also determines the deviation from the Tafel dependence (Fig. 1.5) and so the passivation-process parameters, $\log k_p$ and β_p^*, can be calculated from a linear extrapolation of the $\log A$–E relationship:

$$\log A = \log k_p + \log\left[\frac{RT}{2{\cdot}3\beta_p^*\, FV}\right] + \frac{\beta_p^*\, FE}{2{\cdot}3\, RT} \tag{1.20}$$

Differentiating (Eq. 1.19) with respect to E and equating to zero gives, for the passivation potential

$$E_p = \frac{2{\cdot}3\, RT}{\beta_p^*\, F} \log\left[\frac{\beta_a^*\, FV}{k_p\, RT}\right] \tag{1.21}$$

With $E = E_p$ (Eq. 1.19) the passivation-current density i_p is described by

$$\log i_p = \log k_a + \frac{\beta_a^*}{\beta_p}\left[\log \frac{\beta_a^*\, FV}{k_p\, RT} - \frac{1}{2{\cdot}3}\right] \tag{1.22}$$

In those potential regions for which a Tafel dependence holds, b = 60 mV for all concentrations of the buffer. The value of the logarithm of the rate constant linearly decreases on increasing the total concentration of borate ion $[\Sigma C_B]$, as calculated from the respective dissociation constants of the borates. However, an analysis of the kinetic scheme shows that borates do not directly take part in the iron ionization stage and their influence on $\log k_a$ has to be explained by the shift in the negative direction of the ψ_1 potential.

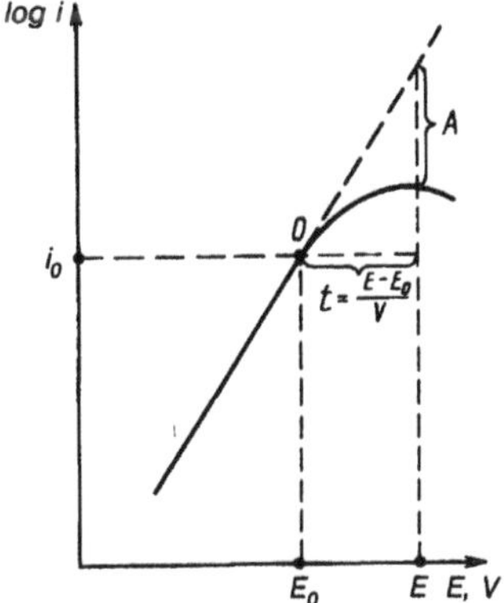

Figure 1.5. Schematic representation of the initial stage of passivation of iron under potentiodynamic polarization of the electrode at various scanning rates ($V_{S^{-1}}$). *A* denotes the departure from Tafel behavior [see Eq. (1.19)].

A different situation exists with the participation of borates in the initial stages of passivation. For the linear sections of the log *A*–*E* dependence (Eq. 1.19), the kinetic constants $\beta_p^* = 1.34 \pm 0.03$ and $\Delta \log k/\Delta \log \Sigma C_B = -0.7$ were calculated. The value of β_p^+ obtained corresponds to the limiting stage of the transfer of a second electron during the equilibrium stage of the transfer of the first. In the studied region of potentials such a reaction can only be $Fe^\circ \rightarrow Fe^{2+}$. It differs from the ionization reaction in the Tafel region which has $\beta_a^* = 0.97 \pm 0.02$ and is characterized by a different dependence of the reaction constants on ΣC_B. According to (Eq. 1.21), for constant β_p^* and β_a^* an increase in log k_p and consequently a decrease in log ΣC_B should lead to a linear decrease in E_p, as observed experimentally. The fact that the value of i_p remains (within the limits of experimental error) essentially unchanged with variation in ΣC_B is also explainable (Eq. 1.22) if the mutual compensating effect of ΣC_B on k_a and k_p is taken into account. Bearing in mind that i_p as a kinetic parameter is determined by the overvoltage of the passivation process, the nonlinearity of the initial section of the log A–E curves can be explained by the fact that at low overvoltages the reverse reaction needs to be taken into consideration; this was disregarded in the derivation of (Eq. 1.20). Actually, a comparison of the potentials of the onset of passivation (−0.4 to −0.35V) with the equilibrium potentials for the formation of oxides (hydroxides) at pH 7.4 leads us to suggest that the onset of passivation of iron is connected with the formation of $[FeO]_{ads}$ during the limiting stage of the transfer of a second electron. Such a process could be as follows:

$$Fe + OH^- = [FeOH]^-_{ads} \quad (1.23a)$$

$$[FeOH]^-_{ads} = [FeOH]_{ads} + e \quad (1.23b)$$

$$Fe[OH]_{ads} + OH^- = [FeO]_{ads} + H_2O + e \quad (1.23c)$$

Borates can hinder passivation as a result of the following reaction which competes with (Eq. 1.23c):

$$[FeOH]_{ads} + HB_4O_7^- = [FeHB_4O_7]_{ads} + OH^- \quad (1.24)$$

Consequently, borates, at least in the initial stage, do not promote the passivation that results from oxide formation. In fact, crystal nuclei of oxides that penetrate into the metal can set up dipole steps in potential which exponentially decrease the rate of the anodic reaction.

The formation of a passive film on iron in a borate buffer is a still more complex process and occurs only with a further displacement of the potential of the electrode. However, even with $E >> E_p$ and if the oxide film is first removed from the surface (by cathodic polarization), low values of i_a characteristic of the passive state of iron are established relatively slowly. Thus, with a stepwise displacement of potential from the cathodic region to $E = 0.2$ V, the passive state requires 15–30 minutes to become established. The rate of film growth continuously decreases with time and follows an inverse logarithmic law, as is valid for the formation of thin passivating films when the limiting stage of the growth is the reaction of metal ionization.(44)

The formation of a passive film can also be followed from electron diffraction patterns (Fig. 1.6) in which the reflections can be divided into two groups: sharp rings characteristic of α Fe and broader rings with diffuse edges corresponding to an oxide structure.(45) Densitometer measurements, which allow for the low (0.5°) angle of incidence of the electron beam on the specimen and the cross section of elastic scattering of electrons in amorphous iron oxides, have been used to determine the ratio of the integral intensities of the α Fe reflections and of the covering film. These measurements have given an estimated 3–5 nm for the thickness of the film. This value is commensurate with the results of coulometric measurements, i.e. 3.4 nm calculated for unhydrated γFe_2O_3 and 5.4 nm for $\gamma Fe_2O_3 \cdot 1.8H_2O$. The inverse-logarithmic growth law for oxide films is usually observed only for film thicknesses of ≤ 4 nm, so the degree of hydration is lower than in $\gamma Fe_2O_3 \cdot 1.8H_2O$.

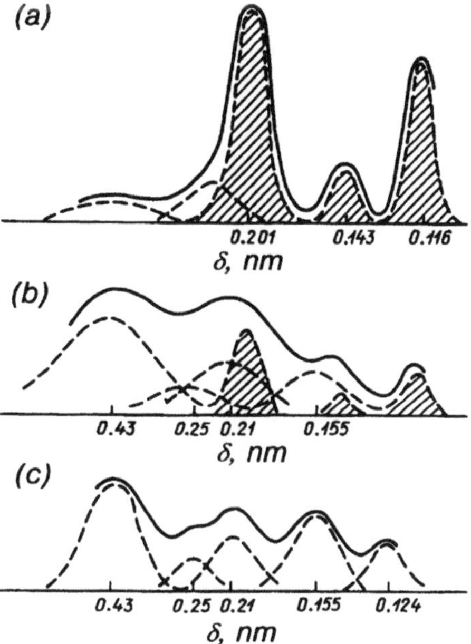

Figure 1.6. Curves for the distribution of intensity along a vertical line passing through the center of an electron diffraction pattern (solid lines) and the diffraction maxima (dashed lines). The shaded maxima correspond to αFe. The patterns were obtained from specimens: (a) with an original αFe surface; (b) after 1 and (c) after 30 minutes passivation in a borate buffer at E=0.2V.

Using measurements of 50 electron-diffraction patterns, a selection was made of the most characteristic series of diffraction lines corresponding to the parameters of the oxides and hydroxides of iron with a cubic structure, e.g. FeO, Fe_3O_4, Fe_2O_3, $Fe[OH]_3$, etc. However, complete agreement of the parameters was not obtained with the interplanar spacings of any of these structures.

A clear diffraction picture of the oxide phase was obtained from the surface of a specimen that had been held in the buffer for 30 minutes. The diffraction pattern for the initial stage of formation of the passive film had a wide central halo. Using model structures, a special computer calculation (for a rigid lattice in a kinematic approximation) was made of the diffraction intensities of iron oxides and showed that in this case the film was either amorphous or consisted of two-dimensionally ordered regions of FeO with diameters of 1 nm containing up to 50 atoms. Such an interpretation is in agreement with the results of the electrochemical studies on the kinetics of passivation (Eq. 1.23) which, although not providing direct proof of the formation of such clusters in the initial stages, does nevertheless allow for their formation. The thermodynamic formation of FeO is already possible at the potentials of active dissolution, since, for the reaction (at 25°C):

$$Fe + H_2O = FeO + 2H^+ + 2e \qquad (1.25)$$

$$\text{the value of } E_{Fe/FeO} = -0.04 - 0.059\text{pH}$$

As the period of holding the electrode at $E = 0.2$ V is increased, crystallites begin to form in the oxide and regions with a three-dimensionally-ordered structure occur. The lower degree of ordering of the oxide in the initial stage of the passivation is explained by the significant heterogeneity of the electrode–solution system in which, from the thermodynamic point of view, many reactions are possible. In particular, along with the reaction in Eq. 1.25, the formation of magnetite is possible:

$$3Fe + 4H_2O \rightarrow Fe_3O_4 + 8H^+ + 8e \qquad (1.26)$$

$$E_{Fe/Fe_3O_4} = -0.09 - 0.059\text{pH}$$

The possibility of hydration of the already nucleated oxide makes the distinction from hydroxides, such as $Fe(OH)_2$, difficult.

On the basis of coulometric and ellipsometric studies, Bockris *et al.*[46] came to the conclusion that in the region of potentials of active dissolution of iron an $Fe(OH)_2$ film forms on the metal surface. Close to E_p, this begins to transform into γFe_2O_3 with this process accelerated by anodic polarization and its completion providing a stable passivation of the metal. Reinhard[47] proposed the following scheme for the transformation of the film on passivation:

$$3Fe(OH)_{2\,ads} \rightarrow Fe_3O_4 + 2H_2O + 2H^+ + 2e \qquad (1.27a)$$

$$2Fe_3O_4 + H_2O \rightarrow 3\,Fe_2O_3 + 2H^+ + 2e \qquad (1.27b)$$

Frankenthal[48] considers that stable passivity of iron arises with the formation of Fe_2O_3 of significant thickness. Of interest in this respect is the analogy between the anodic behavior of iron in borate buffer and of oxide electrodes of FeO and Fe_3O_4. According to the work of Tsuru *et al.*,[49] a thin film of γFe_2O_3 is found in all cases and the authors associate its formation with the passivation of iron.

Nagayama[50] pointed out that the passive film could not be identified as any particular oxide of iron and this has been confirmed by our own studies. Although with the formation of oxide–electrolyte and metal–oxide boundaries, the role of diffusion and recrystallization processes becomes greater, the structure of the film even after some hours remains insufficiently ordered to be described by any definite oxide. To build up a structural model of such imperfect oxide layers, it is convenient to use an octahedral configuration of FeO_6 atoms, which is the basic part of the structure of all cubic oxides of iron. The plane two-dimensional structure of FeO which has been observed on the surface of specimens held in a buffer for several minutes can be constructed by joining the octahedra at the edges in a plane parallel to the surface of the specimen. However, during longer periods of oxidation, hydration of the surface and adsorption of the components of the solution on the outer side of the film will eventually lead to defectiveness of the oxide and a departure from stoichiometry.

The question of the degree of oxidation of iron in the passive film is controversial. Although Evans[20] using chemical analysis found only Fe^{3+} in the film, many authors believe that the internal layer adjacent to the metal corresponds to magnetite. This conclusion is based on coulometric studies and thermodynamic assessments. However, in the opinion of Sato,[51] the passivation of iron in borate buffer is connected with the formation of only the hydrated oxide $Fe_2O_3 \cdot 0.4\ H_2O$. The two waves on the coulometric curves are explained by the successive reduction of this oxide to Fe_3O_4 and then to metal, or by the presence of layers with various degrees of hydration in the film. Water was actually found in the passive film, and surface layer analysis showed a dependence of the film composition on the potential of the iron. In all cases the passive film consisted of a barrier (less hydrated) layer and a precipitated film. However, if the electrode was held in the solution at potentials less than a certain value, sometimes referred to as the Flade potential ($E < E_F$), then the barrier layer was found to contain Fe^{2+} as well as Fe^{3+}, whereas with $E > E_F$ only Fe^{3+} was detected in the oxide. In the latter case only a barrier layer of approximate composition $Fe_2O_3 \cdot 0.2\ H_2O$ formed with a thickness which increased linearly with increase in E and with a degree of hydration that was independent of pH. The precipitated layer was formed by the anodic oxidation of Fe^{2+} passing into solution at an earlier stage of the passivation.[51]

Heterogeneity of the passive film was also confirmed by the results of ellipsometric and X-ray studies. The surface of iron passivated in a neutral borate buffer at E = 0.4V and then argon-ion etched[52] showed that the "chemical shift" of the binding energy E_B of the $Fe2p_{3/2}$ electrons decreased as deeper parts of the film were reached. Experiments showed that this effect could not be explained by any partial reduction of the iron oxides by ion bombardment, and it was concluded that the degree of oxidation of the iron decreased, going from the outer layers (Fe^{3+}) to the metal. The inner layers could be represented as $[(FeO)_n(Fe_2O_3)_m]$, where $n < m$. This model was compared with the ellipsometric data. Mathematical analysis allowed the system to be described as a multilayer oxide film on iron characterized by changes in the refractive index of each layer.

Originally, Fischer,[53] in discussing the heterogeneity across the thickness of the passive film, proposed a more general model for the passivation of iron in aqueous solutions (ranging from acid to alkaline). He suggested that next to the metal there is a barrier layer of an oxide–hydroxide nature, but without definite stoichiometry. This amorphous layer constitutes a basic impediment to the transport of metal cations. These cations are transformed into Fe_3O_4 on the outer side of the amorphous layer. Later it was considered—as did others—that the passive film transforms into Fe_2O_3 with the degree of hydration increasing towards the solution side, and that it is completed by an adsorbed layer of anions from the solution. The formation of oxides initially occurs at dislocations, and the dissolution of iron occurs at the sites on the surface with maximum coordination unsaturation. Fischer, as others, assumes the attachment of a second OH^- to $Fe(OH)_{ads}$ in the initial stages of passivation. The further oligomerization of these compounds and their spatial orientation leads to the formation of a layer of magnetite. As the potential is displaced, magnetite is electrochemically oxidized from the solution side to γFe_2O_3. The latter occurs only with full passivation and suppression of the process of iron dissolution.

Significantly, Fischer ascribes an important role to the adsorption of non-oxidizing components of the solution which can stabilize the oxide structure and prevents the detachment of nuclei from the metal surface. However, as we have already seen, this does not apply to borates. If a passive film is formed over a long period at $E = 0.2V$ in a borate buffer then, as secondary -ion mass spectrometry using argon ion etching has shown (Fig. 1.7),[45] borates can be detected effectively only in the outer layer of the film, whereas both OH^- and H_2O are present at significant depths. In view of the high hydrophilicity of borates, it is unlikely that they could exert any stabilizing action on the oxide layer—particularly when compared with such strongly adsorbed anions as OH^-. It is not unexpected that introduction of very small quantities of halide ions into the buffer leads to depassivation of the iron. On the other hand, the components of this buffer do not possess any noticeable complexing properties—unlike acetate and citrate buffers—and promote the formation of only oxide layers on the metal surface, as compared to phosphate and oxalate solutions. It is for these reasons that the passivation of iron in borate solutions has been so widely studied.

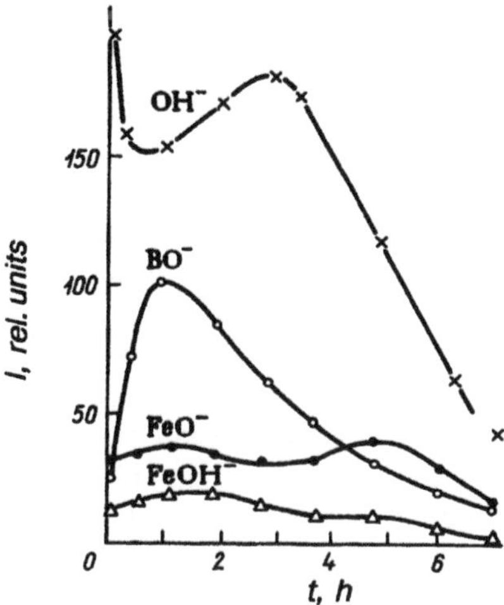

Figure 1.7. Changes in the peak intensities, *I,* in the mass spectra of secondary ions with duration of 3 keV etching after 30 min passivation at $E = 0.2$ V in a borate buffer.

In neutral solutions oxide films of appreciable thicknesses have been found on the surface of passivated iron; therefore, there is believed to be a relationship between the mechanism of the passivation with the growth kinetics and semiconducting properties of these oxides. Thus, to explain the growth law, the concept of a "variable site" has been proposed, with a peculiar rotation of the metal–oxygen dipole. Another model postulates that the controlling stage of the process is either the emission of cations from the metal to the film, or the transport of oxygen anions in conditions such that the field strength in the oxide is independent of its thickness.[44,54]

However, these approaches make it difficult to explain certain laws regarding the formation of passive films on iron in borate buffer or in alkaline solutions. Furthermore, the inverse-logarithmic law for film growth can also be obtained in acid solutions which, according to Chao *et al.,*[55] is difficult to associate with the models of Mott and Cabrera or of Fehlner because these ignore the fact of film dissolution. Because of this, Chao, Lin, and Macdonald put forward another model in which it was postulated that for all potentials more positive than the critical-passivation potential an oxide with a large concentration of point defects will form. By considering that the field strength in the passive film is 10^6V cm^{-1} (of the same order as that required for dielectric breakdown), they proposed that the passive film behaves as an incipient semiconductor because it exists on the verge of dielectric breakdown.

The controlling stage of the film growth is believed to be the transfer of vacancies through the film, and the establishment of an equilibrium condition at both boundaries. Using this approach, the authors deduced equations to explain the growth kinetics and the effect of temperature on the anodic oxidation of iron. The data of Goswani and Staehle[56] on the growth kinetics of the passive film on iron in a borate buffer (pH 8.4), which can be described equally well by either a logarithmic or inverse logarithmic law, are consistent with the theory. However, the most valuable contribution has been an attempt to introduce into the kinetic equation a coefficient which reflects the influence of the solution anions on the sensitivity of film growth rate to pH. As we shall show below, a change in the anionic composition of the solution is capable of not only changing the kinetics of the oxidation of the metal, but also of leading to a change in the actual mechanism of passivation.

Of all metals, the passivation of iron is the most widely studied, not only because its alloys are those most widely used in technology. Many construction materials, such as aluminum and its alloys, do not require passivation in neutral media. Increases in their corrosion resistance are most closely associated with the prevention of pitting or other types of localized corrosion. From the position of film theory, this can be explained by the presence of protective oxide layers on the metal surfaces. Based on this, Brasher divided metals into two groups.[57] One group (Fe, Zn, Al) was characterized by the stability of the air-formed oxide film in solutions not containing aggressive ions and with pH close to neutral. In the presence of oxygen, these metals should passivate by the "healing" of defects in the oxide film. Metals of the second group were either stable only in the alkaline region (Cd), or were unstable at all pH values (Pb) and therefore could not, in her opinion, be passivated by an inhibitor. However, in 1956, Akimov[58] pointed out that if a phase-oxide film is present on a metal surface, the adsorption of oxygen could lead to passivation.

According to Novakovskii,[59] an oxide which does not have protective properties can be enriched by chemisorbed oxygen from the solution side, and this also will lead to passivation of the metal. In this case the oxide film does not replace the initial chemisorbed layer but grows between it and the metal. Moreover, the possibility of adsorption passivity must not be ignored. Thus, the mechanism of passivation of metals by chemisorbed oxygen or water has been independently examined by Kabanov, Kolotyrkin, Skorcheletti, Mikhailovskii, and Uhlig. However, it is necessary to point out the difficulties in reliably separating the effects caused by adsorption and the subsequent formation of the passive-oxide film. Therefore, it is of interest to look for a solution–metal system in which both the transfer of the metal into the passive state, as well as a sufficient amount of time in this condition, would not be accompanied by the formation of an oxide film. This would remove the objection by Skorcheletti that adsorption only facilitates the transition of a metal into the passive state and that the maintenance of that state requires an oxide film of appreciable thickness.[60] It follows that a study of the influence of inhibitors on the passivity of metals, particularly iron, is important not only in practice but also for the theory of corrosion protection.

1.4. THE EFFECT OF INHIBITORS ON THE PASSIVATION OF METALS

1.4.1. The General Situation

The conversion of a metal or alloy into the passive state can be achieved by several methods, the effectiveness of which will depend on many factors, including the nature of the metal, the oxidation–reduction potential of the environment, temperature, the possibility of forming difficultly-soluble compounds of cations of the dissolving metal with components of the solution. Changing the composition of an environment so as to facilitate passivation in a metal–solution system is a classical method that was recognized by Lomonosov who, in 1743, showed that a sufficiently strong "spirit of saltpetre" (nitric acid) could modify the dissolution of a metal. This action of nitric acid, for example in relation to iron, is the result of its strong oxidizing properties. Thus, 50% solutions lead to passivation by forming a thin film of difficultly-soluble oxide on the metal surface. Contemporary analysis based on thermodynamic data and the kinetics of corrosion diagrams[1,22,23] now provide a theoretical basis for the use of inhibitors of the oxidizing type and in practice, experience with passivation treatments for metals has confirmed their effectiveness.

However, it also has been found that the oxidizing agents that are most successful in protecting metals from corrosion often form surface layers which contain reduction products of limited solubility; this points to the possibility of a second type of passivation.

Other corrosion inhibitors do not have any clearly-defined oxidizing properties.* These include several that have been known for a long time,

*The appearance of oxidising properties of chemical compounds in relation to a metal will ultimately depend on the nature and potential of the metal and on the pH and temperature of the solution.

for example, carbonates, silicates, and phosphate, particularly when used with calcium or zinc cations. At low concentrations, these can provide significant inhibition of the corrosion of low-carbon steel; however, they do not provide complete protection. The steel surface becomes covered with a film of (basic) salt, which mainly retards the cathodic reaction. At higher concentrations—above the critical concentration for full inhibition—non-oxidizing anions in air-saturated solutions can stabilize the passivating film on iron to give essentially complete inhibition of corrosion. The passivating film of iron oxide then contains small inclusions of salts or basic salts of the inhibitive anion. These inclusions form at localized anodic areas in the film to block them off and complete the inhibition of the anodic reaction. Thus, as pointed out by Mayne in 1954,[61] the passivating film formed on iron in 0.1M Na_2HPO_4 during natural aeration of the solution includes $FePO_4 \cdot 2H_2O$ as well as iron oxide. Since iron is not passivated in phosphate solutions in the absence of oxygen or other oxidizers, a secondary role must be ascribed to phosphates. However, Kolotyrkin and co-workers[62] established that the dependence of the dissolution rate of iron on potential in solutions of phosphates is characterized by a specific salt-passivation which precedes oxidation. Therefore, the region of active dissolution of iron in these solutions is contracted in comparison with that in the background borate buffer. It is significant that the primary, i.e., the phosphate passivation, is reached with noticeably lower current densities than the oxide type of passivation, although this is no longer the case when the concentration of phosphates is reduced or the pH of the solution is changed. In this region of potentials, compact-phase films consisting of crystals of $Fe_3(PO_4)_2 \cdot 8H_2O$ form on the surface of the metal, whereas at more negative potentials or in the volume of the solution, amorphous deposits form.

Accepting that the passivation of iron in phosphate solutions is the result of a monolayer of a difficultly soluble salt, Kolotyrkin and co-workers[29,63] have compared the experimental and theoretical dependence of anodic-current density on the rate of movement, the potential, and the solution concentration of phosphate. This analysis of the monolayer model of passivation by phosphates showed that, although it described the basic effect of the primary passivation, it did not explain certain aspects of the anodic behavior of iron. This problem is not lessened by the possibility of polymolecular adsorption of $FeHPO_4$. An adsorption model is, on the whole, inadequate, and it is therefore necessary to consider the effect of a porous coating crystallizing from the solution. It must also be recognized that the adsorption of oxygen or other components of the solution can also make a definite contribution, although this manner is not as clear in salt passivation as in normal passivation. Sukhotin and Osipenko came an analogous conclusion on the mechanism of salt passivation in their study of the electrochemical behavior of manganese in phosphate solutions.[64]

The passivation of copper in these solutions occurs by another mechanism. In weakly-alkaline solution (pH 9.2), the passivation initially is connected with the formation of oxide films only, but our studies[65,66] using electron diffraction and mass spectrometric analysis of the surface have shown that the role of the phosphates cannot completely be ignored. By forming soluble compounds with Cu^+, phosphates can accelerate the dissolution of copper, but in the presence of dissolved O_2 they can act as special oxidation catalysts capable of forming a protective film from CuO crystallites. In deaerated solutions, the passivation of a rotating electrode is caused by the formation of $Cu(OH)_2$ on the metal surface, no phosphates are found in the passive film. However, in a quiescent electrolyte with natural aeration, copper passivates most readily, and in this case phosphates are found not only on the surface but also through almost the whole depth of the passive film. Consequently, even in the absence of characteristic signs of salt passivation, the possibility of inclusion of difficultly-soluble compounds formed from the components of the solution and ions of the corroding metal must be considered.

Chelating agents, notably those related to heterocyclic compounds, should also be mentioned in this group of inhibitors. The formation of difficultly-soluble complex compounds of these agents on metals or oxides in the form of continuous or slightly defective, and often quite thin films, can passivate the surface and impart anti-corrosion properties that are unusual even for oxide films. The thicknesses of the films formed by the chelating agents depends on the nature of the metal, its surface condition, the temperature, and the pH of the solutions. On copper, for example, quite good protection by azoles is provided in neutral and weakly-alkaline media with film thicknesses of 5–20 nm; in acid solutions films reaching several hundreds of nanometers in thickness are formed.

It would appear that since effective suppression of corrosion can be achieved by oxidizers, by compounds forming salts of low solubility, or by complexes, the adsorption mechanism of passivation must be put in doubt. However, it is known that anions of organic acids, long chain amines, amides, and certain other compounds which do not possess oxidizing properties can significantly inhibit the anodic dissolution of metals and facilitate their passivation. Furthermore, in these solutions the passivation is not accompanied by the formation of salt or chelate films of noticeable thicknesses. Such inhibitors could be referred to as adsorption passivators. However, it is not easy to answer the question "can such an inhibitor, the concentration of which is relatively small, prevent not only the dissolution of a metal but also its oxidation by water, i.e., is the passivation of iron possible without the formation of passivating films of oxide?" We shall examine this question in detail with the example of the passivation of iron in the presence of organic additives of the nonoxidizing type.

1.4.2. Passivation of Iron and the Adsorption of Organic Inhibitors

Tsai and Hackerman,[67] on finding that passivation of iron in acid-sulfate solutions was encouraged by the presence of certain amides, (A), proposed the following scheme for the anodic reactions:

$$\underset{1-\theta}{Fe(H_2O)} \rightarrow Fe^{2+} + 2e$$

$$\underset{1-\theta}{Fe(H_2O)} \rightarrow \underset{\theta_1}{Fe(OH)_2} + 2H^+ + 2e \quad (1.28)$$

$$\underset{1-\theta}{Fe(H_2O)} \downarrow \underset{\theta_2}{FeOH\ A} + H^+ + e$$

where $\theta = \theta_1 + \theta_2$ is the degree of surface coverage by the reaction products.

The inhibitive action of the amides increased as the potential of the iron was ennobled. However, in the authors' opinion, one could draw no conclusions as to whether this was a consequence of a facilitation of the growth of oxide, or of the formation of adsorption compounds of amides with the oxide which would decrease the ionic conductivity of the oxide. One of the amides used in this work was dimethylformamide. The effect of this on the passivation of iron and other materials merits further attention.

Ogura and Kobayashi[68] found that in 0.1M $NaClO_4$ in mixed dimethyl-formamide (DMFA)–water solvents, iron is in the passive state when $C_{H_2O} < 11.4\%$. This result indicates a passivating action of DMFA impeded by adsorption of ClO_4^-. In our own work[69] with DMFA in borate buffer, we have shown a promotion of passivation by the presence of the DMFA. Of significance in that work was the fact that oxide growth was inhibited in the passive region.

If it is remembered that in a pure borate buffer the passivation of iron has a clearly-expressed oxide character, then the addition of various organic compounds to this solution and the study of the composition of the passivating layers will be important in understanding what properties are required for an inhibitor to provide passivation. As seen from Fig. 1.8, certain cationic-active compounds that are often effective in the inhibition of the

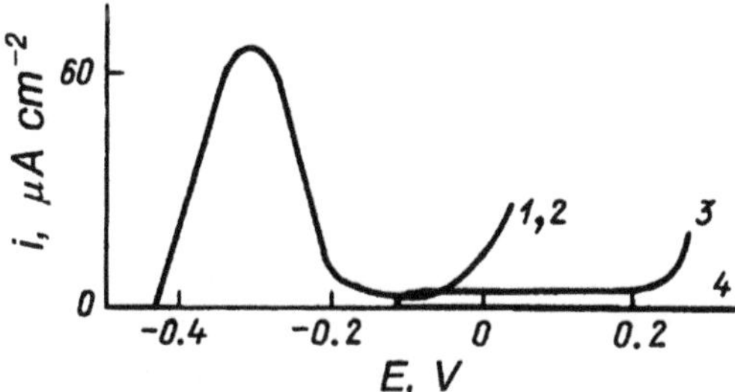

Figure 1.8. Anodic polarization curves of iron in a borate buffer at pH 7.4 containing 0.01M NaCl (1), the oxyethylated alkylphenol OP-7 (2), alkylbenzyldimethyl ammonium chloride (3), and sodium oleate (4).

corrosion of iron in acids are also capable of promoting passivation in neutral solutions. Furthermore, the introduction of simple amines (triethylamine, piperidine, and others), even in conditions of natural aeration but without alkalization of the solution, has been shown to be incapable of converting iron into the passive state if the oxide film was first removed from the surface by cathodic polarization. The effect of aniline on the passivation critical current density i_p (Fig. 1.9) is significant. Thus, in the region of relatively low concentrations, aniline increased i_p so that $\gamma_p = i_p^{background}/i_p^{organic} < 1$ and only slightly facilitated passivation ($\gamma_p > 1$) at higher concentrations. Such a dependence of γ_p on concentration is also found with benzhydroxamic acid, which is practically undissociated in these conditions ($pK_a = 8.89$). Stronger organic acids, which are also benzene derivatives, exist in solution as anions (phenylarsonic, benzoic) and impede the passivation of iron. The polar groups in such anions are very hydrophilic and the hydrophobicity of the benzene nucleus is insufficient for conferring

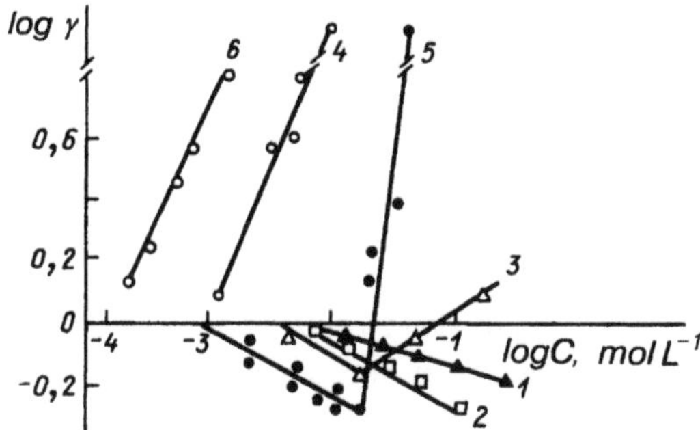

Figure 1.9. Effect of the concentration of substituted benzenes C_6H_5X on the passivation current density of iron in borate buffer at pH 7.4 containing 0.01M NaCl, $\gamma p = i_p^{background}/i_p^{organic}$: (1) benzoate, (2) sodium phenylarsonate, (3) aniline, (4) phenylborate, (5) benzhydroxamate, (6) sodium phenylundecanoate.

a passivating action on the adsorbing particles. In such a case the particles of adsorbate evidently remain outside the dense part of the electrical double layer and will form soluble compounds with the ions of the metal. Passivation can therefore be promoted only with high concentrations of the additive as a result of the supersaturation of the near-electrode layer by soluble complex compounds (salts) or the products of their hydrolysis. On the other hand, the double role of water in the dissolution and passivation of the metal indicates that its displacement by organic molecules or ions can also effect a replacement of a stimulating action ($\gamma_p < 1$) by a passivating action ($\gamma_p > 1$) of the additive. This is observed during the passivation of not only iron but also of other metals. It should be noted that the less hydrophilic additions lead to an increase of γ_p at lower concentrations. It is not surprising that of the substituted benzenes, the sodium salt of phenylundecanoic acid is particularly effective. Close to this in passivating action is the well-known corrosion inhibitor and anionic surfactant—sodium oleate (Fig. 1.8).

It does not follow, however, that imparting sufficient surface activity to an organic compound will make it capable of effective passivation. For example, additions of oxyethylated alkylphenols [corresponding to the USSR type OP-7, see Table 3.5] which are nonionic surface-active compounds do not significantly lower i_p. The reason for this could be the absence of functional groups in the alkylphenols that are capable of effectively interacting with the surface and thereby being strongly adsorbed.

It is worth recalling that even in the work of Frumkin *et al.*[24] the passivation of metals is explained by changes not so much in the structure of the double layer but rather in the chemical nature of the surface. The latter is often achieved with the adsorption of anions, with their bonding to the surface strengthened as the potential moves to more positive values. Consequently, anionic-active inhibitors with tendencies to chemisorption should promote adsorption passivation of iron in neutral media. According to Frumkin,[7] these inhibitors hydrophobize the surface—an action which can also impede the growth of the oxide film. In this respect a significant effect should be expected from anions containing a substantial number of hydrocarbon radicals. This is in agreement with the views of Antropov[5] on the possible blocking effect in the case of adsorption of large, easily-polarizable anions. Nevertheless, in 70 years not one case has been described in the literature where passivation of iron in neutral solution has been achieved without the formation of an oxide film.

It seemed unlikely to us that hydrophilic oxidizers, compounds with high electron donor capabilities, and those with hydrophobizing properties, could encourage oxide growth. Among inhibitors of the last type, sodium phenylanthranilate, $C_6H_5NHC_6H_4COONa$ [PAN] in particular, can more effectively promote passivity of iron in deaerated solution than the well known passivator, sodium nitrite.[70] It is logical to suppose that in this case the determining role will be played by chemisorption of the actual inhibitor, which suppresses the dissolution of iron over a wide range of potentials.

If a potential corresponding to the passive state is applied to a cathodically-reduced surface and the thickness of the oxide that is formed is determined coulometrically at various times, the contribution of the organic compound to the formation of the passivating layer can be estimated. With a galvanostatic reduction of an electrode after 540 sec at $E = 0.2$ V (a usual steady-state value for the complete protection of iron) in a buffer solution, three waves are observed. One of these, beginning at $E = -0.5$ V, corresponds to the discharge of H_3O^+ (Fig. 1.10). The two other waves ($E = -0.05$ to -0.23 V) can, by using the data in Refs. 50, 51, be explained by the reduction of γFe_2O_3 or some oxide close to this in composition. According to Sato *et al.*[71] the reduction is initially of the outer layer, a hydrated oxide, by the reaction $Fe_2O_3{\cdot}nH_2O + 6H^+ + 2e \rightarrow 2Fe^{2+} + (n+3)H_2O$ and then of the inner layer, conditionally described as γFe_2O_3.

Significantly, all the organic compounds we studied did not alter the potentials for the reduction of the surface oxides, although to various extents they affected the quantity of electricity (Q_{ox}) necessary for the

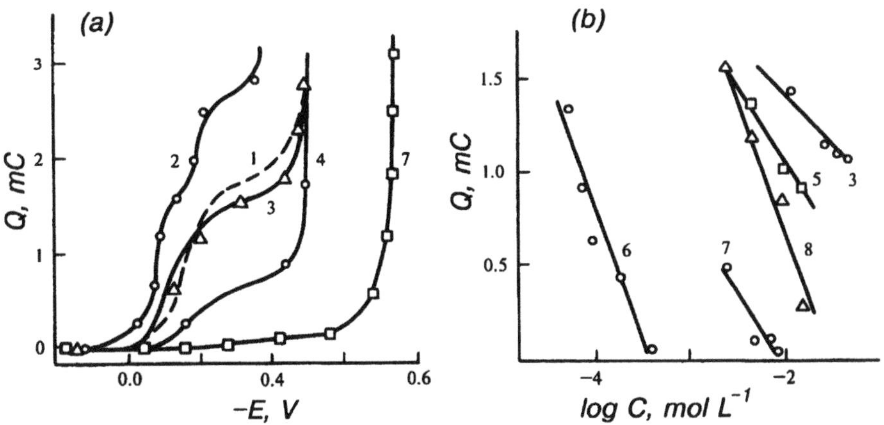

Figure 1.10. The dependence of Q, measured during the galvanostatic reduction of iron passivated for 540 sec at $E = 0.2$ V in a borate buffer at pH 7.4 (1), containing $NaNO_2$ (2), sodium benzoate (3), sodium anthranilate (4), sodium diethylaminopropionate (5), sodium oleate (6), sodium phenylanthranilate (7), and diethylaminopropionic amide (8), on the potential of the electrode (*a*) and the concentration of the additive in mol L^{-1} (*b*).

reduction. Furthermore, in the presence of nitrite an arrest was observed at $E = -0.37$ V which was associated with the reduction of the inhibitor on a surface that would be already free from the passive film.

In nitrite solution both oxide arrests were increased, whereas sodium benzoate and sodium anthranilate decreased Q_{ox} and consequently the thickness of the oxide film. An increase in concentration of, for example, benzoate, had no marked effect on the value of Q_{ox}. The sodium salt of diethylaminopropionic acid (DEAK) had an analogous effect on the passivation of iron. However, for the amide of this acid, the noticeable increase in the anti-oxidizing action with increase in its concentration led to a more effective reduction in the passivation current. Moreover, in the absence of primary oxide films, all these organic compounds were unable to suppress the corrosion of iron, with benzoate even stimulating the active dissolution, particularly in deaerated solutions. This would have pointed to an oxide nature of passivity if in solutions of sodium phenylanthranilate there had not been a more abrupt decrease in Q_{ox} as the achievement of passivity was facilitated. Thus, with the concentration of phenylanthranilate, $C_{PAN} > 0.02$ M, all the signs of the passive state are present even in deaerated solutions: an independence of anodic reaction rate on potential, extremely low values of current, and $Q_{ox} \to 0$.

It should be noted that sodium benzoate—which is significantly below phenylanthranilate in hydrophobicity and chemisorption tendencies—has no noticeable effect on the oxidation of iron. The introduction of an $-NH_2$ group (sodium anthranilate) increases the electron-donor properties of the adsorbate. This impedes the release of metal cations into the solution, but not to the extent that the oxidation of iron is suppressed, since its surface remains hydrophilic. An analogous effect is shown by derivatives of DEAK, although its amide, being a more hydrophobic addition, strongly inhibits the growth of the oxide film.

Phenylanthranilate even at low concentrations affects the iron-oxidation kinetics. In a deaerated buffer, Q_{ox} has a logarithmic dependence on the time of oxidation (Fig. 1.11). However, in a 2.5 mM solution of phenylanthranilate, after a short period of film growth (900 sec) its thickness does not change further, that is, the phenylanthranilate stabilizes the thickness of the oxide at a certain level.

The value of the potential in the passive region also affects the thickness of the passive film to differing extents depending on the presence of the inhibitor. In a pure buffer it increases linearly with increase in anodic potential E ($t = 900$ sec), whereas in an inhibited solution the potential has practically no effect on the quantity of electricity Q_{ox} expended in the reduction of the oxide. In agreement with ideas being developed, the curtailment of oxide growth upon reaching a certain thickness is the result

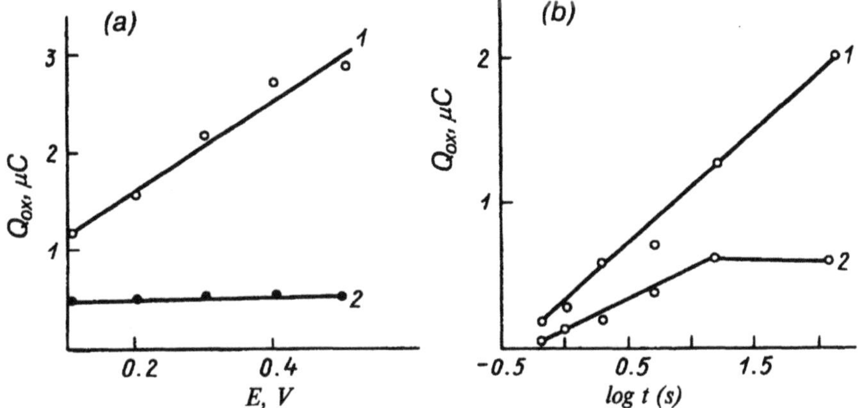

Figure 1.11. The effect of the potential (*a*) and time of passivation (*b*) at $E = 0.2$ V on Q_{ox} in a borate buffer at pH 7.4 (1) and containing 2.5mM phenylanthranilate (2).

of the accumulation of hydrophobic complexes of iron in the passivating layer with phenylanthranilate anions thereby preventing the penetration of water molecules and the formation of an oxide film. The limiting thickness of the oxide should be determined by the concentration of phenylanthranilate and is weakly dependent on potential.

The important role of the hydrophobicity of organic anions in the adsorption passivity of iron should include recognition of the even more effective suppression of oxidation of the metal by sodium oleate since this can be achieved at concentrations even lower than the critical concentration for micelle formation. In fact, the decrease in the rate of oxide formation brought about by an increase in concentration of these anions above a certain value is such that the oxide cannot be detected coulometrically.

This situation is also true for solutions with natural aeration, which often facilitates the passivation of iron by inhibitors of the nonoxidizing type. In a simple borate buffer at pH 7.4, an iron surface free from oxide can be passivated only upon displacing the potential to more positive values. However, with a sharp change in potential from the region of active dissolution up to $E = 0.2$ V, the passive state is reached in 30 min, by which time, according to coulometric measurements, a layer of oxide of 3–5 nm thickness has been formed (Fig. 1.12). With time the rate of film growth continuously decreases and follows an inverse logarithmic law, as does the formation of thin passivating films, in which the reaction of metal ionization is the rate-determining step.[44]

The presence of phenylanthranilate in a borate-buffer solution accelerates attainment of the passive state on switching the potential from the

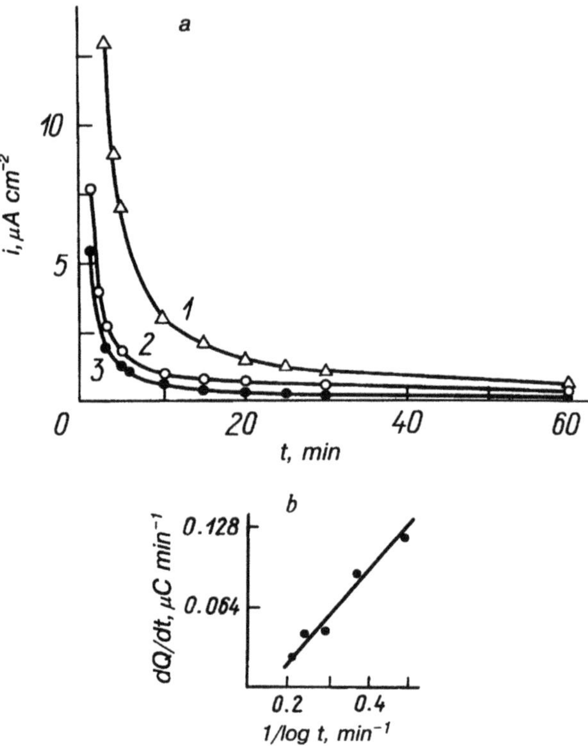

Figure 1.12. Changes in the anodic current density on switching from $E = -0.42$ to $+0.2$ V (a), and the rate of oxide growth on iron with time (b), in a borate buffer without (1) and with addition of 0.05M (2) and 0.2M (3) phenylanthranilate.

active to the passive region.[45] Thus, the introduction of 0.05 M and 0.2 M phenylanthranilate leads to the attainment of a current density of $i = 1$ μA cm^{-2} after 10 min and 6 min, respectively. Coulometric measurements at $E = -0.05$ to $-$ 0.23 V showed no reduction wave even after holding the electrode in a phenylanthranilate solution of $C \geqslant 0.05$ M at $E = 0.2$ V for 15 min. This wave appeared only at significantly longer times, and only after $t = 120$ min did Q_{ox} correspond to a coating of thickness 0.1–0.2 nm.

Electron diffraction analysis of specimens exposed to solutions of phenylanthranilate failed to show any lines that were not characteristic of structures of oxides obtained on specimens from solutions without phenylanthranilate, even when the analysis was conducted after only a few minutes of immersion. This indicates that the passivation of iron is not connected with the formation of a phase film of complex compounds or salts of an organic inhibitor. Furthermore, even after oxidation of the

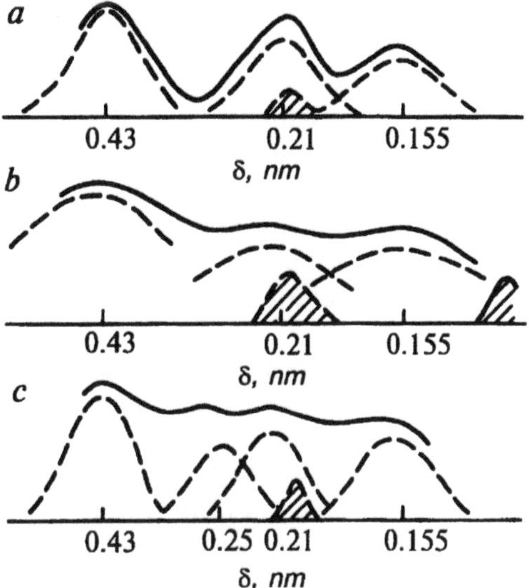

Figure 1.13. Electron diffraction intensity curves as in Fig. 1.6. Diffraction pictures were obtained from specimens after passivation of iron for (a) 180 min in 0.5 M phenylanthranilate; (b) 120 min in 0.2 M phenylanthranilate; and (c) 120 min in 0.0025 M phenylanthranilate.

electrode for 180 min in 0.05 M phenylanthranilate solution, the electron diffraction picture (Fig. 1.13) showed α-Fe reflections and the oxide film thickness did not exceed several atomic layers (< 1 nm). The oxide line for 0.25 nm, which is characteristic of a three-dimensional ordered oxide structure, was absent and the intensity distribution of the diffraction signal was best described by diffraction in a two-dimensional defect structure with a characteristic dimension for the ordered region of ~1 nm. A defect-free structure should have a 0.3 nm line, which was absent in these diffraction patterns.

The electron diffraction data also confirmed the dependence of the thickness of the passive film on the phenylanthranilate concentration, C_{PAN}. Thus, at 120 min, the intensity of the αFe reflections decreased with a decrease in solution concentration from 0.2 to 0.0025 M. A characteristic feature was that in dilute phenylanthranilate solution, where its passivating action was relatively small, the structure of the film was very close to that of the oxide formed in the pure borate buffer. With higher concentrations of phenylanthranilate, the oxide formation was of a more isolated nature, in the form of islands, which with time slowly increased the surface coverage. This behavior, together with phenylanthranilate adsorption, weakened the αFe reflections in the diffraction patterns.

X-ray photo–electron studies of the surface layers of passivated iron also show significant differences in their degrees of oxidation.[72] A broad peak on the spectrum obtained from a specimen held in a buffer solution (Fig. 1.14) is characteristic of that for Fe^{3+} oxides, but the absence of any distortion in its shape in the low-energy region indicates a significant (> 3nm) oxide thickness. The spectra obtained from a specimen passivated in the presence of phenylanthranilate showed a dip at $E_B = 707$ eV representing a signal from Fe°, despite the fact that after passivation the electrode had been removed from the solution, rinsed in 100 ml distilled water, and dried.

The ratios determined from the integrals of the oxygen and iron peaks, which also provide a measure of the relative contents of these elements in the layer (Table 1.1), noticeably exceed the values calculated from the stoichiometry of compounds such as Fe_2O_3, Fe_3O_4, or FeOOH. Taking into account the exponential fall in the intensity of the signal with depth, the explanation lies in the enrichment by oxygen-containing compounds of the outer layers of the film. In fact, the rather high O/Fe ratios exceeding stoichiometry are not unusual since the thin surface layers always contain adsorbed water and oxygen species that are not removed in the high vacuum used in the XPS studies. Even after 24 h oxidation in 0.05 and 0.2 M solutions of phenylanthranilate, the signal from Fe° and the low values of O/Fe—compared with the uninhibited solution—are still retained. This finding is consistent not only with a decreased oxidation of the surface but also with surface dehydration and dehydroxylation by the hydrophobic phenylanthranilate anions.

From the ratios of the metallic (E_B = 706.9 eV) and oxidized iron peaks the thickness of the film can be estimated to be 1 nm. However, whereas E_B for the $Fe2p_{3/2}$ line in the spectra of specimens treated in the pure buffer corresponded to that for Fe^{3+} oxides, in the case of a deaerated solution of phenylanthranilate it was 710.2 eV (710.6 eV with natural aeration). These values do not correspond to binding energies in any known oxide of iron and cannot be assigned to a complex of phenylanthranilate with Fe^{3+} in which the binding energy is found to be 713 eV. It is probable that the observed values of E_B are characteristic of adsorption complexes of phenylanthranilate with iron in a lower state of oxidation.

The high adsorbability of phenylanthranilate is confirmed by the presence in the spectra of an N1s peak of electrons with a binding energy of $E_B \approx 400$ eV, which can be associated with a nitrogen atom in the anion. The basis for this suggestion is that the N/Fe ratio in this case is significantly greater than that in the oxidation of iron in a solution without an inhibitor (0.24–0.38 and 0.1–0.15 respectively). Unfortunately, the quantitative assessment of the adsorption of phenylanthranilate from the intensity of the

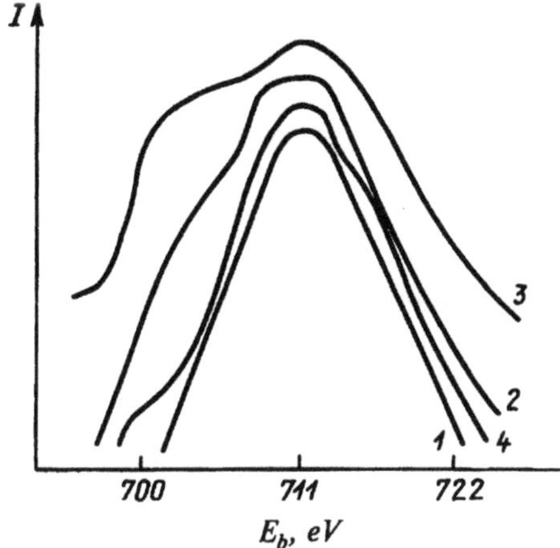

Figure 1.14. Photoelectron spectra $Fe2p_{3/2}$ lines of specimens passivated in a borate buffer solution at pH 7.4 (1), in the presence of 0.005 M phenylanthranilate (2), 0.004 M sodium oleate (3), and in 50% dimethylformamide (4).

N1s peak was difficult because of its coincidence with the lines for nitrogen adsorbed from the air. To explore this further, an assessment of the irreversible part of the adsorption was carried out in the presence of a substituted phenylanthranilate, sodium N-(m-difluormethylthiophenyl)-anthranilate (isodifluorant).

Passivation in a 0.025 M solution of this inhibitor led to a similar retardation of oxide growth on iron as in the case of phenylanthranilate; the form of the electron spectra due to Fe2p electrons in both cases was identical (Fig. 1.14, curve 2). The S2p and F1s peaks present in the spectra belonged to sulphur and fluorine atoms present in the inhibitor and could

Table 1.1. The O/Fe Atomic Ratio in X-Ray Electron Spectra of an Fe Surface after Various Treatments

	Air formed film	Treatment (oxidation at $E = +0.2V$)									
		Background solution pH 7.4	Deaerated 0.2M PAN	Naturally aerated PAN							
				0.02M		0.05M		0.2M			
Treatment time, h		0.25	2	0.25	1	2	24	0.5	1	2	24
O/Fe	4.5	7.2	2.7	3.4	3.7	3.6	3.6	3.3	3.8	3.6	3.6

act as markers. Their intensities remained effectively unchanged on washing the specimen with water and alcohol, thus indicating a strong bond of the inhibitor with the surface and confirming that the N 1s peak observed in the case of phenylanthranilate was also related to the inhibitor. The quantity of chemisorbed inhibitor was estimated from the values of the F/Fe, S/Fe, and N/Fe ratios and found to be sufficient for the formation of at least one monolayer. The same conclusion was reached from a mass spectrometric analysis of the surface layers of iron passivated for 30 min in a 0.02 M buffer solution of phenylanthranilate on argon ion etching of the specimens.[45]

Attention must be drawn to one particular observation made during the study of passive films by this method. This is the marked reduction in the intensity of the OH^- peak and the almost complete absence of an $FeOH^-$ peak in the mass spectra. This points to hydrophobizing and dehydroxylizing actions of the inhibitor. A further characteristic feature was a sharp reduction in the content of borates in the surface layer due to displacement by the inhibitor. Thus, according to the X-ray electron studies, the B:Fe ratio of 2.4 obtained on passivation in the pure buffer was reduced to 0.1 in an 0.02 M phenylanthranilate solution, and in solutions of less than 0.05 M concentration, the B1s peak effectively disappeared.

The results of these studies support our proposed adsorption mechanism for the passivation of iron in relatively high concentrations of phenylanthranilate ($C > 0.05$ M). Unfortunately, the methods used (apart from coulometry) were applied to electrodes that had been removed from the solution. Therefore, despite the valuable information they provide, it would have been more appropriate to conduct further studies using "in situ" methods. Such a system using an automatic ellipsometer[73] has enabled 100 measurements a minute to be made in which the changes in optical parameters with time were recorded as an electrode was electrochemically polarized. To study the kinetics of film growth, the potential of the electrode was stepped from $E = -0.65$ V to $E = 0.2$ V. The differences in the phase angle shift Δ and the polarization restoration angle ψ with time in the pure and the inhibited buffer indicated a difference in the kinetics of formation of the passivating layer on iron. In both cases film formation occurred in the first second. With the inhibitor present, the film thickness reached 1.2 nm in 0.5 sec. Since the size of the molecule is 1.3 nm, it can be assumed that instantaneous adsorption of the inhibitor had occurred. With the oxidation of the iron electrode in a borate buffer in these conditions, the "thickness" of the oxide after 0.5 sec was 0.1 nm. Even clearer differences in the thickness and nature of these layers is seen from a comparison of the nomograms obtained in the plane $\delta\Delta = \Delta - \Delta_0$ .... $\delta\psi = \psi - \psi_0$ where Δ_0, ψ_0 are the initial values of the ellipsometric parameters of the surface [the complex index of reflection of the "pure" surface of iron $n_K = 3.07$–

3.63 (Fig. 1.15)]. In constructing the nomograms in the potentiodynamic regime, potential is taken as a parameter for each point; however, in the potentiostatic regime, time is used. Curves 1 and 2, which refer to iron in phenylanthranilate solution and the pure buffer, respectively, were obtained as a function of potential. Curves 3 and 4 were obtained by switching the potential from $E = -0.65$ to 0.2 V in the inhibited and pure buffer solutions, respectively. Curve 5 was calculated and related to $n_K = 2$. The inflections in curve 2 are associated with various oxide phases on the surface of the iron. The inflection on curve 1 at $E = -0.09$ V (where the light intensity begins to increase) is probably connected with polymeric adsorption. There are no obvious inflections on curve 3 and it is best described by the nomogram for $n_K = 2$. The increase in value of n_K in comparison with that for the phenylanthranilate molecule, calculated through refraction elements and equal to 1.7, is connected with the formation of a chemisorbed complex of iron with the inhibitor, which is in agreement with our conclusions drawn from the X-ray electron studies. The difference in these curves in the $\delta\Delta - \delta\psi$ plane demonstrates the presence of different compositions of films formed on the iron surface.

A shift of potential in the positive direction increases the quantity of adsorbed phenylanthranilate. Bearing in mind that for $n_K = 2$, and the coefficient of proportionality in the Drude equation, $\alpha = 0.66$ nm deg^{-1}, the maximum thickness of film which grows in a linear manner with potential can be calculated as 7 nm. In fact, it reaches 5 nm after only 15 min. Such a thick film was not found in any of the methods described above. The discrepancy in thicknesses can be explained by the washing off of the inhibitor before the specimen was placed in the spectrometer chamber. To confirm the presence of a reversible part of phenylanthranilate adsorption, the electrode was allowed to reach adsorption equilibrium in the solution at $E = 0.2$ V and then withdrawn from the solution, rinsed with water, dried with filter paper, and again placed in a cell, but this time a pure buffer solution. As well as a decrease in $|\delta\Delta|$ at the steady-state potential, there was a change in $|\delta\Delta|$ of 4° upon switching the potential to $E = -0.65$ V, which was almost half that which would be obtained if the switching were in the phenylanthranilate solution ($|\delta\Delta| \approx 8°$). Such a decrease in $|\delta\Delta|$ points to a partial washing away of the inhibitor from the iron surface.

Film formation begins in the first seconds of immersion. This rapid stage can be associated with an initial fast adsorption of inhibitor which, even after only 0.5 sec, leads to a 1.2 nm thickness. Since the size of the organic anion is ~1.3 nm it can be assumed that rapid adsorption of the inhibitor has occurred. On oxidation of iron in the pure buffer, the "thickness" of the oxide was estimated as 0.1 nm. The formation of films with different compositions during oxidation in borate buffer and in the inhibited

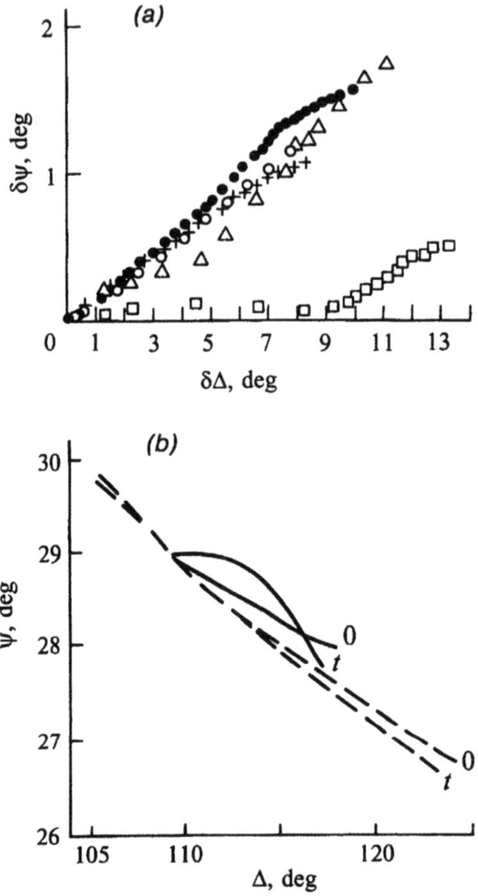

Figure 1.15. The dependence of δΔ–δψ for films on the surface of iron on potentiodynamic polarization (●,□) and after a stepwise change in potential (+,Δ) in a borate buffer (□,Δ) in 0.5M phenylanthranilate (●,+) and theoretically for n_K=2 (○), (a), and also the changes in Δ and ψ during growth of the film and its removal (b). Solid line, 0.05 M phenylanthranilate; dashed line, borate buffer. Curve 1—●; 2—□; 3—+; 4—Δ; 5—○.

buffer is also confirmed by the $\Delta - \psi$ curves, obtained on removal of these films. Fig. 1.15b shows that on a $\Delta - \psi$ plane the changes in Δ and ψ with film growth (stepped to E = 0.2 V) and during its subsequent removal from the surface (stepped to E = −0.65 V). The index 0 refers to the initial values of Δ and ψ and the index t to the final values established at E = −0.65 V. The full curve describes the growth of the film and its removal in phenylanthranilate solutions. In the removal process a hysteresis loop appears and this, together with the thinning of the film in the background solution at the steady-state potential, confirms the presence of a polymolecular layer on the solution side of the film. Figure 1.15 shows a similar curve for the growth and reduction of the oxide in the pure borate buffer (broken line). The electrode was held at $E = 0.2$ V for about 40 min. The oxide was reduced without a hysteresis loop appearing.

Thus, studies of the iron surface by several independent methods have confirmed the possibility of adsorption passivation of metals in neutral solutions by organic compounds possessing high electron-donor and hydrophobizing properties. Not only phenylanthranilate but also other compounds with similar properties, besides facilitating passivation, also act as antioxidizers of iron, e.g., sodium oleate with its known hydrophobizing properties. Sodium oleate passivates iron and prevents its oxidation at concentrations even lower than those of phenylanthranilate. The spectra from a specimen passivated in an oleate solution clearly separates the low energy component of the $Fe2p_{3/2}$ electrons caused by the signal from Fe° (Fig. 1.14). As in the case of phenylanthranilate, it may be assumed that the oxidized iron peak in the spectra corresponds not to an oxide but to a compound of oleate with iron. This proposition was confirmed by a study of the reflection–adsorption infrared spectra of an iron surface.[74] This showed that difficultly-soluble compounds of the inhibitor with iron can form by either precipitation from solution (in the case of Fe^{2+} accumulation) or interaction of oleate directly with the metal surface. It is of interest that the role of the adsorbed oleate layers in the passivation of iron is large enough for the presence of a primary oxide on the metal, which usually increases the effectiveness of inhibitors, to have no beneficial effect in oleate solutions. Furthermore, the ability of adsorbed layers of oleate to polymerize on storage of specimens in air increases the irreversibility of the adsorption, which thereby increases the effectiveness of these layers in the protection of metals from atmospheric corrosion.

1.4.3. Passivation of Metals by Chelating Reagents

As shown above, the adsorption of organic compounds has an especially beneficial effect on the passivation of a metal when the anions involved are sufficiently hydrophobic. This is understandable because, on the one hand, there is a strong "squeezing out" of the anions from the solution onto the surface and, on the other hand, they are pinned to the surface as a result of their significant nucleophilic properties. Moreover, the interaction of such inhibitors with ions of the dissolving metal is more likely to take place on the surface of the metal, where the concentration of ions will be well in excess of that in the solution. The products of such a reaction are usually of low solubility and provide only a supplementary blocking action; it is therefore difficult to expect the formation of thick films of salts or complexes from these anions.

A different situation exists with hydrophilic compounds even when these are capable of forming stable complexes with ions of the dissolving metal. First, it is more difficult for these to form directly on the metal surface because of the high affinity of the organic additive for water, which thus prevents its entry into the dense part of the electrical double layer. Second, such complexes or salts cannot exert a blocking action because they are insufficiently insoluble. In this case, if there is a sufficiently high concentration of the organic reagent and a low pH (since the competitive formation of extremely stable hydroxy complexes must be taken into account) it is possible to form complexes (salts) in the near-electrode layer which, on reaching a certain limiting solubility, will precipitate on the surface—thus leading to a salt passivation mechanism.

An example is the well-known oxalate treatment of steel as used in industry. The production of oxalate coatings is used even on stainless steels for improving cold forming. As in the case of phosphate coatings, they provide strong adhesion to the metal and act as a good sublayer for lubricants, as well as facilitating wire drawing, tube forming, and so on. The oxalate anion is a well-known complexing reagent (for $Fe^{2+}L$, $\log K_s = 3.05$ and for $Fe^{3+}L$, 9.4), and provides salts with Fe^{2+} of relatively low solubility, $SP = 2 \times 10^{-7}$.(75) In 0.5M $H_2C_2O_4$ iron has two regions of passivity.(76) The primary passivation potential is 0.5 V more negative than that for the passivation of iron in sulphuric acid solutions at the same pH (Fig. 1.16). On passivation, the surface is covered by a dark grey layer of oxalate, $FeC_2O_4 \cdot 2H_2O$, with a thickness of several microns. The protective properties of this film are to a large extent determined by the conditions of its crystallization. Thus, a high rate of entry of Fe^{2+} into the solution, which will provide rapid anodic polarization, leads to a considerable reduction of currents in the passive state. On the other hand, factors which lower the near-electrode concentration of Fe^{2+}, such as the stirring of the solution or the presence of oxidizers or anions forming soluble salts with Fe^{2+}, will impede passivation. A long retention of the electrode at primary passivation potentials lowers the protective effect because of recrystallization of the deposit or reduction of oxalate. An analysis of data obtained from polarization and impedance measurements leads us to the conclusion that the oxalate layer is formed by precipitation from solution. It is of interest that on holding the electrode in the secondary passivation region, that is, at more positive potentials, signs of salt passivation are absent and the content of oxalate in the film then does not exceed 0.01 mg cm^{-2}.

Sukhotin and Bereskin(77) constructed a Pourbaix diagram for the Fe–H_2O–0.21M $H_2C_2O_4$ system and compared it with the diagram for this

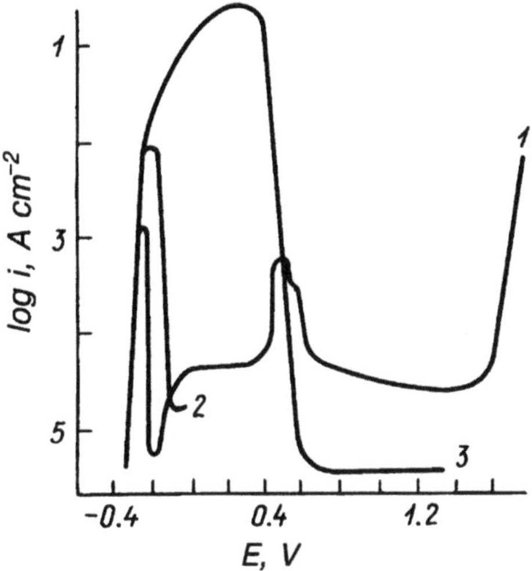

Figure 1.16. Potentiodynamic anodic curves for iron in 0.5 M $H_2C_2O_4$ (1,2) and in 1.0 N H_2SO_4 (3). Scanning rate (1) 1.2 V h^{-1} and (2) 12 V h^{-1}.[76]

system without oxalate. They found that in the former there was not only a wide range of potentials touching both the acid and neutral solution regions and characterized by $FeC_2O_{4(solid)}$ stability (salt passivity), but also a change in the stability regions of the oxides. Thus, the cathodic boundary for γFe_2O_3 stability:

$$\gamma Fe_2O_3 + 6H^+ + 2e = 2Fe^{2+} + 3H_2O \tag{1.30}$$

in the presence of oxalate was displaced to more positive potentials and the corresponding line on the diagram intersected the line for the transpassive reaction of iron at pH 2.37:

$$\gamma Fe_2O_3 = 2Fe^{3+} + 1.5\ O_2 + 6e \tag{1.31}$$

At pH < 2.37, an electrochemically stable condition for $\gamma Fe_2O_{3(solid)}$ was not possible in oxalate solution.

In carrying out a further electrochemical study with a special electrode of powdered FeC_2O_4 in an aerated solution of oxalic acid, the authors concluded that the increase in anodic current on curve 1 of Fig. 1.16 was caused by the breakdown of the salt film as a result of its intensive oxidation. In this region of potentials there is the possibility of formation of a passivating oxide:

$$2Fe + 3H_2O = \gamma Fe_2O_3 + 6H^+ + 6e \quad (1.32a)$$

$$2Fe_3O_4 + H_2O = 3\gamma Fe_2O_3 + 2H^+ + 2e \quad (1.32b)$$

At $E > 0.99$ V the oxide becomes resistant to reduction by reaction (1.30) and on the polarization curve a horizontal section of passivity is traced out, with oxygen evolution beginning at E 1.5 V.

Unfortunately, we do not have this much information for inhibitor-chelating agents, which also form difficultly-soluble compounds with metal ions. Among such inhibitors, there is no doubt that passivation is connected with the formation of complex compounds and less frequently—oxides on the metal surface. Thus, Sastri and Packwood[78] in studies using ESCA, Auger-, and Mossbauer spectroscopies of the composition of the surface films on steel during its inhibition by 8-hydroxyquinoline showed that the passive condition was caused by the formation of a two-layer protective film consisting of an outer layer of adsorbed inhibitor and an inner layer of Fe^{3+} 8-hydroxyquinolate. Practically no oxides of iron were detected under this film. In neutral and weakly-alkaline aqueous solutions, the inhibiting action of 8-hydroxyquinoline was more marked than in weakly-acid conditions—an effect associated with the dependence of the stability of the complex on pH. Another point of view on the mechanism for the occurrence of the passive state is given by Voshida *et al.*[79] and Klyuchnikov *et al.*[80]. According to them, a two-layer structure of the passivating layer is not necessary; achievement of the passive condition requires only the presence on the metal of complex compounds of the inhibitor with Fe^{3+} or Zn^{2+}. The minimum solubilities of the hydroxyquinolate complexes of copper, zinc, and aluminum (but not of iron) are reached at pH 4.0.

It is known that the presence of a sulphur atom in heterocylic chelating reagents often intensifies interactions with metals as a result of the formation of coordination bonds with the metal ions. The best known of these inhibitors is 2-mercaptobenzothiazole [MBT] or 2-mercaptobenzimidazole [MBI]. Studies using Auger, ultra-violet, infra-red, and X-ray electron spectroscopy, laser mass spectrometry, nuclear magnetic resonance, as well as crystallographic investigations of the protective films on metals after contact with aqueous solutions of MBT and MBI at pH 6–10 have shown that the inhibitors form insoluble complex compounds with Fe^{2+},[81] Zn^{2+},[82] or Cu^+,[83] which inhibit the diffusion of O_2 to the surface. Chadwick and Hashemi[83] suggest that in the complex that is formed, monovalent copper is bound to the pyrrole N since in solution both inhibitors exist in the tautomeric form.

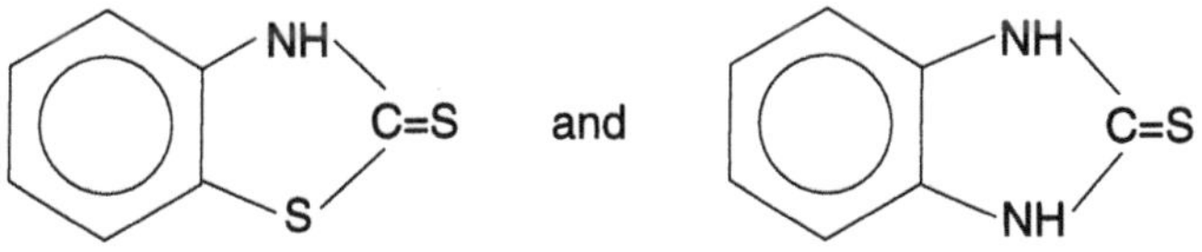

However, in air these complexes are oxidized and dimerized, e.g. Cu^{2+} and

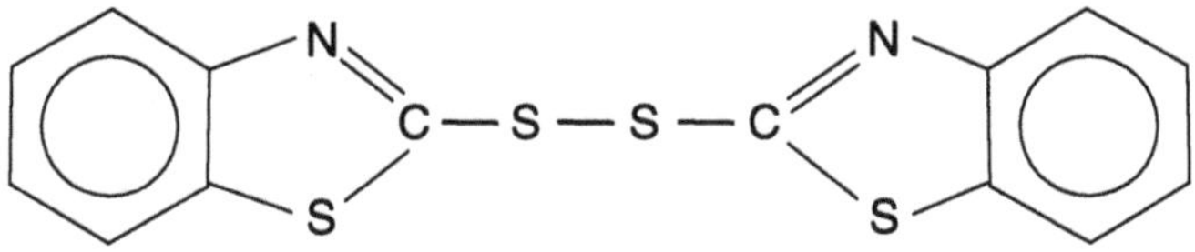

dibenzthiazolyldisulphide [DBDS] are formed from the MBT complex. According to Boffardi[84] and Ohsawa and Suëtaka,[85] the DBDS bands in the infra-red spectra disappear after ultrasonic treatment of the copper specimens in distilled water, although the intensity of the Cu^+MBT remains unchanged. It is suggested that at pH 5–10, DBDS forms an outer layer on the Cu^+MBT film by deposition from the solution. The protective film of Cu^+MBT forms on the Cu_2O surface during the first stages of corrosion and grows by reaction of Cu^+ diffusing through the film of Cu^+MBT or Cu_2O. Even in quite acidic solutions (pH 1–3), MBT forms a protective layer on copper, which according to infra-red studies, is composed of Cu^+MBT and DBDS, with the latter always disappearing after washing. Despite the fact that it is difficult to associate such protection in acid solutions with passivity (only the cathodic process is effectively inhibited), attention must be drawn to the wide range of pH in which a film of a complex of MBT with Cu(I) can exist. The latter is insoluble not only in water but in the majority of organic solvents. This, together with its stability at elevated temperatures (~200°C), has led some investigators to propose that the complex film on copper is of a polymeric nature.

In general, the point of view which is extremely widely held maintains that the passivating layer on metals in solutions of heterocyclic chelating agents is a polymeric complex. A demonstration of this comes from the results of studies of one of the best known inhibitors of the corrosion of copper, benzotriazole, BTA. According to Lynch *et al.*,[86] the initial stages of the interaction of BTA with copper are formally described by the Langmuir isotherm which assumes a linear correlation between the degree of surface coverage and the effectiveness of inhibition. The energy of adsorption of

BTA on copper is 2.5 and that of desorption 41.8 kcal mol^{-1} which points to a high degree of irreversibility of the adsorption. Data from infra-red and X-ray photoelectron spectroscopy, coulometry, and impedance measurements[87–90] of copper or brass electrodes in 1–3% solutions of NaCl at pH 3.3–13.0, $t = 25–130°C$, containing 0.1–25mM BTA, provide a basis for suggesting that the protective film is a linear polymer in which the molecules of BTA and ions of copper occur alternately. In the polymer, the copper forms covalent bonds through the sp-orbitals by substitution of the hydrogen of the NH groups, as well as a coordination bond with the free electron pair of the nitrogen atom of another molecule of BTA.

By using electron paramagnetic resonance, Vysokovskaya and Belousov[91] attempted to explain why copper was in the univalent state in polymeric films of complex $(CuBTA)_n$ compounds. It was found that during the interaction of cations with BTA anions, a one-electron oxidation of the latter occurred, leading to the formation of the corresponding azolyl-azo-centered radicals. In solutions of Cu^+ salts, the oxidation of BTA anions does not occur and coordination polymers are formed. Consequently, BTA anions are reducing agents for Cu^{2+} cations.

According to Youda *et al.*[92] the polymer complex $[Cu^+(4\text{-hydroxy BTA})]_n$, as distinct from $[Cu^+(BTA)]_n$, does not have a linear but rather a 2- or 3-dimensional structure as a result of the occurrence of additional hydroxyl bonding of the inhibitor with water molecules. ESCA (electron spectroscopy for chemical analysis) and laser mass-spectrometer data indicate that the covalent bonding in the $[Cu^+\text{-(mercapto-BTA)}]$ complex is formed by substitution of the hydrogen of the SH group and the coordination bond by the interaction of copper with nitrogen atoms.[93]

In spite of significant differences in the composition of the passive films, it remains difficult to separate the passivating effects arising from adsorption of the inhibitor, the formation of difficultly soluble complexes, and the oxidation of the metal. In view of this, the protection of copper by chelating agents deserves special attention. Copper oxides are electron conducting and the formation of Cu_2O, for example, does not guarantee passivation of the metal, even in neutral media. Many data showing that copper is in the monovalent condition in passivating films suggest that its protection by chelating agents can be considered as a special variant of salt passivity. In the majority of cases, such protection is associated not so much with an increase in the energy of activation for metal dissolution, but rather with a significant blocking action. The passivating properties of films formed by chelating agents can exceed those of oxide–hydroxide types. Nevertheless, it is evident that, irrespective of the mechanism of passivation, the stability of the surface layers towards local depassivation must play an important role in the corrosion behavior of a metal.

1.5. BREAKDOWN OF THE PASSIVE STATE OF A METAL AND ITS PREVENTION BY INHIBITORS

It is widely held that with the possible exception of conditions involving high temperatures and pressures, the local depassivation of metals and alloys takes place only in the presence of anions and at potentials more positive than a certain critical value for the particular conditions, E_{pit}. The metal surface condition, the presence in the metal of various inclusions, and the temperature of the solution may play an important part, but overall the most important factor is the presence of activators in the solution. Various anions can behave as these activators—even OH^- in the case of some amphoteric metals[94] although halides have been the most widely studied.

For example, no pitting is observed on aluminum in pure water (resistivity 1 Mohm cm) even with anodic polarization up to 10 V, whereas in the presence of only $1 \cdot 10^{-4}$M NaCl pitting occurs at $E=-0.7$ V.[95] According to the work of Marshakov *et al.*,[96] in water rigorously purified from electrolytes, not only nickel, tin, 18:10 Cr–Ni (Ti stabilized) steel, and German silver, but also zinc—which forms a protective film of ZnO—are passivated. However, in the presence of salts in neutral solutions, all these materials can be subject to localized corrosion. Thus, aluminum immersed in neutral solution is effectively passivated and with time its potential moves in the positive direction; this behavior can be explained by the formation of a protective film. However, this situation persists only up to a certain value of potential, which depends on the composition of the solution. Above this value the potential will begin to debase (Fig. 1.17). When this happens it is possible to observe breakdown of the surface at several points, which suggests that this potential corresponds to a steady-state potential of pitting

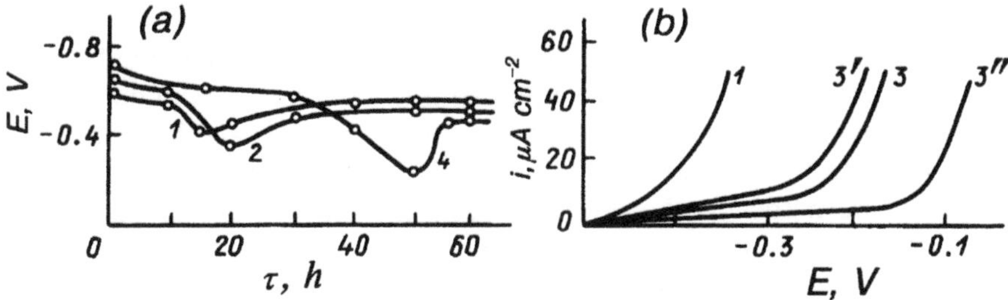

Figure 1.17. Changes in potential with time (a) and polarization curves (b) for aluminum in a borate buffer (pH 8.08), containing 10 mM NaCl (1) 5 mM phenylanthranilate (2), 10 mM phenylanthranilate (3,3′,3″) and 20 mM phenylanthranilate (4). Curves 3′ and 3″ were taken after holding the electrode for 1 h in solutions without phenylanthranilate or NaCl, respectively.

formation E_{pit}^{st}—the most positive potential in Fig. 1.17a. When the concentration of the activator is increased, E_{pit}^{st} becomes more negative and the time τ' to achieve E_{pit}^{st} decreases. That is, there is an increase in the rate of the depassivation, $V'_{pit} = 1/\tau'$, which evidently takes place at defects in the primary oxide.

Pryor[97] associated these initial stages of depassivation with the replacement of oxygen ions by chloride ions in the aluminum oxide lattice. This occurs preferentially at emerging dislocations, grain boundaries, or impurities, and leads to a nonhomogeneous distribution of cationic vacancies. In regions where the concentrations of vacanies are high, there is an increase in transport of Al^{3+} through the oxide. Bearing in mind that when the pH is higher than 3, the specific adsorption of hydroxyl ions predominates over the protonization of the surface, the reaction $Al^{3+} + OH^{-}_{ads} \rightarrow Al(OH)^{2+}$ can be considered to be the rate-controlling step with the induction period being the time necessary for the migration of the activator through the passive film. However, analyses of passive films have given results which contradict this approach in that they do not confirm the entry of aggressive ions into the oxide lattice. Furthermore, it would seem unlikely that such large anions as ClO_4^-, CH_3COO^-, and $AlkCOO^-$ could act by the proposed mechanism. Moreover, it is important to note the important role of defects, which as a term is not just restricted to mechanical damage of the oxide. According to Richardson and Wood,[98] depassivation also occurs as a result of the interaction of an activator with the surface of the metal at defects in the passive film, which although relatively inert in itself, breaks down because of some undermining as a result of pit development. Finally, in the general case the oxide film should not be considered to be inert and the possibility of changes in its structure, if not of changes in thickness, needs to be taken into account. Nevertheless, in the initial stages of depassivation it is the defect condition of the film that plays the important role. This is also indicated in studies[99] of E_{pit} of various microareas of the surface using a special capillary. It appears that on an appreciable part of the surface, E_{pit} is more positive than the potential of the previously formed oxide film, E_f, and since with increase in E_f the film can thicken, then E_{pit} can depend on the thickness of the oxide. However, on aluminum, for example, there were areas with lower values of E_{pit} which on polarization of the whole electrode could establish the usual value of E_{pit}. A feature of such areas is their increased electron conductivity, as a result of which current can pass through them on anodic polarization without leading to oxide thickening. If on the remainder of the surface the potential drop occurs in the growing layer of oxide as the potential is raised, then in the defects it will be partly localized in the electrical double layer. This establishes a pre-condition for the adsorptive displacement of passivating oxygen-containing particles by aggressive anions with the subsequent dissolution of the electrode.[100] Strong support for such a mechanism is provided by the fact that prevention of depassivation of niobium can be achieved by a special treatment of its surface to provide a defect-free oxide film.[99]

Kolotyrkin and co-workers[101] point out that depassivation should not necessarily be connected with the random occurrence of defects since for a number of metals and alloys, nonmetallic inclusions provide the most reproducible and active centres for pit formation. A reduction in the amount of inclusions in alloys is a prospective method for combating pitting corrosion, although pits can also nucleate at particles of intermetalloids.[94] Thus, various factors which can cause heterogeneities in the energy of a passivated surface can be the cause of its depassivation; adsorption phenomena are the basis for this.

Sato[102] has suggested that the actual adsorption of an activator can lead to the formation of defects in the film as a result of a decrease in surface tension and an increase in internal film pressure. How often this occurs remains unclear but undoubtedly adsorption of anions is always only the first step in their interaction with the surface which eventually will lead to either the formation of soluble compounds upon reaction with the metal or to the metal remaining passive.

In this connection, E_{pit} is sometimes considered to be the sum of the standard electrode potential for the formation of a soluble compound of the metal with the anion and some kinetic factor. This approach was apparently first used by Kaesche[103] in the case of the pitting of aluminum in halide solutions. It is based on the fact that a saturated solution of aluminum halide forms in the pit by the reaction:

$$\mathrm{Al} + 3\mathrm{Hal}^- + x\mathrm{H_2O} \rightarrow \mathrm{AlHal_3}\cdot x\mathrm{H_2O} + 3e$$

and he proposed the equation:

$$E_{pit} = E^{\circ}_{\mathrm{Al/AlHal_3 xH_2O}} + \frac{RT}{F}\ln(a_{\mathrm{Hal}^-})_{pit} + \eta_{\mathrm{Al}} + E_{dif} \tag{1.33}$$

where $(a_{\mathrm{Hal}^-})_{pit}$ is the Hal^- activity in the saturated $\mathrm{AlHal_3}$ solution, η_{Al} is the overvoltage of the electrode reaction at a current density I_{pit} for dissolution at the bottom of a pit, and E_{dif} is the diffusion potential at the mouth of the pit. Later, Vermilyea[104] analyzed in detail the views on E_{pit} as a potential at which the formation of a salt becomes preferred to the oxidation of the metal and noted, first, that these were inappropriate for describing the depassivation of some metals (e.g. zirconium) and second, that it was not possible to predict which anions could be considered to be activators in this case. It should be emphasized that later, Kaesche also recognized that there was insufficient basis for this "quasi-thermodynamical approach" in which E_{pit} is treated as a potential characterizing a functional "initiation" of pitting. The initial stages of depassivation can be distinguished in their nature from subsequent processes of the dissolution of the metal. For the inhibition, or for prevention of depassivation, the initial stages are particularly important since once the pit is actively functioning, there will be additional obstacles to the adsorption of the inhibitor (changes in pH, formation of corrosion products, concentration of the activator, etc).

A model of competitive adsorption between the activator and passivating species of the solvent has been particularly widely used for the treatment of E_{pit}. Schwabe[105] suggested that the adsorption of Cl^- follows the Temkin isotherm and depends on potential with a Boltzman distribution, but that adsorption of the competing OH^- ions is independent of potential.

From the equation he obtained:

$$E_{pit} = K_1 \ln \frac{a_{OH^-} + a_{Cl^-}}{a_{Cl^-}} + (K_1 - K_2) \ln a_{Cl^-} + \text{const} \tag{1.34}$$

it follows for the particular case for acid and neutral media, i.e. with $a_{Cl^-} >> a_{OH}$ that

$$E_{pit} = a - b \ln a_{Cl^-} \tag{1.35}$$

where a and b are values which remain constant for varying a_{Cl^-} but dependent on the nature of the metal, the temperature of the medium, and so on.

Strehblow and Titze[106] interpreted E_{pit} as the potential which, for a given bulk concentration of activator, would provide a critical degree of coverage, θ_{crit}, of the prospective pitted surface. They proposed that with Langmuir-type adsorption the activator enters the double layer, performing only the electrical work $\gamma zF\Delta\phi$, where γ is the electrosorption valency which takes into account the extent of the possible transfer of charge (in practice the anion can be considered to be a component part of the metal side of the double layer). $\Delta\phi$ is the potential drop in the double layer, and z the charge of the anion. Assuming that an adsorption equilibrium is established at E_{pit} and that the composition of electrolytes in the pit and in the bulk solution are approximately the same, the authors obtained the following dependence for the case of monovalent anions:

$$E_{pit} = \frac{2.3\,RT}{\gamma F} \log \left[\frac{\theta_{cr}}{1-\theta_{cr}}\right] + \frac{\bar{\mu}_{ads} - \bar{\mu}_s}{\gamma F} + \Delta\phi_h - \frac{2.3\,RT}{\gamma F} \log a_{Cl^-} \tag{1.36}$$

where $\Delta\phi_h$ is the potential drop in the double layer of the standard hydrogen electrode, $\bar{\mu}_s$ and $\bar{\mu}_{ads}$ are the chemical potentials of the anions in the electrolyte with activity a_{Cl^-} = 1M and in the adsorbed state with θ = 0.5, respectively. By combining the first three terms of (Eq. 1.36) we obtain the form of (Eq. 1.35). The literature provides many examples satisfying (Eq. 1.35) but, as will be shown in the following chapters, it does not provide a general law for E_{pit}. Moreover, ideas on the competing adsorption of activator and passivator have played an important role in developing theories of inhibition and in effect form the basis for the use of inhibitors in practice.

In fact, the introduction into a solution of an anion which does not cause depassivation but which can compete with Cl^- in adsorption, for example phenylanthranilate, decreases the rate of nucleation of pits and ennobles E_{pit} (Fig. 1.17).[107] The protective effect can be expressed either as $\gamma_{pit} = V'^{,0}_{pit}/V'^{,inh}_{pit}$ or as the difference in the respective values of E_{pit}. In the first case (with C_{inh} = 3mM):

$$\log \gamma_{pit} = 0.10 - 0.05 \log C_{KCl} \tag{1.37}$$

it can be seen that the inhibiting effect decreases by a power law as the concentration of the activator increases. An increase in phenylanthranilate concentration can lead not only to an increase in γ_{pit} but also to a prevention of depassivation in general. In this case it is more convenient to use a criterion such as the minimum concentration of inhibitor necessary to prevent corrosion, C_{min}. The dependence of this on C_{KCl} also has a power form with, for example, the protection by phenylanthranilate of aluminum alloys as expressed by:

$$\log C_{min} = a + b\, C_{KCl}$$

The assessment of the effectiveness of an inhibitor based on the positive shift in E_{pit} has been widely used, particularly since the necessary polarization measurements are easily automated. The value of E_{pit} as usually obtained from anodic potentiodynamic curves is somewhat more positive than E^{st}_{pit}, but the difference between the values with slow rates of potential scanning is small. This, together with the better reproducibility of E_{pit} leads to the wide use of the potentiodynamic method for the comparative evaluation of the effectiveness of inhibitors despite the fact that it is a non-steady-state method. In many cases, to improve the electrical conductivity and the buffering properties (without which it would not be possible to assess the influence of pH and other thermodynamic aspects of depassivating processes), the background electrolyte contains salts of boric acid.

It is important that the components of the buffer should not have an activating action themselves, at least in relation to iron, nickel, cobalt, tin, copper, aluminum, bismuth, or zinc. The components should also not show any significant inhibiting properties. For example, aluminum in borate buffer as in pure water is not subject to localized corrosion for a long period and remains passive over a wide potential range. In the presence of NaCl, its potential even after one hour in the solution is more negative than E_{pit}, but on anodic polarization the passive state of aluminum is broken down. If the aluminum is held in a solution also containing phenylanthranilate, then depassivation would set in at more positive values of E_{pit}. It is immaterial whether the inhibitor is present before immersion of the electrode or added immediately before the polarization: the effect of Cl^- appears abruptly only when the potential is moved in the positive direction. Probably, the adsorption of Cl^- depends to a greater extent on potential than does the adsorption of the organic anion, which is why in the initial period after the immersion of the electrode in the solution, when the potential is more negative, conditions are more favorable for the preferential adsorption of phenylanthranilate.

An increase in the protective effect observed after a previous one hour immersion in the pure inhibitor solution points to an important kinetic factor in combating the depassivation of a metal. To understand the significance of this it is necessary to consider the basic concepts of the theory of reactivity of chemical compounds.

2

The Role of the Chemical Composition of the Medium in the Protection of Metals from Corrosion

2.1. THE EFFECT OF pH

The role of pH is important because it characterizes the activity of the dissociation products of the most widely occurring solvent of all corrosive media, water, and consequently, to a first approximation, its chemical composition. A knowledge of the potential, *E*, of the metal and of the pH of the environment may often be sufficient for assessing the aggressiveness of the environment. Thus, by changing the pH with the value of *E* held constant it is possible to suppress corrosion by means of either passivation or cathodic protection. This can be seen from Pourbaix diagrams which show the pH–*E* dependence of the electrochemical reactions that cause corrosion. As a rule, the kinetics of these reactions depends on pH since H_3O^+ and OH^- either participate directly in them or are their products. According to Antropov[(108)] the rate of corrosion of a metal, *K*, in acid is associated with pH by the equation:

$$\log K = a - b\text{pH} \tag{2.1}$$

in which *b* is a constant dependent on the nature of the metal. Analysis of the values of *b* has shown that for transition metals, such as iron, it is possible to inhibit corrosion by the screening that results from both adsorption of an additive as well as from changes in the bonding energy of hydrogen with the metal. Inhibitors of the cationic type are even capable of affecting the nature of the controlling stage of the cathodic reaction. For non-transition metals, the choice of inhibitors is restricted because many anions accelerate the discharge of H_3O^+. Grigor'ev and Ekilik[33] used Eq. (2.1) to explain the reduction in the role of the chemical structure of a series of organic inhibitors with decrease in pH of hydrochloric acid solutions. However, for the corrosion of metals in neutral media, Eq. (2.1) is, as a rule, not applicable because of the diffusion limitations of the cathodic reaction, among other reasons.

Consideration of the effect of pH on the corrosion of metals is complicated by the change in kinetics of anodic reactions that can occur with increase in OH^- concentration, since OH^- can not only participate in the dissolution of metals but can also affect the inhibiting effect of other anions. Thus, in the dissolution of nickel in sulfate solutions, I^- ion becomes effective as an inhibitor only at $pH > 2$, an effect which is associated with its adsorption and with the displacement of aggressive anions from the surface.[29] Another indication of this effect is provided by the dissolution of iron in phosphate solutions. At $pH < 4$ this takes place with participation of OH^-; in less acid solutions, in which adsorption of phosphates increases, a mechanism of complex formation with $H_2PO_4^-$ operates.[109] If OH^- is not participating directly in the dissolution of a metal, then the effect of pH on the mechanism of the anodic reaction may be associated with changes in the anionic composition of the solution. Thus, at pH 4.5 only $H_2PO_4^-$ ions are found in a phosphate solution, with the dissolution of iron,[109] copper,[110] and zinc[38] having a first-order dependence on these. In a weakly alkaline solution (pH 9.5), $H_2PO_4^-$ anions are absent and the anodic reaction is second order with respect to HPO_4^{2-}.

Often the importance of pH is associated not just with the bulk solution but also with the near-electrode layer. Alkalization of this layer often arises with cathodic protection in neutral solutions, and can have serious effects in overprotection of aluminum alloys. This occurrence provides a possibility for increasing the effectiveness of cathodic protection by adding small amounts of inhibitors, which can form difficultly-soluble deposits when the pH of the medium increases (such protection will be examined in Chapter 5). Significant changes in pH of the near-electrode layer can also occur in the presence of strong oxidizing agents. Even in buffer solutions at potentials corresponding to the active dissolution of iron, the introduction of piperidine 3,5-dinitrobenzoate raises the pH from 7.4 to 9.5.[11] This reduces the critical potential and current density for passivation of iron, thereby increasing the ability of the iron to passivate. Mikhailovskii and Spitsyn[111] made a quantitative assessment of the effect of oxidizing agents on the corrosion of a metal on the basis of the ability to generate OH^- in the reduction process:

$$Ox + nH^+ + me^- \rightarrow Red + nOH^- \quad (2.2)$$

If $m \approx n$, an acceleration of the anodic process will occur as a result of the reaction

$$Me + nOH^- \rightarrow Me[OH]_n + ne + nH^+ \rightarrow Me^{n+} + nH_2O \qquad (2.3)$$

which usually takes place at potentials more negative than the corrosion potential in the base electrolyte. The authors associate the stimulation of corrosion in acid solutions by O_2, H_2O_2, MnO_4^-, NO_2^-, and $S_2O_8^{2-}$ with this effect. When $m/n >> 1$, the oxidizing agent does not directly affect the rate of the anodic process but can move the metal into the passive state by accelerating the cathodic reaction. When $m/n < 1$, an excess of OH^- will be established on the surface which, in our opinion, will change the conditions so as to make the formation of an oxide–hydroxide layer thermodynamically possible, which, if not causing passivation, at least, will facilitate it. Mikhailovskii suggested that this excess of OH^- can also intensify the adsorption of cations and change the sign of the ψ_1 potential (the drop in potential in the diffuse double layer). As a result, an ionic blocking layer will form with an additional difference of potentials which will provide a sharp increase in the overpotentials for the reduction reaction and the stabilization of the metal surface. Of course, in a solution in real conditions, the picture can become significantly more complicated since, apart from the oxidizing agents and the solvent, compounds or ions may be present that can (to some degree or other) interact with the surface or with the products of electrode reactions. Nevertheless, it is important to note that the effect of the oxidizing agent on the corrosion of a metal cannot be examined without taking into account the possibility of changes in the activity of OH^- on the metal surface or in the near-electrode layer of the solution.

Thus, hydroxyl ions, depending on their concentration, as well as on other factors including the nature and potential of the metal, can fulfill the role of either activators or inhibitors of corrosion. Therefore, the corrosion behavior of metals in neutral solutions differs significantly from that in acids and alkalis. In neutral solutions in the pH range 5–9, the corrosion rate of the majority of metals has a weak dependence on pH. This is due to the establishment in the near-electrode layer of a certain value of pH which is sustained by the formation of difficultly-soluble corrosion products. In many respects, this aspect of the corrosion of metals in neutral media is connected with the increased role of oxygen depolarization, which is usually under diffusion or diffusion–kinetic control. Bearing in mind that the retardation of such a process is made difficult by the lateral approach of the depolarizer, and that the blocking of a significant part of the surface is required, i.e. $\theta \rightarrow 1$, it can be understood that the kinetic retardation of the cathodic reaction that operates in the inhibition of acid corrosion will be of little effect in neutral media.

Furthermore, the corrosion of metals in neutral media takes place in conditions where there is a significantly lower solubility of oxides and hydroxides than in acid solutions. This situation facilitates the transition of the metal into the passive state and provides the prerequisite for the use of inhibitors of the passivating type. However, as a rule, oxidized electrodes have a surface that is energetically heterogeneous, and in the presence of certain anions this condition will lead to the nucleation of a dangerous form of corrosion damage—pitting. For some metals and alloys, pitting is the principal form of corrosion in neutral media, whereas for others, corrosion will begin with the local destruction of primary oxide films and then extend over the whole surface. In both cases, the onset of corrosion is associated with local damage of the surface, although the probability and rate of the depassivation will in many respects be determined by the pH. With appropriate change in pH, complete suppression, or at least a significant reduction, of corrosion can be achieved. We shall return later to an examination of the role of OH^- in corrosion processes. However, a deeper understanding of its importance can be gained if we consider OH^- along with other chemical compounds and ions. This may then reveal the particular properties of this anion which lead to its high reactivity in many physico–chemical processes.

2.2. REACTIVITY OF CHEMICAL COMPOUNDS AND THE EFFECT OF THIS IN THE CORROSION OF METALS

The concept of reactivity is of prime importance in various branches of chemistry, but until recently it has not been used in corrosion studies. Thus, the existence of widely-differing corrosive media and the considerable influence of their composition on the corrosion rate or on the stabilization of the passive state indicate that it is not possible to establish a scientific basis for corrosion protection, particularly inhibition, without the use of quantitative criteria for the chemical activity of the components of the corrosive media. Difficulties in the application of a generalized understanding of this topic in corrosion science are not only associated with the extreme complexities of the processes under study. As pointed out in Taub's handbook,[112] any arguments about the reactivity of compounds without an indication of the specific type of reaction are meaningless. In fact, it is difficult to expect that the reactivity, which—according to classical ideas—includes both kinetic and thermodynamic characteristics, will change in the same way in various processes, for instance, in electrophilic substitution or in oxidation of reagents. For this reason, the rule that is well known to all chemists—similar changes in structure lead to similar changes in reactivity—cannot be regarded as unconditional. According to Hammett,[113] "application of the rule requires so great an exercise of judgement, offers so wide an opportunity for the wisdom that comes only with experience and for the genius that seems almost intuition." For the researcher dealing with metallic corrosion, the way out from this difficult position can be found in the phenomenological approach, which points to certain analogies between complex heterogeneous processes and "more simple" homogeneous reactions or equilibria. However, it must be remembered that the electrochemical reactions that cause corrosion differ from chemical reactions in that their energetics depend on the difference in potentials between the metal and the contacting phase, e.g., an electrolyte solution.

Frumkin considered an electrode reaction taking place at different potentials as a series of uniform reactions differing only in the value of the free energy of the reaction $\Delta G = -nFE$.[114] In this case, the Brønsted relationship, according to which the rate constant, k, for the reaction in acid or base catalysis is determined by the dissociation constant, K, of the catalyzing acid or base:

$$k/k_0 = [K/K_0]^\alpha \tag{2.4a}$$

where α is the Brønsted transfer coefficient, can be described in the form

$$i = K' \exp[-\alpha nF\Delta E/RT] \tag{2.4b}$$

Here K' is a constant that includes all the values that are independent of ΔE. The equation obtained for proton transfer agrees in form with the empirical equation of Tafel, which has an exponential dependence of the electrode reaction rate on the overpotential, η_H, i.e., the difference between the equilibrium potential, at which the rates of the forward and reverse reactions equal the exchange current i_0, and the potential at the current i. This work by Frumkin is noteworthy not only because it explains the physico-chemical sense of the Tafel equation, which plays such an important role in the theory of electrochemical processes, including the corrosion of metals, but also because it was written before Hammett formulated the principle of linear free-energy relations (LFER)[113] of which the Brønsted relationship is a partial case. It is unfortunate that despite so much initial success, it has taken a long time for the LFER, or other principles providing quantitative "composition-property" dependencies, to become recognized in the electrochemistry and science of the corrosion of metals.

Finally, one can note that not only the logarithm of the rate constant (and for a reversible reaction, the equilibrium constant) but also the relative change can provide a measure of the reactivity of chemical compounds. This, as we shall see, lies at the base of the energetic assessment of the reactivity.

On the whole, the development of the study of reactivity is related to concepts of the electron interaction of reactants and with the displacement and distribution of electron density in their molecules or ions. An important role is played by the theory of the transition state (absolute reaction rates) according to which, in the reaction between reactants A and B, only those molecules will participate which have energy equal to or greater than a certain critical condition, i.e., the system must pass over a potential energy barrier, the peak of which will characterize the energy of the transition state. The rate of reaction is determined only by the difference in free energy between the ground and transition states (Fig. 2.1). This means that the transition state can be viewed as a normal molecule AB for which the enthalpy can be calculated using statistical thermodynamics. However, it is impossible to examine accurately the potential energy surface because of difficulties in the quantum-chemical calculation of the values of the intrinsic energy of multielectron systems.[115] Therefore empirical methods for determining reaction rates are mainly used.

It follows that there are two important scientific tasks for chemists. One is to examine the various reaction centres as a function of the molecular structure in order to determine their relative activity. The other is to study the reactivity of separate classes of compounds possessing the same reaction centre (or functional group). In the latter case, the properties of a specific compound can be predicted from data on other compounds of this class. Such a comparative analysis of, for example, rate constants also provides a powerful conception of the chemical reactivity expressed in terms of the structure of the molecule, and brings us closer to an understanding of reaction mechanisms.[115] Organic compounds provide a wide field of action for coping with the second task and, in fact, it is with classes of such compounds that a correlation analysis based on the LFER principle has been developed in depth.

The principle of LFER can be described as follows:

Let compounds XYR, differing only in the nature of R which itself remains unchanged during the course of reaction, react with a reagent A. Then for any such reaction, one can write

$$\Delta G_R = -2.3RT \log K_R \tag{2.5a}$$

where K_R is the equilibrium constant.

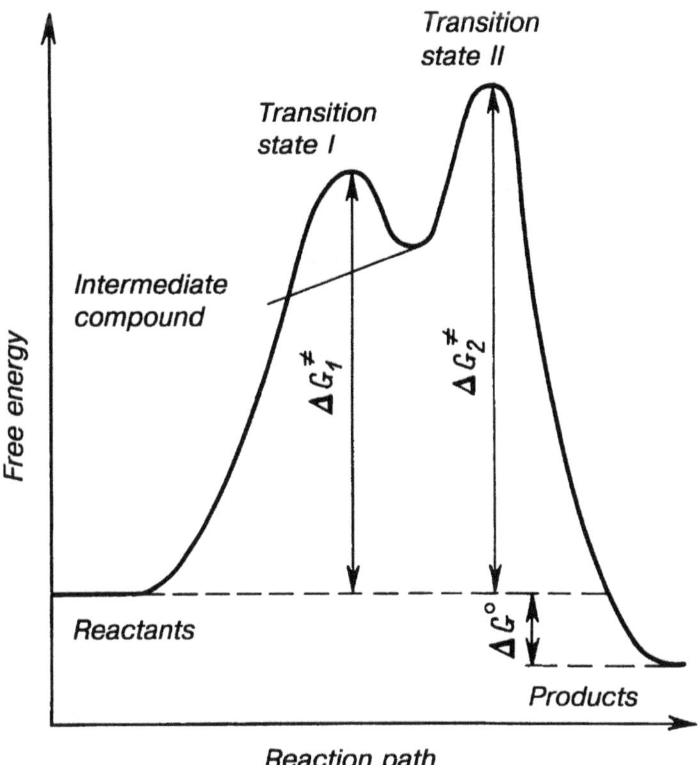

Figure 2.1. Changes in the free energy of a reaction associated with the changes in free energies of the initial and final states.

In the case of irreversible reactions

$$\Delta G_R^{\neq} = -2.3RT\left[\log k_R - \log \frac{\kappa\, kT}{h}\right] \tag{2.5b}$$

where k_R is the rate constant, κ is a coefficient characterizing the probability of the reaction reaching a transition state, k is Boltzmann's constant, and h is Planck's constant. If hydrogen, as is usually the case, is taken as the standard substituent, i.e., H is written for R, then for any other substituent in the series the relative change in the energetic characteristics of a reaction can be expressed in the form:

$$\log K_R - \log K_{\mathrm{H}} = -\frac{1}{2{\cdot}3RT}\left[\Delta G_R - \Delta G_{\mathrm{H}}\right] = -\frac{1}{2.3RT}\Delta\Delta G \tag{2.6a}$$

$$\log k_R - \log k_H = -\frac{1}{2.3RT}\Delta\Delta G^{\neq} \tag{2.6b}$$

Let all the set of the reactions of these compounds with A be taken as a standard reaction series, with each pair of differences of free energy or free energy of activation given the index of the respective substituents, $\Delta\Delta G_R$ or $\Delta\Delta G_R^{\neq}$. Then, according to the LFER principle, those values caused by changes in the structure of the compound are linearly connected with the analogous values for other reaction series (for example, for reaction with reagent B) in conditions in which the same parameter is varied, for example, the substituent in an aromatic nucleus in the para position relative to the reaction centre X. In principle, it does not matter whether the standard reaction series is reversible or whether we are dealing with a constant reaction rate. For this reason, Hammett, to whom credit must be given for realizing the possibility of formalizing the dependence of the reactivity of organic compounds on their structure, selected the dissociation of meta- and para-substituted benzoic acids as a standard series:

$$R\text{-}C_6H_4\text{-}COOH \underset{25°C}{\overset{H_2O}{\rightleftharpoons}} R\text{-}C_6H_4\text{-}COO^- + H_3O^+ \tag{2.7}$$

since the values of K are known with some accuracy. Whence the characteristic of the substituent R can be written as:

$$\log K_R - \log K_H = pK_a^H - pK_a^R = \sigma \tag{2.8}$$

where σ is the substituent or Hammett constants.

For the reaction of alkaline hydrolysis of the ethyl esters of the corresponding benzoic acids in 87.8% aqueous ethanol the correlation

$$\log[k_R/k_H] = \rho\sigma \tag{2.9a}$$

was first obtained. Similarly for a reversible process

$$\log[K_R/K_H] = \rho\sigma, \tag{2.9b}$$

and both these equations represent the well known Hammett equation. Here, ρ is a reaction parameter which is taken as 1.0 for the standard series of the dissociation of substituted benzoic acids. The value of ρ depends on the nature of the reaction under study and also on the conditions for its occurrence. Since the value of ρ increases with increase in the degree of ionic nature of the reaction, that is, with the polarity of the transition state, it can be assumed that electrochemical processes taking place with charge transfer will be extremely sensitive to the electron effects of the substituent. Actually, the first polarographic studies had confirmed this proposal and Zuman modified Eq. (2.9) for application to the half-wave potential:

$$\Delta E_{\frac{1}{2}} = E_{\frac{1}{2}}^{R} - E_{\frac{1}{2}}^{H} = \rho_{\pi}\sigma \tag{2.10}$$

This relationship characterizes the disturbance introduced by the electron effect of the substituent on ΔG in the transition from the oxidized to the reduced form or, for the irreversible process, on $\Delta G^{\neq}$ in the transition to the activated complex. Despite the success of this approach in the study of electrochemical reactions, it was much later before it began to be used in corrosion studies. The main reason for this is that the inhibitor may not take part directly in the electrode reaction, but may influence it only through changes in the ψ_1 potential or by blocking part of the active surface.

It should be emphasized that the LFER principle is realized in various ways which depend both on the reaction mechanism and on the selection of the standard series. For example, for compounds of an aliphatic series the effect of the substituent cannot be reflected by the value of the Hammett constant; other σ constants are often used—sometimes even in aromatic series—which take into account, for instance, only the inductive effects, σ_I, of the substituent.

Without going too deeply into the problem (which has been considered in detail in specialist literature), one can point out that the various electronic and spatial effects within LFER can be allowed for by bringing together the principle of additivity and the independence of the effect of structure factors on the reactivity. According to this principle, which was put forward by Taft, the values of the free energy of the reaction or of the activation are approximately represented by the sum of the components—independently of each other—that are associated with polar, resonance, and other interactions. Among the last, special attention should be given to hydrophobic interactions as first introduced by Hansch[116] for assessing the biological activity of organic compounds. The hydrophobicity of a compound is reflected by the logarithm of the coefficient of distribution (log P) between water and an immiscible liquid—n-octanol. For substituents, the relative hydrophobicity is reflected by the π constant which, analogously to the electronic constant σ, is proportional to the difference:

$$\log P_R - \log P_H = K[-\Delta\Delta G] = K_1\pi$$

where for the selected standard conditions $K_1 = 1.0$. We have shown that the use of the LFER principle for describing many cases of inhibition of corrosion-electrochemical reactions requires account to be taken simultaneously of the electronic and hydrophobic influences of the substituent.

Whatever the features are that affect the effectiveness of inhibitors—the nature of the metal, the pH of the medium, the nature of the solvent, and the composition of the corrosive environment—it is clear that the type of dissolution and the region of potentials in which this occurs are also important. Therefore we should first examine the protection of metals by organic compounds under conditions in which the passivating phenomena do not lead to any abrupt ennoblement of potential, and where the dissolution of the metal is not localized. One can find many examples of such systems among cases of acid corrosion; it was with these that the first studies were conducted that showed the quantitative connection between the chemical structure of an organic compound and its effectiveness as a corrosion inhibitor. The work of Hackerman played a large part in this, and he repeatedly emphasized the importance of taking into account the electron density at the reaction centre of the inhibitor, for example, on the nitrogen in various amines.[117] In a distinct way, the LFER principle had begun to be applied to the problem in the mid-1960s when Donahue and Nobe,[118] starting from the blocking mechanism of the action of acid-corrosion inhibitors, obtained the equation

$$\log [1 - \gamma_R]/[1 - \gamma_H] = -\rho\sigma \tag{2.11a}$$

where γ_R and γ_H are coefficients of corrosion inhibition calculated from the rates of corrosion in the background solution $[K_0]$ and in the presence of the inhibitor $[K_{inh}]$, i.e., $\gamma = K_0/K_{inh}$. However, it was shown later by Grigor'ev and Ekilik[33] that the LFER principle for describing the inhibition of acid corrosion by organic compounds and the limiting of the electrochemical evolution of hydrogen can be used successfully even without the assumption of a blocking action of the inhibitor.

It is well known that the LFER principle for organic compounds is valid in the case of isoentropy (ΔS = const), isoenthalpy (ΔH = const), or isokinetic series ($\Delta H \sim \Delta S$). For the case where the inhibitors act only by a blocking mechanism, i.e., they only reduce the active surface area of the corroding metal without altering $\Delta G^{\neq}_{corr}$ of the remaining area, such processes can, to a first approximation, be considered isoenthalpic. According to Grigor'ev and Ekilik,[33] this is the case for the inhibition of corrosion of nickel in 1 N HCl at T = 323–353 K by 2,4-diphenyl-6-(R-styryl)pyrilium cations. The blocking action of the inhibitor is associated with its specific adsorption caused by π electron interactions of this organic cation, which contains four aromatic rings, with the transition metal, nickel. This agrees with the increase of γ when the electron-donor properties of the substituent are increased.

Another case concerns the inhibition of corrosion by small molecules or ions in which their adsorption does not cause any noticeable blocking of the surface but does change $\Delta G^{\neq}_{corr}$ as a result of a redistribution of charges in the electrical double layer. Using several reaction series as examples, Grigor'ev and Ekilik[33] showed that for inhibitors of such a type the relation

$$\log \gamma_R = \log \gamma_H + \rho\sigma \tag{2.11b}$$

will hold. For example, for substituted anilines with iron in HCl solutions, they observed a linear change in $\Delta G^{\neq}$ with increase in the electronic influence of the substituent (as reflected by its σ constant). It was noticed that there was a close connection between the value of $\Delta G^{\neq}_{corr}$ and the activation energy for the evolution of hydrogen on iron which controls the corrosion process. A simple example which illustrates such results is provided by studies of the inhibition of corrosion by substituted benzaldehydes. The introduction of nucleophilic substituents which increase the basicity of benzaldehydes facilitates their adsorption and increases both $\Delta G^{\neq}_{corr}$ and their protective properties in relation to iron (Fig. 2.2). This diagram also shows that analogous changes occur in the potential E_c of iron polarized at a constant cathodic current density i_c for hydrogen evolution.

Finally, it should be stressed that the two mechanisms of corrosion inhibition that have been examined are strongly idealized and this is why inhibitors are often found that act by both mechanisms simultaneously. Such reaction series when the LFER principle is applied are isokinetic and the equation takes the form[33]

$$\log[\gamma_R/\gamma_H] = \rho\sigma - \log[\theta_R/\theta_H] \tag{2.11c}$$

where θ is the degree of coverage of the metal surface by the inhibitor. In a manner distinct from their inhibition of nickel corrosion, the 2,4-diphenyl-6(R-styryl)pyrilium perchlorates in protecting iron in 1 N HCl at $T < 313$ K not only restrict the active surface of the metal but also strongly increase $\Delta G^{\neq}_{corr}$, with the effect being more marked the higher the electron-donor properties of the substituent. This serves as a good illustration of the undisputed fact that the nature of the metal can introduce significant amendments to the mechanism of action of organic inhibitors.

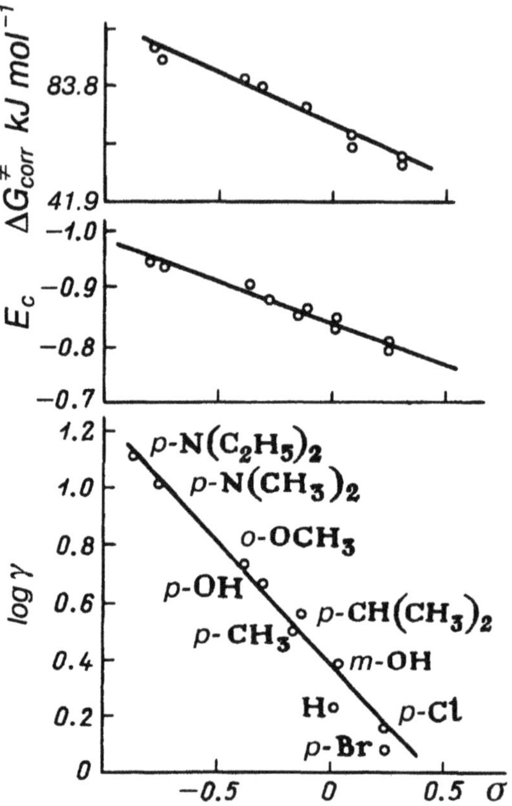

Figure 2.2. The dependence of $\Delta G^{\neq}_{corr}$, log γ, and E_c for iron during corrosion in 1N HCl on the value of σ for the substituent in benzaldehydes [Reproduced by permission of the authors[33]].

In this last connection, attention must be drawn to the study of the effects of 22 substituted benzimidazoles (BI) on the anodic dissolution of copper and zinc in phosphate solution that have been reported by Kuznetsov and Podgornova (Fig. 2.3).[119] Electron-accepting substituents improved the protection of copper but decreased that of zinc. On the other hand, an increase in electron-donor properties of the substituent weakened the inhibition of copper dissolution but increased it for zinc. However, in both cases log γ_{an} (where $\gamma_{an} = i_0/i_{inh}$ for E = const) correlated well with the inductive constants of the substituents.

The different influences of the chemical structure of benzimidazoles on the protection of these metals can be explained by the ability of copper and zinc to form chemical bonds in their complexes. In a covalent σ bond,

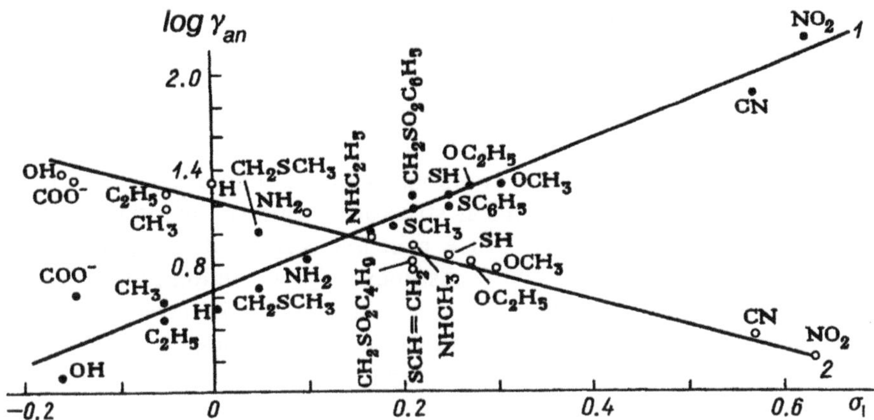

Figure 2.3. The dependence of logγ on the σ_I constant of the substituent R in benzimidazole on the anodic dissolution of copper (1) and zinc (2) in 0.5 M phosphate solution at pH 11.4. Inhibitor concentration = 5×10^{-4} M (1); 5×10^{-3} M, (2); E=0.14 V (1) and −1.0 V (2).

only ligands are donors of electrons and the metal cation is the acceptor. An increase in electron density at the reaction centre of benzimidazole, in which the "pyridine" and "pyrrole" nitrogens are mutually equivalent because of a rapid proton exchange, should be capable of strengthening the σ bonds and increasing the effectiveness of inhibition. Such is the case for the protection of zinc. Copper, on the other hand, as a consequence of its low ionization potential and high electronegativity, is capable of acting as an electron donor and forming an overall molecular orbital with π-like electrons of the ligand. Thus, a decrease in electron density at the reaction centre of the benzimidazole improves the protection of copper. The value of the minimum concentration, for inhibition, in mol L^{-1}, C_{min}, as determined from long-term corrosion tests, is also found to depend on the nature of the substituent

$$\text{Cu:} \quad \log C_{min} = -2.55 - 2.21\,\sigma_I \quad r = 0.92 \tag{2.12a}$$

$$\text{Zn:} \quad \log C_{min} = -2.65 + 3.30\,\sigma_I \quad r = 0.92 \tag{2.12b}$$

(where r is the correlation coefficient). Interestingly, distinct from the results shown in Fig. 2.3, here we have a situation in which passivation of the metal is brought about by the formation of relatively thick layers. Evidently, in both cases the effectiveness of the inhibitors depends on the stability of the complexes that are formed, which will be determined by the nature of the metal–ligand bond, which will be different for zinc and copper complexes. Unfortunately, the stability constants, K_s, of the appropriate complexes are not known (apart from those of benzimidazole), and therefore it remains uncertain whether a still more precise prediction of the inhibitive action of azoles could be made, i.e., if K_s values were known. This example shows the need for an all round study of the mechanism of corrosion inhibition by organic compounds and the important place in this of investigations of the "chemical structure–property" association. It is important that in this case the LFER principle is fulfilled in a quite simple form: one type of interaction of a substituent with a reaction centre (induction) not being complicated by effects leading to a breakdown of the linear type of dependence (Eq. 2.9). Very often this dependence has a more complex character and the reaction parameter inside one and the same reaction series will change sign. The type of dependence that may be considered to be the most usual is that given in Fig. 2.4 and has a V-shaped character.

From these considerations, it follows that the classical Hammett equation can be divided into two linear dependencies distinguished both by the sign and the value of ρ. Unsubstituted compounds ($\sigma = 0$) obey both these dependencies, but an increase in protection of iron by a single absolute value of the σ-constant is more easily achieved by the introduction of electron-donor substituents, this being reflected in a steeper slope of the left branch of these dependencies. The reasons for the V-shape differ, but in the majority of cases they are associated with a change in the controlling stage of the process. Taking into account the heterogeneous nature of the inhibition of corrosion and the obvious changes in ρ with increase in C_{inh}, Grigor'ev and Ekilik[33] examined the nature of these dependencies, but in conditions such that θ, not C_{inh}, was constant. It was found that neither log γ nor $\Delta G^{\neq}_{corr}$ were dependent on the value of the σ constant for the case of electron-acceptor substituents i.e., inhibition of corrosion was caused only by blocking of the surface, although an analysis of the data in Fig. 2.4 points to an error in the conclusions concerning the influence of such substituents on $\Delta G^{\neq}_{corr}$. Unfortunately, the determination of θ for a corroding metal presents considerable difficulties. Because of this, dependencies of the (Eq. 2.11) type are more usually set up for conditions such that C_{inh} = constant. Moreover, even with θ = constant, the effect of the polarity of electron-donor substituents on $\Delta G^{\neq}_{corr}$ is retained, which suggests differences in the mechanism of inhibitors that are related in one reaction series. Finally, one should not confuse the formal and the true mechanisms of the process; the latter involves an understanding of the consecutive stages in the complex reactions and the identification among these of the controlling stages. However, in many cases, changes in these brought about by variations in the chemical structure of the reagent take place in a similar manner. This

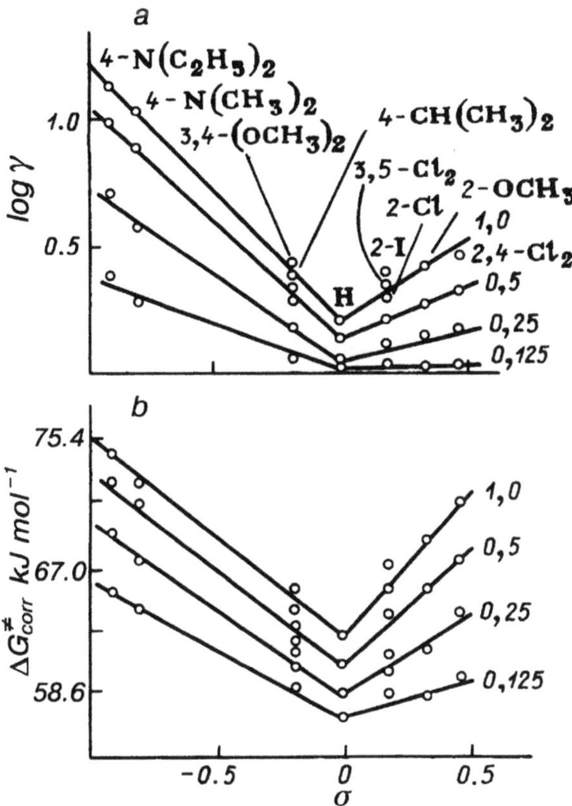

Figure 2.4. The dependence of log γ (a) and $\Delta G^{\neq}_{corr}$ (b) on the σ-constant of substituents in the perchlorates of 8-(R-benzyl)-2-(4-methoxyphenyl)-4-phenyl-5,6,7,8-tetrahydrochromilium for the corrosion of iron in 1.0 N HCl. (The numbers on the lines refer to the concentration $C_{inh} \times 10^{-5}$ M.) [Reproduced by permission of the authors[33]].

allows one, within the scope of a formal approach, to predict the optimum composition of an inhibitor and also with suitable precautions to construct a hypothesis for the true mechanism of the process.

Not only the sign but also the value of ρ, which reflects the sensitivity of the inhibitive effect to the polarity of the substituent, is of considerable value in interpreting the mechanism of the protection and, particularly, for predicting the possibility of protection. Thus, in applying the Hammett equation to chemical processes the effect of temperature on their kinetics is described by the Leffler[120] relationship:

$$\rho = \text{const}\left[1 - \frac{T_0}{T}\right] \tag{2.13}$$

where T_0 is a constant, the so-called isokinetic temperature, at which the effect of the polarity of the substituent on the rate of the process disappears. When applied to inhibition, this means that at $T = T_0$ all the compounds of a given series inhibit corrosion equally, but at $T > T_0$ the sign of the effect of substituent polarity will change. Grigor'ev and Ekilik[33] having analyzed Eq. (2.13) for the inhibition of corrosion of iron in acids by perchlorates of substituted 5,6,7,8 tetrahydrochromilium, found that at $T > T_0$ the increase in the σ-constant of electron-accepting substituents led not to an increase in protection but to a decrease. They also noted a lower value of T_0 than would be the case for the majority of organic reactions. Characteristically, with the V-shaped form of the correlation dependence the improvement in inhibiting properties brought about by the introduction of more nucleophilic substituents is retained up to temperatures that are higher than for electrophilic substituents. Consequently, if we wish to have an inhibitor which will retain high effectiveness over a wide temperature range, compounds should be sought which are not simply adsorbed on the surface, for example by electrostatic forces, but which interact with it in a deeper chemical process.

By analogy with Eq. (2.13) and using Eq. (2.1), Grigor'ev and Ekilik introduced the concept of isokinetic acidity of the medium, pH_0, at which $\rho = 0$:

$$\rho = \text{const}\,[1 - \text{pH}/\text{pH}_0] \tag{2.14}$$

Using the examples of substituted anilines (with $\sigma > 0$), benzaldehydes and perchlorates of 2,4-diphenyl-6-(R-styryl)pyrilium, they showed that for iron in hydrochloric acid the increase in acidity of the medium, in conditions of similar screening effects of the substituents, effectively led to the inhibition of corrosion being independent of their polarity. It was of interest that here also the introduction of nucleophilic substituents (in the aniline molecule) was found to have little sensitivity to changes in conditions.

The anionic composition of the medium has no less a significant effect on ρ. For example, in the corrosion of iron in sulfuric acid, ρ is always lower than in hydrochloric acid solutions, which in the latter case is associated with an increase of adsorption of the inhibitor in the cationic form.[33] This explanation does not conflict with ideas about ρ as a characteristic of the degree of ionicity of the transition state, or with the fact that chemisorption of Cl^-, by displacing the potential of the uncharged surface, facilitates the adsorption of cations. However, it is difficult to associate with changes in surface charge those differences in ρ that are caused by additions of Cl^-, I^-, or NO_3^- in the inhibition of iron corrosion in neutral buffer solution by N-substituted sodium anthranilate (Fig. 2.5). In this case, these anions ful-

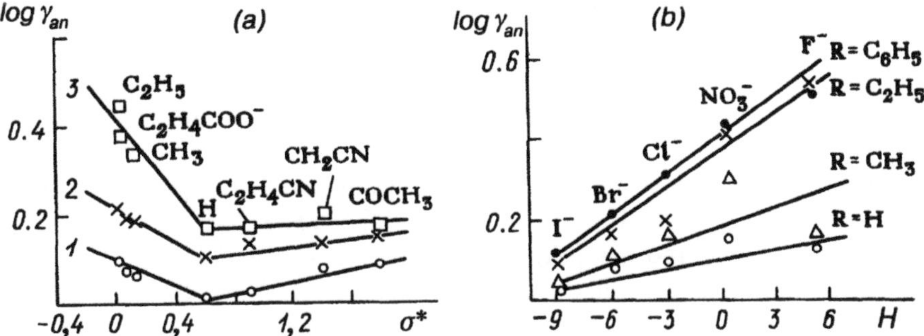

Figure 2.5. The dependence of log γ_{an} in the anodic dissolution at $E = -0.03$ V of iron in a borate buffer containing 0.03M NaI, NaCl, $NaNO_3$, NaBr, or NaF and 0.01 M N-substituted anthranilate on: (a) the Taft sigma constant, σ^*, of R in NaI (1) NaCl(2), or $NaNO_3$ (3) solution; (b) the Edwards basicity, H, of the anions.

fill the role of activators of anodic dissolution and, depending on their nature, will to various degrees compete with the inhibitors (anthranilates). However, to understand the role of the chemical structure here, the LFER principle alone is not sufficient and new assumptions are required. These would include, for example, a constancy of the types of interaction in the studied reaction series since variations in the nature of the activator are accompanied by a change in the nature of the reaction centre.

This task is, of course, more complex. In its resolution, the concept of hard and soft acids and bases (HSAB)[121] will provide help. While not being strictly quantitative, this concept is widely applied to various fields of chemistry to which Klopman[115] has contributed significantly. According to Pearson, the author of the HSAB concept, all reactions take place in such a way that hard acids prefer to be combined with hard bases and soft acids with soft bases. Hard acids and bases have low polarizability, small sizes, and high values of the electronegativity of the donor and acceptor atoms of the base and acid, respectively. Soft acids and bases have opposite properties, with the presence of deep vacant orbitals for the formation of π-dative bonds providing another characteristic sign. Hard bases are difficult to oxidize (high ionization potential); their donor atoms have high electronegativity with considerable localized negative charge. Hard acids are difficult to reduce, they have a high degree of oxidation, and their acceptor atom usually carries a significant positive charge. It follows from these properties that hard–hard interactions should take place mainly as a result of the availability of long-range electrostatic forces between the polar and charged molecules. Examples of hard bases, according to Pearson, are: OH^-, F^-, CH_3COO-, SO_4^-, NO_3^-, NH_3, CH_3NH_2, and $C_2H_5NH_2$. Examples of hard acids are: H^+, cations of alkali metals, Mg^{2+}, Ca^{2+}, Al^{3+}, Cr^{3+}, Fe^{3+}, BF_3, and $AlCl_3$. Soft–soft interactions are associated mainly with the formation of covalent bonds. The ease of these interactions is often associated with the fact that in the transition state of the reaction the coordination number of the acid atom is increased and the transfer of negative charge to it is enhanced, with the high polarizability of the base facilitating this process. Pearson relates I^-, SCN^-, R_2S, R_3P, RS^-, CN^-, and C_6H_6 to soft bases and Cu^+, Ag^+, Pd^{2+}, Co^{2+}, uncharged atoms of metals, $InCl_3$, quinones and trinitrobenzene to soft acids. Nevertheless, not all reagents can be assigned to hard or soft classes; others are classified as intermediate acids and bases. Furthermore, the concept does not exclude the possibility of hard–soft interactions and refers to "supplementary stabilization" of complexes that are formed by acids and bases of the same kind.

The advantages of a simultaneous consideration of two types of interaction—ionic and covalent—are obvious, and it is not surprising that such attempts are often made in chemistry. For example, mention may be made of the Drago–Ulinda equation for the heats of formation of various acid–base complexes (without metal ion participation). However, the equation proposed by Edwards[122] is of more interest because it is applicable not only to organic reactions but also to the formation of complexes involving cations of various metals. This equation, subsequently known as the "oxy-base scale," has been widely used for interpreting kinetic and equilibrium data and especially for assessing the reactivity of nucleophiles with substrates which contain an electrophilic centre.

$$\log K/K_0 = aE_{OK} + bH \qquad (2.15a)$$

where K_0 is the equilibrium constant of the reaction with participation of an arbitrarily-selected standard nucleophile, $E_{OK} = E_O + 2.60$, the relative value of the oxidation potential of the nucleophile (in V), and $H = pK_a + 1.74$, its relative basicity, and a and b are substrate constants. Since $E_{OK} = a_1 \log[MR/MR_{H_2O}] + b_1 H$ where MR and MR_{H_2O} are the molar refractions of the nucleophile* and water respectively, reflecting their polarizability, then (2.15a) can be rewritten in the form

$$\log K = \text{const} + a_2 \log MR + b_2 H \qquad (2.15b)$$

in which a_2 and b_2 are parameters determined by the nature of the electrophile.

*Pal'm[123] has pointed out that since the nucleophiles under consideration are usually small, then the total value of MR for the whole molecule may be used.

The Edwards equation is of further interest in that by its nature it embodies the HSAB concept since the generalized basicity (nucleophilic nature) and acidity (electrophilic nature) are determined by the "hard" (bH) and "soft" ($a\log MR$) terms. Other known equations—Brønsted or Sven–Skotta, can be considered as partial cases of Eq. (2.15).[123]

The consideration in one equation of the polarizability of the reagent, along with ionic and covalent interactions, broadens the sphere of its applicability, since any bond by its nature is not purely ionic or covalent. Success or failure in the use of the Edwards equation is evidently connected with features of the use of the HSAB principle and with difficulties in including the role of a variable reaction centre in the thermodynamics and kinetics of the various processes.

The complexity of the latter can be illustrated by the example given earlier for the inhibition of the dissolution of iron by N-substituted sodium anthranilates. In these solutions, the unsubstituted anthranilate inhibits the dissolution but there is only a qualitative dependence of its effectiveness on the basicity of the activator (Fig. 2.5b). The low correlation coefficient ($r = 0.87$) shows only a tendency for $\log \gamma_{an}$ to increase with increase in the H parameter (or decrease in polarizability). However, with increase in the hydrophobicity of the substituent, as reflected by the constants of Hansch, which are negative for substituents more hydrophilic than hydrogen and positive for more hydrophobic substituents, there is an increase not only in the inhibiting properties of N-substituted anthranilates but also in the accuracy of the quantitative description of the $\log \gamma_{an} - H$ function. Thus, in solutions of methyl-($\pi = 0.56$), ethyl-($\pi = 1.02$), and phenyl-($\pi = 1.37$) anthranilates of sodium, a linear dependence is found with $r =$ 0.92, 0.94, and 0.96, respectively. Bearing in mind that in aqueous solutions the order of electron donating ability of the aggressive anions is the reverse of their H values, it can be postulated that the effectiveness of inhibition of corrosion with competitive adsorption of anions is determined by the electronic characteristics not only of the amino acid (inhibitor), but also of the aggressive anion. In view of the important role of chemisorption, it becomes clear why electron-donating substituents, which increase the electron density on the reaction centre, increase the effectiveness of inhibitors, whereas electron acceptors have practically no effect in chloride or nitrate solution. Furthermore, these data show that the hydrophobicity of the inhibitor as well as that of the activator can play an important role in such processes. Thus, the extremely hydrophilic acetylanthranilate for which no linear dependence of $\log \gamma_{an}$ on H is found, is not inhibitive for iron in a pure borate buffer and actually accelerates the dissolution. For sufficiently hydrophobic inhibitors, e.g., phenylanthranilate, γ_{an} decreases in the halide series from F^- to I^- in the same sequence as the change in hydration energy.

The role of screening, which includes the hydrophobic factor, in the inhibition of corrosion by organic compounds has been discussed in the literature for more than 50 years. However, discussions have been concerned only with quantitative considerations of the role of the hydrocarbon chains of inhibitors, particularly those containing nitrogen. In recent years, two approaches have been more intensively developed. One of these, the more traditional, but expanded by including perfluoralkyl chains, associates high effectiveness of an organic inhibitor with the critical concentration of micelle formation of its solutions. Its logical consequence is the establishment of a quantitative connection between the *f*-constants of the hydrophobicity of the inhibitor (determined by the method of Nys and Rekker from the partition coefficients)[124] with the criteria of the protective action of the inhibitor. However, the second approach is more general and is based on the simultaneous consideration of the hydrophobic and electronic characteristics of the various atoms, their groups, and in a number of cases, the whole inhibitor. The predominance of the role of the electronic factor (coefficient a) or the hydrophobic factor (coefficient b) can be determined either from the formal theoretical approach which uses the principles of LFER and the particular structural contributions, or from the actual mechanism of the dissolution of the metal and its inhibition. The universality and achievement of such an approach is seen most clearly in studies of the initiation and inhibition of pit formation with various metals and will be examined in the appropriate section of this book.

2.3. THE ROLE OF THE NATURE OF THE ANION IN THE CORROSION OF METALS

The important role of OH^- has already been pointed out in Section 2.1 where the effect of pH on the corrosion of metals was discussed. However, the role of pH is not limited to its effect on the thermodynamic conditions (as assessed, for example, from the Pourbaix diagram) since it can also directly affect the kinetics of the reactions.

Kabanov and Leikis[125] established that in alkaline solutions the dissolution of iron proceeds by the scheme:

$$Fe + OH^- = Fe(OH)_{ads} + e \qquad (2.16a)$$

$$Fe(OH)_{ads} + OH^- \rightarrow FeO_{ads} + H_2O + e \qquad (2.16b)$$

$$FeO_{ads} + OH^- = HFeO_2^- \qquad (2.16c)$$

with (2.16b) as the rate-controlling stage. Even in acid solutions in which the activity of OH^- ions is low, there are reasons for believing that they participate directly in the anodic dissolution of metals. Heusler, Bockris, Christiansen, and Antropov have put forward possibilities for the successive stages of the dissolution of iron involving OH^- participation. As a rule, these approaches differ in the nature of the controlling stage of the process and their merits and demerits are frequently discussed in the literature. The scheme proposed by Zytner and Rotinyan[39] deserves attention. This proposes that one can obtain, depending on the pH region and the extent of surface coverage by intermediate compounds, differing kinetic characteristics of the anodic process, i.e., with valves of b_a between 0.130 and 0.029V and $n = 0 - 2$ (where n is the number of electrons taking part in the reaction):

$$Fe + OH^- = FeOH + e \quad (2.17a)$$

$$Fe + FeOH + OH^- = Fe_2(OH)_2 + e \quad (2.17b)$$

$$Fe_2(OH)_2 \rightarrow Fe_2(OH)_2^+ + e \quad (2.17c)$$

$$Fe_2(OH)_2^+ = 2Fe^{2+} + 2OH^- + e \quad (2.17d)$$

Hydroxyl ions are not unusual in this respect. Even as early as 1928, Tsentnershver had pointed out that the rate of dissolution of cobalt increased with HCl concentration and explained this by complex formation:

$$Co + 4Cl^- = (CoCl_4)^{2-} + 2e \quad (2.18)$$

If one were to accept such a simple role for anions in the anodic process then it would be necessary to consider only the value of the stability constant, K_s, of the complex that is formed. However, an analysis of the available data does not allow such a conclusion to be drawn.

A new step in the understanding of the mechanisms of the anodic process was taken by Florianovich *et al.*[126] They showed that the dissolution of iron in H_2SO_4 can be described by the following kinetic scheme:

$$Fe + H_2O = Fe(OH^-)_{ads} + H^+ \quad (2.19a)$$

$$Fe(OH^-)_{ads} = Fe(OH)_{ads} + e \quad (2.19b)$$

$$Fe(OH)_{ads} + HSO_4^- \rightarrow FeSO_4 + H_2O + e\text{, or}$$

$$Fe(OH)_{ads} + SO_4^{2-} \rightarrow FeSO_4 + OH^- + e \quad (2.19c)$$

$$FeSO_4 = Fe^{2+} + SO_4^{2-} \quad (2.19d)$$

As can be seen, the controlling stage (2.19c) involves the displacement from the adsorption complex of one anion, OH^-, by another. To understand the significance of this step in assessing the reactivity of anions taking part—in some measure or other—in the corrosion process, we need to examine the classification of reagents from the standpoint of contemporary physical chemistry.

Every reaction, even when purely chemical, is an electrical process in which the reagent displays some constitutional affinity either for an atomic nucleus, i.e. for a positive charge, or for electrons (in principle, a reagent having an odd number of electrons can have the same affinity for the nucleus and for electrons, that is an affinity for an atom, although this is a relatively rare case).[127] However, it would appear that a direct classification of reagents into "anionoid" and "cationoid" (according to Lapworth) can be justifiably criticized for its excessive emphasis of the role of charges, without considering the fact that the same reactivity can be observed even with different charges.[127] In fact, the separation of reagents into nucleophilic and electrophilic is the classification which is most widely used. According to Ingold, a reagent that participates in a reaction by donating electrons to, or sharing them with, other atomic nuclei, is described as nucleophilic. One should not, however, confuse nucleophilicity with basicity, which is affinity towards a proton, since the former is a more general property which is dependent on the electronic constitution of the reagents. A reagent acquiring, during the course of a reaction, electrons or a share in electrons which had earlier belonged to another molecule, is referred to as electrophilic. In relation to this definition, acidity is a special case of electrophilicity, which is the affinity for external electrons in general, although acidity is not identical to electrophilicity.

Bearing in mind that there is a conditional distinction between substances reacting between themselves on a metal substrate and reagents reacting with the substrate, we may propose the former to be represented by some form of complex with the metal, MeL. Under the influence of a nucleophilic reagent, rupture of the bond can occur by homolysis or heterolysis. In homolysis, according to Taub,[112] each of the atoms forming the rupturing bond retains an electron

$$\mathrm{Me{:}L} \rightarrow \mathrm{Me}\cdot + \mathrm{L}\cdot \tag{2.20a}$$

For transition metals, this involves a transfer from the ligand to the metal of one electron from the pair providing the bond in the complex; homolysis can be considered an oxidation–reduction process. In the case of non-transition elements, unstable free radicals can be formed.

During heterolysis, the electrons participating in the formation of the bond with the metal remain with the detaching group

$$Me{:}L \rightarrow Me^+ + {:}L^- \qquad (2.20b)$$

An attacking nucleophile, $:Y^-$, can act as a donor of electrons, usually by forming a new complex, and the process itself is referred to as a nucleophilic substitution of ligands, S_N.* It can be synchronous or non-synchronous. In the latter case the rupture of the old bond precedes (dissociative S_N^1 mechanism) or follows (associative S_N^2 mechanism) the formation of the new bond.

It should be remembered that the S_N process can occur in stages, with the formation of an intermediate compound. In such a case, the dissociative process will include successive reactions such as:

$$MeL \rightarrow Me^+ + L^- \qquad (2.21a)$$

$$Me^+ + {:}Y^- = MeY \qquad (2.21b)$$

and the associative will include:

$$MeL + {:}Y^- \rightarrow MeLY^- \qquad (2.22a)$$

$$MeLY^- = MeY + L^- \qquad (2.22b)$$

It is clear that the designation of S_N processes reflects the nature of the controlling stage, which in turn also determines the dependence of their rates on the concentration and nature of $:Y^-$.

From the viewpoint of such classification of reagents and types of reaction, the substitution of Eq. (2.19c) can be considered as a nucleophilic substitution of ligands. If it is rate controlling, then the nature of the attacking nucleophile (SO_4^{2-} or HSO_4^-) can have a significant effect on this S_N process and on the dissolution of the metal. Thus, the strong influence of the anionic composition of the solution on complex dissolution–passivation processes can be understood since these involve competition between the solution components in the formation of compounds of various stabilities and solubilities on the metal surface. To a significant extent these

*If in the attack by a reagent on the substrate the electrons remain at the reaction center of the substrate, then the group entering into combination with it will be an acceptor of electrons. Such reactions (electrophilic substitution) are most often caused by transfer from the metal of π electrons to the ligand.[112]

properties determine whether the metal will corrode and at what rate—or whether it will be passivated. Anions, in particular, are strong nucleophiles and their capacity to participate in complex formation on the surface of a metal is quite well known. In view of this, we have proposed that the dissolution of an electrode and its inhibition in the presence of surface active anions should be looked at as a complex process involving stages that are analogous to the reactions of nucleophilic substitution of ligands.*

An understanding of the role of the chemical nature of anions and of their reactivity in the dissolution of a metal using such an approach, requires addressing the problem of the effect of a variable reaction centre on the rate of the process. As seen from Fig. 2.6, the introduction into a borate buffer of 0.05 M sodium salts of monobasic acids significantly changes the critical passivation current density, i_p, the value of which can, to a first approximation, be considered as a measure of the activating or passivating capability of the anion.

However, the use in this case of equation (2.15b) does not provide an adequate description of i_p. Thus, we have allowed for the fact that one of the deficiencies of the Edwards criteria is the insufficient consideration of solvation effects. This is repeatedly noticed in an analysis of its applicability to homogeneous reactions.[123] To overcome this deficiency and broaden the possibilities for the quantitative consideration of the nucleophilicity of anions, we propose that their solvation effects are proportional to the π constants (which reflect the hydrophobicity, see section 2.2) of the atoms or groups of atoms making up the anion.

For a rotating iron disc electrode, such a modification of the Edwards equation as given in Eq. (2.15b) takes the form

$$\log i_p = 2.642 - 0.292 \log MR + 0.013H + 0.088\pi \qquad (2.23)$$

In 0.05 M solutions of inorganic anions, only cyanide fails to meet this relationship (Eq. 2.23) and is evidently associated with some other mechanism of passivation. The equation is also not valid for solutions of organic anions of such high hydrophobicity that they can fulfill the role of a passivator. Thus, passivation occurs in a solution of a benzoate ($\pi_{OCOC_6H_5} = 1.46$) in agreement with Eq. (2.23), but the passivation is considerably facilitated in presence of phenylanthranilate ($\pi_{C_6H_5NHC_6H_4COO} = 2.86$) and the composi-

*The author realizes that energetic heterogeneity and the presence on the surface of Me—Me, Me—O—Me, etc., bonds, introduce significant difficulties and, at least at present, cannot be directly taken into account in such an approach. Therefore, the proposals that are being developed should be considered only as models and the corresponding reactions as approximate schemes for the process.

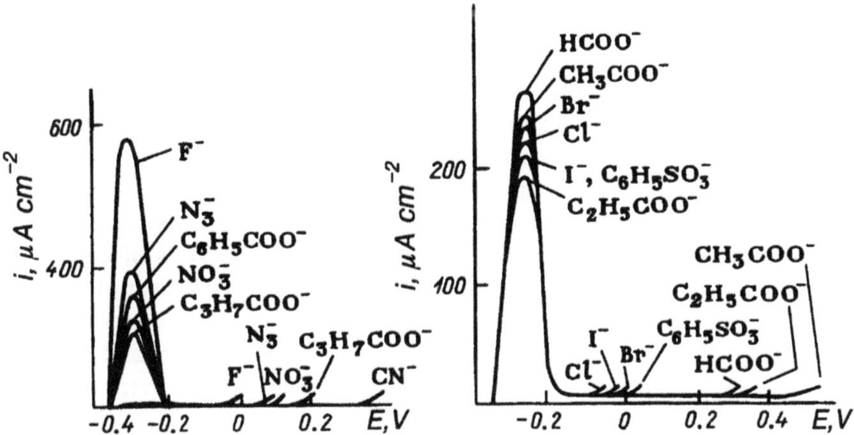

Figure 2.6. Anodic polarization curves for a rotating iron disc in a borate buffer containing 0.05 M sodium salts of monocarboxylic acids.

tion and formation kinetics of the passivating films, as shown in Chapter 1, differ one from the other. Consequently, inhibitors of metal dissolution which encourage passivation should in the first place be those anions whose complexes or salts are of low solubility in the given conditions. The effect of the anionic composition of the solution appears even more clearly in the depassivation of the metal and, in particular, on the characteristic value of the critical potential for pit formation, E_{pit}. The concept of complex formation has been developed particularly intensively for this situation and, in fact, the first attempts to uncover the role of the nature of anions in pit formation were directly or indirectly connected with this concept. Lorking and Mayne,[128] following the hypothesis of Piontelli on the determining role of the polarizability in the adsorbability of anions, associated the aggressive action of Cl^-, Br^-, and I^- with the polarizability, since its increase facilitated the introduction of the anion into the double layer and the formation of soluble compounds with the metal. In their view, F^-, NO_3^-, and ClO_4^- could not act in this manner since they were of low polarizability. Thus, according to Foroulis,[129] SO_3^- and NO_3^- should concentrate in the diffuse part of the double layer and inhibit formation of pits on aluminum in chloride solutions. However, such an assertion is in error since, depending on the anion concentration, the potential and the nature of a metal, depassivation can be initiated by various anions including NO_3^-, SO_4^{2-}, and ClO_4^-.

It is clear that the effect of anions on, for example, E_{pit} cannot be explained solely by the value of their polarizability. Novakovskii[130] identified the following properties which aggressive anions should possess: a high mobility and solubility (or a tendency to supersaturation) of compounds of the anions with cations of the corroding metal; a low affinity for protons; the ability, relatively easily and reversibly, to replace water in the solvation shell of the cation; and an absence of the ability to split off oxygen in strongly acid solutions. Anions having the above properties (Cl^-, Br^-) will support pitting, but in the initial stages of depassivation, which takes place with the displacement of passivating species, an important role, in our opinion, must be played by the nucleophilic properties of anions. In this case, an increase in basicity raises the affinity of particles not only for the metal but also for water, which is the solvent. It is of interest that Rahner *et al.*[(131)] proposed to examine pitting as a consequence of ions and neutral molecules present in the electrolyte influencing proton transfer into the surface layer and thus, the activity of the passivating particles (OH^- and O^{2-}). In the adsorption of an anion close to a hydroxyl ion, the forces of detachment which arise between species with the same charge are compensated by the transfer of a proton to the surface. The neutralization by this proton of the OH^- disturbs the local hydroxide surface layer, thus facilitating the emergence of a metal cation into the solution at this site. This event is then compensated by the supply to the surface of a proton which leaves a defect in the water structure and so facilitates the migration of the activator. The localized nature of the breakdown of the surface oxide will be aggravated with varying rates of protonation of parts of its surface. The authors propose that the effect of the anion type is associated with the differences in the peripheral intensities of the electrical field that they set up, i.e. in their ability to attract cations. Such an approach still does not provide a quantitative assessment of the effect of the nature of the anion, for example, on the value of E_{pit}, or of the predominating electrostatic role, that is, the "hard" factor is also exaggerated and requires further substantiation.

It is more logical to suggest that in the preceding model the effect of the nature of the anion lies in its ability to act as a ligand in complex formation. Here, two approaches are possible. One of these, used by Foley *et al.*,[132–135] was developed in studying the pitting of aluminum alloys, and suggests that there is a rapid electrochemical stage in the oxidation of a metal and that complex formation occurs with a metal cation by the reaction

$$Al^{3+} + xAn^- + yOH^- + (6-x-y)H_2O \rightarrow [AlAn_x(OH)_y(H_2O)_{6-x-y}]^{3-x-y}. \quad (2.24)$$

According to Foley, pitting results from the formation of a soluble complex of the $AlBr_4^-$ type (with an aluminum cation from the lattice) and the passage of this complex into solution with consequent thinning of the passive film. The other mechanism for pitting corrosion involving complex formation was put forward earlier[14] and developed[136] by Kolotyrkin and his co-workers. According to this mechanism, Cl^- forms a complex, $[MeCl_\rho]^{\sigma-\rho}$ which, as a result of its negative charge, orientates the water dipoles with their positive ends towards the complex. This leads to the formation of strong bonds between them and the transfer of the mixed complexes into solution:

$$(MeCl_\rho)^{\sigma-\rho} + nH_2O = [MeCl_\rho\,(H_2O)_n]^{\sigma-\rho} \quad (2.25a)$$

$$[MeCl_\rho\,(H_2O)_n]^{\sigma-\rho} \rightarrow Me^{z+} + \rho Cl^- + nH_2O + (z-\sigma)e \quad (2.25b)$$

The distinguishing features of this type of mechanism are the participation of aggressive anions in the stage of metal ionization and also the labile nature of the formed complexes. The localized nature of the corrosion is associated with the heterogeneity and defect nature of the film, which results in a non-uniform distribution of current across the surface. In this situation, the electrochemical stage is rate determining and E_{pit} is treated as the potential which must be reached for the formation and subsequent ionization of the adsorbed complex.

It should be pointed out that the real surface of a metal is often hydroxylated, with the complex $(MeCl_\rho)^{\sigma-\rho}$ evidently being formed as a result of the displacement of OH^- from the surface by chloride. This stage will have an electrochemical nature, thus it will take place with a partial ionization of the metal ($\sigma > 1$). Such an approach is undoubtedly broader than that described above and points to the possibility of pitting to occur even in those cases when the anion is capable of complex formation in the bulk of the solution. It was on this basis that Kolotyrkin *et al.*[136] developed the physical theory of functional pitting which has shown a good agreement between calculated and experimental data obtained on iron and nickel in chloride solutions.

In our work on complex formation, we introduced the concept of nucleophilic substitution of ligands in a surface complex. This considered the anion as the attacking nucleophile in the reaction:*

$$[Me(OH)_kS_l]^{z-k}_{ads} + mAn^- \rightarrow [Me(OH)_{k-m}S_{l-n}An_m]^{z_1-k}_{ads} + mOH^- + nS + (z - z_1)e \quad (2.26)$$

*Here the anion is shown as singly-charged and the process as a single step, which is not necessarily always the case and is used here only for simplification.

From which it follows that An^- displaces from the surface complex not only oxygen containing particles such as OH^-, but also molecules of solvent S (water or an organic compound). The behavior of a metal is thus determined by the reactivity of An^- and by the properties of the complex that is formed. If the latter is sufficiently stable and difficultly soluble, then the passive condition is stabilized, but the formation of a soluble complex will initiate the depassivation of the metal.

The experimentally measured value of E_{pit} reflects not only the thermodynamics of complex formation but also the kinetic limitations for its occurrence. Whence

$$E_{pit} = E_{eq} + \eta_{pit} \tag{2.27}$$

where $E_{eq} = E^{o}_{Me/Me^{n+}} - RT/nF \ln(K_s a_{An^-}/a_{[MeAn]})$ is the equilibrium potential for the complex formation and $\eta_{pit} = \eta'_{pit} + \Delta$ is the overpotential of the process which includes the potential drop Δ in the passive film—more precisely, in its defects—and the overpotential η'_{pit} which depends on the nature of the metal and the activator. The nature of η'_{pit} is not completely clear and it should be considered to be some effective overpotential connected with the kinetics of the process. It follows from Eq. (2.26) that η'_{pit} depends on at least two factors, the nature of the metal and the affinity of An^- for water. The latter can be viewed as the hydration energy ΔG_{hydr} or the coefficient of hydrophobicity of the atoms or groups making up An^-.

It was found that for the series of metals that was studied, η_{pit} increased linearly with increase in ΔG_{hydr} of the anion and that with a constant anionic composition it was least for copper and greatest for nickel (Fig. 2.7). Hence, one can understand the earlier observation of Rozenfel'd and Maksimchuk[137] that SO_4^{2-}, although a depassivator for iron, does not cause pitting of nickel, since in the latter case η_{pit} must be very large because of the high hydrophilicity of this anion. In view of the limited data available for ΔG_{hydr}, the linear dependence of η_{pit} on iron on the π-constant of the hydrophobicity of various carboxylates merits attention (Fig. 2.8). These include, as well as alkyl and heteroalkyl carboxylates, aminoacids, oxalate, and complexing agents, that is, differently-charged ligands.

The dependencies that are obtained suggest that Eq. (2.26) represents the limiting stage for the initiation of pitting, this being supported by the fact that this reaction is accelerated as the affinity of An^- for water decreases (in this situation there is a reduction in the energy required for the transfer of a nucleophile from the solution to the metal surface). Despite its simplicity and the information this approach gives for understanding whether anions will be inhibitive or aggressive, it suffers from a lack of information

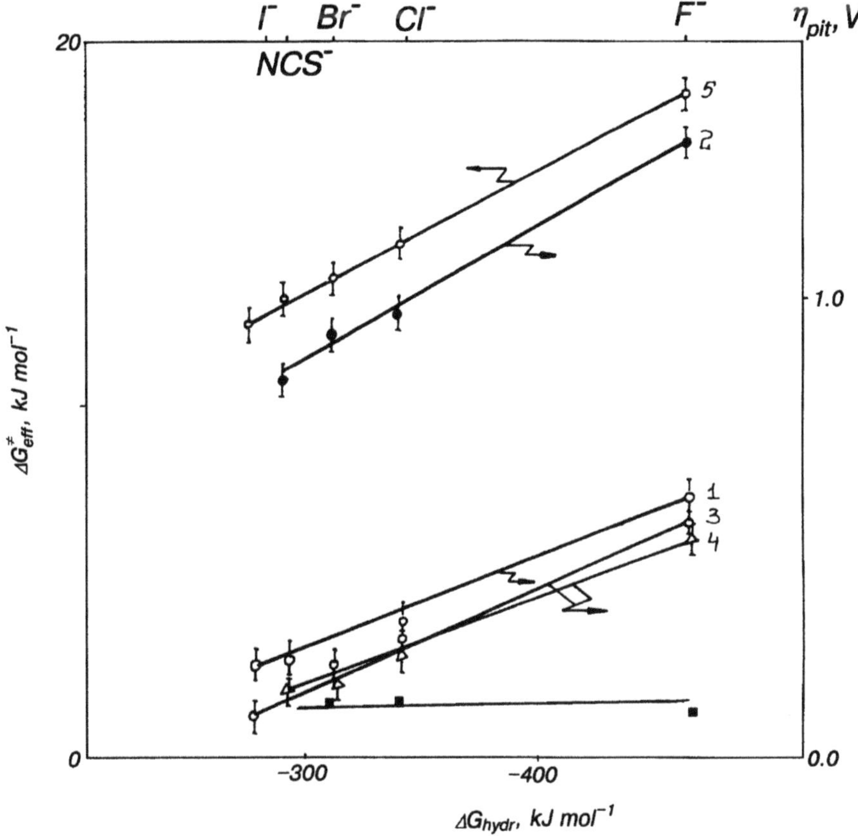

Figure 2.7. The dependence of the apparent activation energy for pitting $\Delta G^{\neq}_{\text{eff}}$ and η_{pit} in a borate buffer containing 10^{-4} M activators on the free energy of hydration (ΔG_{hydr}) for iron (1,5); nickel (2); bismuth (3); tin (4); and copper (6).

of K_s values of the relevant complexes.† Therefore, we have assumed that changes in the value of E_{pit} brought about by changes in solution composition are proportional to the corresponding change in the free energy of activation of the depassivation process, $\Delta G^{\neq}$ (see pp. 117–118). In other words, the higher the value of $\Delta G^{\neq}$ the more positive will be the potential required to initiate the depassivation of the metal. Then, according to the LFER principle, one could expect that for a reaction series represented by

†The values of η_{pit} represented in Figs. 2.7 and 2.8 were obtained by using values of K_s of the complexes of monodentate ligands $[\text{Me}^{n+}\text{L}^{m-}]^{(n-m)}$ provided that $a_{[\text{MeL}]^{n-m}} = 1.0$.

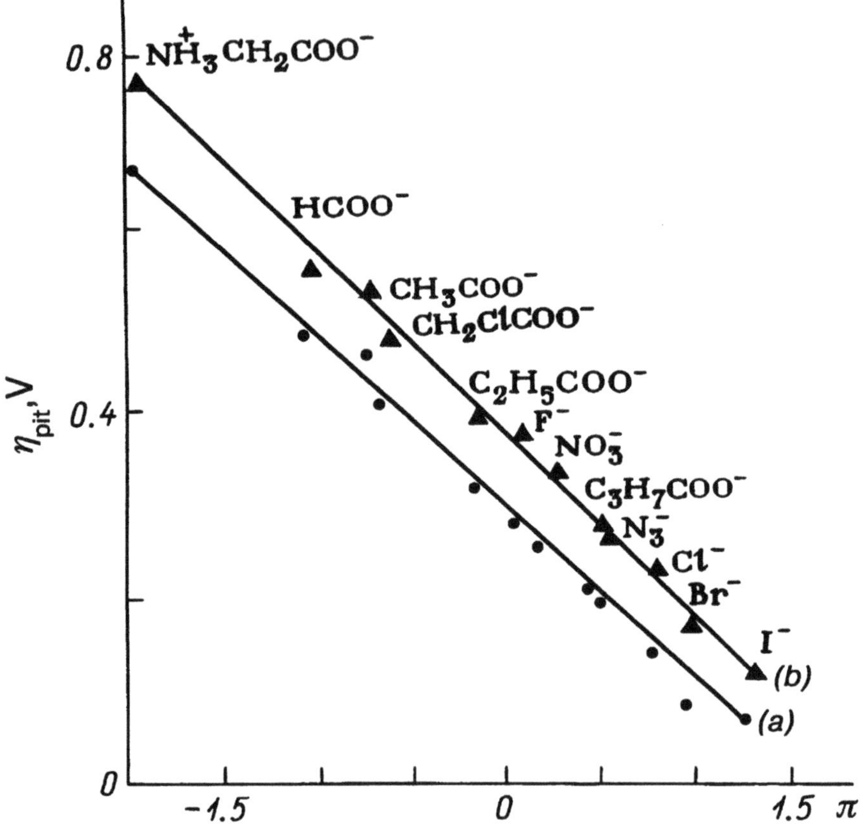

Figure 2.8. The dependence of η_{pit} of iron in a borate buffer containing 10^{-3} M activator without (a) and with (b) addition of 10^{-4} M sodium phenylundecanoate on the π constant of atoms or groups of atoms contained in the activating anions as shown.

carboxylates of the general formula $RCOO^-$, E_{pit} for the initiation of pitting on any metal would be a simple function of a constant of the substituent. In practice, it has been found that E_{pit} does not correlate with the induction Taft constant σ^*, although by taking into account the hydrophobicity, f, and the steric effects, E_s,[(113)] of the substituent it could be successfully described by the relation:[(138)]

$$E_{pit} = a - bf - c(2.6 + 0.6\sigma^* - 2.4E_s) \qquad (2.28)$$

The relationship (2.28), first found for the case of nickel in a borate buffer with pH = 7.4 and C_{an} = 0.05M, has also been confirmed for the pitting of iron, cobalt, and aluminum. From this, it is seen that the basicity and also the more clearly expressed electron-donor properties as described by the electronic constants of the substituent, do not uniquely determine the reactivity of the anions and it is necessary to consider changes in the solvating effects.

The suggested treatment of changes in E_{pit} as an energetic characteristic of the process allows us to use the concept of nucleophilic substitution of ligands in a surface complex also for the initiation of pitting. In this case, the weak effect of the basicity of the anions shows up even more clearly, and E_{pit} can be represented as functions of only π and log MR[139]

$$E_{pit} = a - b\pi + c \log MR \tag{2.29}$$

The validity of Eq. (2.29) has been checked with many metal oxide-solution systems. It holds not only for the depassivation of metals (Fe, Ni, Co, Al, and Sn), but also for several alloys that have been studied (Fe—Ni, Fe—B) both in aqueous and mixed solutions. Although this equation is applicable only to singly-charged anions, it does not require knowledge of K_s or ΔG_{hydr} and the values of π and log MR of the respective atoms and groups are widely quoted.[116] Furthermore, differing from Eq (2.28), it extends to anions of differing chemical types: hal^-, CNS^-, N_3^-, CN^-, NO_3^-, carboxylates, and sulfonates.

The coefficients in Eq. (2.28) and (2.29), are empirical and depend on the nature of the metal, the solvent, pH, etc. Moreover, they provide interesting information on the mechanism of the process. Thus, increases in the π- or f-constants, (the hydrophobicity of the anion) are accompanied by decreases in the value of E_{pit}, consequently, depassivation is facilitated, which effect—according to our views—is associated with a decrease in the energy barrier of the process. However, the effect is not without its limits, since further increase in the hydrophobicity of the ligand can lower the solubility of the complex and eventually lead to a replacement of the controlling stage which then becomes the transfer of the complex into the solution. Depassivation is then retarded when the hydrophobicity exceeds critical values of f_{cr} or π_{cr} so that Eq. (2.28) is found to transpose in this way completely when $f > f_{cr}$, and Eq. (2.29) to transpose when $\pi > \pi_{cr}$. Thus, caproate ($\pi_{C_5H_{11}COO^-} = 2.65$) in normal conditions in a borate buffer at pH 7.4 does not initiate pitting of iron in contrast to chloride ($\pi_{Cl} = 0.71$) and iodide ($\pi_I = 1.12$). The value of π_{cr} depends on the nature not only of the ligand but also on the complex-former. In this situation, the range of aggressive ions differs significantly for different metals. Thus, for copper and nickel it is significantly wider than for zinc and especially tin, and this, at least in part, can be associated with an increase in the absolute values of the energy of hydration, i.e. the affinity for water of cations of these metals in the order: $Sn^{2+} < Zn^{2+} < Ni^{2+} < Cu^{2+}$.

Anions with $\pi > \pi_{cr}$ are usually the classical inhibitors of localized corrosion and include the higher alkylcarboxylates, aminoacids, and substituted benzoates. Of course, this approach does not exhaust the ways of inhibiting localized corrosion, although it is one of the most widespread methods of achieving this. An assessment of the reactivity of anions in inhibiting solutions is still more complex, since in the scheme of Eq. (2.26) there is the possibility that not only a hydrocomplex but also the complex formed by the inhibitor $[Me(OH)_{k-m}In_mS_l]^{z_1-k}$ can act as the substrate. We shall examine the inhibition of depassivation from this position in the next chapter, only pointing out here that until recently the selection of an inhibitor was mainly dictated by considerations of the particular mechanism of its action. Thus, having in mind the increase in the resistance of steels to pitting corrosion brought about by alloying with molybdenum, special attention has been given to the study of Na_2MoO_4 as an inhibitor of the corrosion of steel[140,141] and aluminum.[142] Sugimoto and Sawada[143] compared the protection of 20Cr–5Ni type steel produced either by alloying with molybdenum $\leqslant 5\%$ or by introducing Na_2MoO_4 into the solution. The minimum temperature at which pitting formation was observed increased linearly with increase in content of Mo and MoO_4^{2-}, with the greatest effect obtained with a combination of both methods. Furthermore, photoelectron analysis of the passive film formed on steel in a solution of 0.1M NaCl + 0.2M Na_2MoO_4 indicated the presence of only Mo^{6+}, therefore the inhibiting action of molybdate must be associated with its adsorption. This is in agreement with the power dependence of the protective concentration of MoO_4^{2-} on Cl^-, which follows from the assumption that adsorption of both anions obeys the Freundlich isotherm.

The protective action of chromate is, on the other hand, associated with inclusions of difficultly soluble Cr^{3+} compounds in the passive film. It is generally considered that chromates protect stainless steel from pitting corrosion by a mechanism which differs from that in the inhibition provided by nitrate or sulfate. In the first case, the passive region of potentials is broadened mainly as a result of displacement of the steady-state potential, E_{st}, in the negative direction, and in the second case by an ennoblement of E_{pit}.[144] The inhibiting action of phosphates is also explained by the formation of difficultly soluble compounds which, according to Batroff and Tostmann,[145] can be enhanced by the combined introduction of phosphates and MoO_4^{2-} or WO_4^{2-} as a result of the formation of heteropolyacids in the solution. Evidently, the effectiveness of phosphates depends strongly on the composition of the metal that is to be protected and on the conditions in which the pitting occurs. Thus, according to Rajagopolan and Venkatachari,[146] HPO_4^{2-} is more effective than CrO_4^{2-} in preventing pit formation on steel, but CrO_4^{2-} is the more effective with aluminum.

Until recently, organic anions were considered to be less effective in this respect than inorganic anions which, as proposed,[147] should form stronger bonds with a metal because of their better polarizability. However, it has been known for a long time that some salts of organic acids are capable of fully suppressing the corrosion of low carbon steel in neutral salt solutions. Since an ennoblement of E_{st} is observed in such cases, it is evident that the inhibitor also prevents pit formation. Thus, a power dependence exists between the concentration of various aggressive anions and the minimum concentration necessary for protection not only by nitrite and chromate but also by benzoate.[148] This relationship is often considered to be the result of competition between the two types of anions, the adsorption of which is described by the Freundlich isotherm. The power law also holds in solutions of various organic compounds, including those used in the protection of aluminum alloys where corrosion is usually of a localized nature.

The role of the chemical structure of organic anions in inhibition of pitting corrosion has been studied in detail in our own work and is one of the basic questions addressed in the next chapter.

2.4. THE INFLUENCE OF THE SOLVENT ON THE EFFECTIVENESS OF CORROSION INHIBITION

Changing the solvent can have a significant effect on the nature and behavior of other components of a solution. According to the concepts of complex formation, the energy of adsorption of an anion or other nucleophile, $:Y^-$, will be greater the higher the energy of its bond with the metal ΔG_{Y-Me}, the higher the energy of the mutual interaction of solvent molecules ΔG_{S-S}, the lower the energy of the bond of the metal with the solvent ΔG_{Me-S}, and the lower the energy of the bond of the nucleophile with the solvent ΔG_{Y-S}.[149] Hence, the gain in ΔG_{ad}^{Y} on replacing solvent S_1 by S_2 can be described by the equation:

$$\Delta\,(\Delta G_{ad}^{Y}) = (\Delta G_{Y-S_2} - \Delta G_{Y-S_1}) + (n_2\Delta G_{Me-S_2} - n_1\Delta G_{Me-S_1}) - (n_2\Delta G_{S_2-S_2} - n_1\Delta G_{S_1-S_1}) \quad (2.30)$$

The last term can be neglected for weakly-associated solvents, but in any case, changes in the adsorption of the nucleophile, and consequently its activating or inhibiting action, will depend to a significant extent on the known that the change in the equilibrium configuration brought about by solvent replacement is caused by the difference in the solvation energy of the species that exist in equilibrium. The influence of the solvent on the reaction rate will depend on changes in stability of the initial and transition states. If the solvation energy of the initial compounds increases only slightly (or if it decreases), and that of the transition state increases, then the rate of the reaction will be increased.

In considering the importance of solvation and the large contribution made to it by electrostatic effects, it must be pointed out that anions are best solvated by solvents of the polar type $\overset{+}{H}-\overrightarrow{X}$ (where X=halogen, O, or N), for example, chloroform or formamide, which can form hydrogen bonds with the anions. However, the best solvents for cations are usually compounds possessing free pairs of electrons in hybrid atomic orbitals, i.e. $\overset{+\rightarrow}{X}$: . It can then be understood that it is no coincidence that formamide, which also satisfies the latter requirement, is, together with water and methanol, the strongest solvent. Polar aprotic solvents [dimethylformamide (DMFA), dimethylsulfoxide (DMSO)] are poor solvators of anions but good solvators of cations: acetonitrile generally has a weak solvating action towards ions. Solvation usually increases with an increase in the charge of the particle being solvated and decreases with dispersion of the charge, however the latter effect is less than it would be in the complete absence of the charge. Hence, the greater hydrophilicity of F^- or SO_4^{2-} than I^- and CNS^- can be understood.

It is tempting to estimate the solvating ability of the solvent on the basis of electrostatic considerations using the dielectric constant ε as the characteristic property. However, it was found that its influence on the solvation of ions in the majority of cases is only significant at small values ($\varepsilon \leqslant 30$).[113] With high values of ε, specific solvation predominates when the molecules of solvent enter into complex formation with particles of the dissolving substance. In practice, there are many estimates made of solvating power in which specific and non-specific solvations are considered simultaneously and in which the solvent is represented as an isotropic uniform medium characterized by a particular macroscopic constant.

The empirical approach of Alexander and Parker in Ref. 113 merits attention. Having determined the solubility of difficultly-soluble electrolytes in solutions with $\varepsilon > 30$, they calculated the logarithm of the standard activity of anions in relation to a DMFA solution, since the latter provides almost no solvation of anions. Some interesting observations were made using conditional values for the activities as shown in Table 2.1. For example, it can be seen that in almost every sequence the changes in the calculated activities of the anions or of the influence of the solvent on these were the same. Therefore, the average value of the logarithm of the conditional

Table 2.1. Logarithms of Conditional Standard Activities of Anions at 25° C

Anion	H_2O	CH_3OH	$HCONH_2$	CH_3SOCH_3	CH_3CN	$CH_3CON(CH_3)_2$
CH_3COO^-	−18.4	−11.9		−3.0	−2.4	1.0
Cl^-	−15.3	−9.2	−8.8	−1.3	−1.2	1.3
Br^-	−13.3	−7.6	−7.3	−1.6	−1.7	1.0
N_3^-	−13.0	−7.6		−1.7	−1.2	1.3
I^-	−10.4	−5.3		−1.6	−1.2	0.4
SCN^-	−6.7	−5.4		−1.6	−1.1	0.5
ClO_4^-	−6.7	−2.0		0.3		
S-parameter	1.76	1.0	0.92	0.24	0.19	−0.11

activity when related to the corresponding value in methanol (using this as a standard solvent) can be taken as a parameter S which reflects the strength of the solvent. It should be noted that water solvates anions more strongly than other solvents and various types of interactions can result from this effect. An important example is that of donor–acceptor interactions which can in a broad sense represent acid–base reactions between a solvent and a dissolved substance. Water, as is the case with many other solvents, can act as a donor or acceptor depending on the nature of the dissolving substance. For characterizing these properties of the solvent, Gutman has introduced donor, DN, and acceptor, AN, numbers. DN is determined from the enthalpy of formation of the adduct between a molecule of the investigated solvent and a standard acid ($SbCl_5$) in a solution of 1,2-dichlorethane. AN is found from the chemical shifts of phosphorus in NMR spectra of $(C_2H_5)_3PO$ in the respective solvents. The dependence of the rate of the chemical reaction on the nature of the solvent can be expressed in this case by the equation

$$\ln k = a + bAN + cDN \tag{2.31}$$

According to Lewis, AN and DN are often correlated with other parameters, for example, basicity (nucleophilicity) or acidity (electrophilicity) by the Dimroth–Reichardt and other parameters. This means that certain solvating effects connected with covalent bonding, ion–dipole and acid–base interactions, formation of hydrogen bonds, and so on have been taken into account. Serious deviations from Eq. (2.31) occur in two cases, that is, when using highly structured solvents or solvents with $\varepsilon \leqslant 10$. The former include solvents in which the molecules show both donor and acceptor properties which mutually interact. Naturally, the formation of hydrogen bonds has a considerable influence on the structuring of a liquid, and protonated solvents (particularly water, alcohols, and formamide) are typical examples of this group. In analyzing deviations from Eq. (2.31) in the second group of solvents, Schmid and Sapunov[150] came to the conclusion that ε is a measure of the amphoteric character of the solvent and that it could be described by the linear relation:

$$\log \varepsilon = 0.0711\,AN + 0.054\,DN + 0.2581 \qquad (2.32)$$

Consequently, the addition to Eq. (2.31) of a third parameter, ε, does not make sense. This is the case with any attempt to use universal characteristics of the solvent to find a general approach to all situations in which the nature of the solvent affects reaction rates. Evidently a dependence of the type of Eq. (2.31) is fulfilled only where steric hindrances—as well as other more subtle effects not yet fully understood—can be neglected.

One can see that the difficulties in considering the effect of the solvent are much the same as those that arise in examining the reactivity of other components of the solution. This can be further appreciated when examining the reactions of complex formation in which the solvent, particularly when it has high nucleophilic properties, can only conditionally (and far from always!) be taken to be chemically inert. In general, reactions of complex formation in donor solvents, as characterized by high values of DN, must be understood as reactions of ligand substitution of the solvent molecules coordinated with the central atom.

It is significant that the most interesting solvent for corrosionists, water, has an amphoteric character. Its strong solvation of anions lowers their adsorption which, depending on the An^{n-} function (inhibiting or activating), is capable of preventing or improving the protection of a metal. Furthermore, specific solvation in protonated solvents of highly basic nucleophiles lowers their reactivity to the extent that the nucleophilicity of halide ions is usually in the following order: $I^- > Br^- > Cl^- > F^-$. On the other hand, the aprotic solvents, which have effectively no solvating action on anions, the sequence is in the reverse order. When considering the role of anions in affecting the dissolution of a metal through complex formation, it is useful to remember that the stability of complexes in a solution usually decreases with increase in the donor strength of the solvent.[151]

Thus, solvating effects of the solvent are capable of significantly complicating the prediction of the effectiveness of inhibitors. If there are activators and inhibitors present in a solution, then changes in their solvation resulting from a change in the solvent can lead to different effects, depending on the nature of the change. Usually, large organic anions are less sensitive to the nature of the solvent, although their complex forming capabilities may be changed and furthermore, one cannot exclude the probability of the formation of soluble complexes in the new solvent.

It must also be kept in mind that the solvation of metal cations is usually more marked than that of anions and this fact will in some complex manner affect the stability and solubility of complexes.[152] Furthermore, since in the formation of complexes by, for example, the scheme of Eq. (2.26), an important role is played by the adsorption capability of the solvent molecule, it can be understood why the true mechanism of the effect of the solvent on inhibitor effectiveness often remains unclear. Even hydrocarbon solvents, which do not contain polar groups, often take part in the formation of protective layers on a metal and cannot be considered to be inert. In many respects, this property has contributed to the wide use of inhibitors in the oil industry. Bregman[21] has expressed the opinion that "the advent of organic long chain, high molecular weight corrosion inhibitors revolutionised the industry" The films that are formed by these inhibitors are usually of multilayer thickness. Apart from adsorption of an inhibitor, in which the interaction of its polar group with the metal plays an important role, although it can also be chemisorbed, the screening action can be further enhanced by the attraction of its alkyl groups to the hydrophobic molecules of the hydrocarbons. Such blocking of the surface prevents the emergence of metal cations and the penetration to the metal of the water. In the absence of this, corrosion of steel would not normally develop in hydrocarbon solutions. The probability of the formation of a "sandwich" is higher the longer the hydrocarbon part of the inhibitor, and this is particularly the case with colloidal solutions in water, i.e., fully developed surfactants. On reaching the critical micelle concentration, CMC, they form a new colloidal phase—micelles—aggregations of several tens or even hundreds of molecules. In water, a minimum free energy is associated with the formation of these micelles, which have alkyl groups on the inside and hydrophilic polar groups on the outside. In hydrocarbon solutions, the micelle is constructed in the opposite way and not surprisingly, the adsorption of a surface active inhibitor from such solvents is different.

According to Rebinder, the best conditions for the adsorption of amphiphilic surfactants occurs at the interphase boundary where there is the greatest difference in polarities. This provides the possibility for hydrophobization of the metal surface by hydrocarbon-soluble surfactants and particularly those with polar groups which will then guarantee a chemisorption bond with the metal. In this case, the difference in polarities at the interphase boundary is decreased as a result of the orientation of the hydrophobic alkyls towards the hydrocarbon medium and of the hydrophilic groups towards the metal. This is a situation usually made use of in the inhibition of corrosion in hydrocarbon media where the surface of the metal is hydrophilized by a layer of FeS.[152] The alkyl chains of the inhibitor adsorbed through polar groups in the FeS layer can interact with the hydrocarbon and dissolved inhibitor, so as to form a 'sandwich' consisting of an inner layer of adsorbed inhibitor, then a hydrocarbon layer, and then an outer layer of inhibitor oriented with the alkyl chains towards the metal and the polar groups towards the bulk hydrocarbon medium. The triple layer corresponds to a laminated micelle which solubilizes, that is, contains within itself, the hydrocarbon. Obviously, it will be very difficult for aggressive components to penetrate through such a barrier film to the metal surface. If the adsorption of these colloidal surfactants from water takes place, then after the attainment of a CMC by the first chemisorbed layer, the surfactant can form a second layer with the reverse orientation of its groupings, hydrophobic to the surface, polar to the solution, although such a film will not contain hydrocarbon solvent between the layers. In Gonik's view,[152] the absence of the stabilizing influence of the hydrocarbon interlayer in the protective surfactant film is responsible for the reduction in inhibitive properties.

In the case of cationic surfactants in oil–water systems, the micelles with hydrocarbon inclusions can be considered as polycharged cations which will be easily adsorbed on a negatively-charged metal surface. When rearranged into the "sandwich" as described above, the protective film will, as a result of its structure, provide a noticeable post-treatment protection, it will retain its effectiveness in a corrosive medium even in the absence of an inhibitor. Clearly, the screening by the hydrophobic film of a chemisorbed or electrostatically-adsorbed first layer of inhibitor on the metal will play a positive role here. The problem of the stabilization of such a structure, which is particularly important for two-phase systems of the oil–water type, has been discussed in the literature.[21,152] However, it should be noted that in a number of situations there are significant difficulties in achieving stabilization. For example, problems arise when outer layers are removed as a result of movement of the environment, where the effectiveness of the inhibitor film formation is markedly dependent on the composition of the corroding surface, or when various pollutants, deposits, etc. are present. Despite the considerable number of inhibitors of this type in use, the tasks of increasing their effectiveness, developing better inhibitors, and improving the technology of their application remain, as always, real problems.

Water, which is the main solvent among corrosion media and which usually stimulates the active dissolution of a metal has, nevertheless, no influence on the rate of dissolution of chromium in methanolic HCl.[153] This example indicates that protonated solvents, in a way similar to water, can act as corrosive agents and solvate cations of the metal or its complexes. According to Shatma *et al.*,[154] the lower fatty alcohols are especially aggressive, but ethylene glycol (ethanediol) and glycerol do not cause corrosion of low carbon steel. In ethanol solutions, the corrosion rate increases pro-

Table 2.2. Critical Concentrations of Water in Alcoholic Solutions for the Passivation of Metals[149]

Metal	Medium	Critical concentration of water, mass %
Iron	(0.2–2) N H_2SO_4, methanol	>10
Chromium	1 N HCl, methanol	10
Nickel	0.1 N H_2SO_4, methanol	1
Titanium	0.1 M NaCl, methanol	4
Titanium	1.0 N HCl, methanol	>4.4
Titanium	$NaClO_4$, methanol	0.5
Zirconium	HCl, ethanol	>10

portionately with the concentration of aggressive alcohol which, like chloride, here fulfills the function of an activator. It is significant that in the absence of water, or more precisely, below a certain critical quantity of water, metals will not become passivated in alcoholic solutions. For example, the introduction of even a small quantity (0.6%) of water into a methanolic solution of rhodamine 6G* can cause inhibition of the corrosion of aluminum.[155] In examining such systems, it is necessary to bear in mind that water is the stronger donor of passivating oxygen, whereas alcohols (methanol) provide a better competition with aggressive anions in adsorption on a passivated metal, as has been shown in the case of niobium.[156] As a result of this, the metal will remain passive in certain ratios of water and alcohol over a wide range of potentials. The critical concentration of water necessary for holding metals in the passive state depends on many factors which can be associated either with the nature of the metal or with the nature of the solute (Table 2.2).

Unfortunately, there is relatively little data on the effectiveness of corrosion inhibitors in solutions of alcohols not containing water, or with $C_{H_2O} < C_{crit}$. The majority of the available data relate to acid solutions and have been considered by Ekilik and Grigor'ev.[157] It is generally considered that the inhibiting action of organic compounds towards any metal is lower in alcoholic than in aqueous media. Two factors may be operating in alcoholic media. Thus, the adsorbability and hence the blocking effect of the inhibitor will be lowered, and there will be an increase in the rate of the

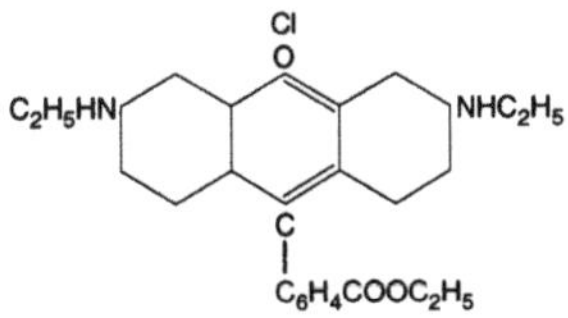

* Cl, O, C_2H_5HN, NHC_2H_5, C, $C_6H_4COOC_2H_5$

recombination stage of hydrogen depolarization control of the corrosion process. This is particularly the case with iron for which organic cations are often used to inhibit corrosion in acid media. If such inhibitors or those of the cationic–molecular type do not possess a strong screening action, then they will be practically without any protective effect on the metal in alcoholic media (Table 2.3).

An extremely effective route for inhibiting the corrosion of iron in alcohols can be proposed by applying our knowledge of the various features of the mechanism of the cathodic process. It is based on the use of sulfur-, selenium-, tellurium-, or antimony-containing compounds which even in small quantities are capable of retarding the recombination of H-atoms, by acting as catalyst poisons. Among these, antimony chloride merits special attention since it stimulates the corrosion of iron in aqueous solutions while effectively suppressing it in various alcoholic solutions of HCl. Unfortunately, such a route to inhibition is promising only for transition metals; however, even in this case the possibility of accelerating electrolytic hydrogen charging should be taken into account, especially in conditions of prolonged protection or with insufficiently high concentration of the corrosion inhibitor.

Evidently, a more universal approach that would be less dependent on the nature of the metal would be one associated with compounds having high chemisorptive properties which, providing the molecular structures were appropriate, could lead to significant blocking of the surface. In a

Table 2.3. Inhibition Coefficients for the Corrosion of Iron in Aqueous and Alcoholic Solutions of HCl [157] by Organic Compounds

Inhibitor	Concentration mol L^{-1}	Inhibition coefficient, $\gamma = K_{blank}/K_{inh}$				
		Water	Methanol	Butanol	Hexanol	$C_8H_{17}OH$
Anthranilic acid	0.01	20	1.1	1.8	—	—
p-tolylthiourea	0.01	140	4.6	1.2	—	—
p-dimethylamino-benzaldehyde	0.002	9.6	1.1	1.1	—	—
1-phenyltetrazole	0.001	220	1.4	1.8	—	—
1-ethyl-2-amino-5-decylimidazolin-bromide	0.001	160	1.1	1.9	—	—
Pyridine	0.001	1.3	1.6	1.2	1.3	1.1
Antimony chloride	0.001	0.4	120	45	53	115
Sodium selenide	0.001	2.8	9	3	2	1.5
$(R\text{-}Te)_2$	0.001	—	140	70	75	75
$R\text{-}TeCl_3$	0.001	—	105	50	60	66

manner distinct from catalyst poisons, inhibitors of the blocking type are effective in protecting iron in hydrochloric acid solutions of alcohols having more than six carbon atoms by retarding the anodic reaction and chemical dissolution.[157] For example, Ekilik and Grigor'ev examined the role of the solvent type using tetrahydrobenzo-pyrrole derivatives as inhibitors of iron corrosion in hydrochloric acid solutions. They found that in aqueous solutions the contribution of the blocking action increased on going from electron-donating to electron-accepting substituents, producing a V-shaped log γ–σ dependence. In alcoholic media, beginning from propanol, the V-shaped log γ–σ dependence was replaced by a linear dependence.

This was explained by the fact that an increase in polarity of electron-donating substituents ($\sigma > 0$) increased the number of chemisorbed particles and the strength of the bond of the inhibitor with the metal. In the case of electron-accepting substituents, the improved protection obtained with an increase in σ is the result of increased physical adsorption, which decreases going from water to alcohols in accordance with the reduction in their ε values and with the increase in n_c, the number of carbon atoms.

As shown below, inhibitors of the chemisorption type are the more universal even in aqueous–organic media, including neutral media. However, it must be kept in mind that complex-forming reagents usually tend towards chemisorption and a change of solvent can facilitate the transfer of the surface complex into the corrosive environment. Furthermore, the molecules of the organic solvent can themselves have a high chemisorptive capacity and can provide a significant competition to the inhibitor. This is particularly the case with aprotic solvents, although a passivating action has been reported with some metals in anhydrous acetic acid.[149] The ability of such solvents to form a strong donor-acceptor bond with a transition metal should increase with increase in their *DN* or nucleophilic constant *B**.[123] Such an inhibiting (passivating) action of solvents is no random event and can in fact be predicted from the value of i_p (critical passivation current density) for iron in a borate buffer,[69] and will increase symbiotically with *B* in the series: acetonitrile, AN(160)—methylethylketone, MEK(209)—dimethylformamide, DMFA(291)—dimethylsulfoxide, DMSO(362).

In this series, the aprotic solvents impede oxide film growth as a result of the chemisorption of their molecules. It is significant, for example, that using coulometric methods it has not been possible to detect an oxide film

*The constant *B* is calculated from the shift in frequency of infra-red absorption by the OH group in phenol in solution in carbon tetrachloride (the solvent in this case) on the addition of bases which form hydrogen bonds, $B = \upsilon^{CCl_4}_{C_6H_5OH} - \upsilon^{CCl_4}_{C_6H_5OH} ......S$.

on a passive iron surface in the H_2O–DMSO (1:1) system. A passivating action of DMSO has also been found with nickel.[149]

DMFA has noticeably lower inhibiting and antioxidizing actions. However, in aqueous–DMFA media (1:1) a sharp decrease in the adsorbability of inhibiting anions (phenylanthranilate, phthalate, and sodium chlorobenzoate) has been found on iron surfaces. This is difficult to associate with solvation effects since the mixed solvent is quite rich in water, which strongly solvates anions. It is more probable that nucleophilic substitution in the surface complexes formed by the DMFA inhibitor is difficult because of the high electron-donor properties of the actual aprotic solvent.

The chemical interaction of an organic solvent with a metal can also be the cause of corrosion; this is particularly the case in chlorinated hydrocarbons which occur widely in industrial practice. Thus, according to Winterton *et al.*,[158] pitting of aluminum in 1,1,1-trichlorethane occurs as a result of the release of Cl^- from the organic solvent that is weakly adsorbed on the metal surface,

$$Al + Al_2O_3 + 3CH_3CCl_3 \rightarrow 3AlOCl + 3CH_3CCl_2\cdot \qquad (2.33)$$

The next chemical step involves the dimerization of the 1,1-dichlorethane radicals, a disproportionation reaction and, at the bottom of the forming pit, an interaction between the metal and the solvent which then acquires an autocatalytic character leading to breakdown of the latter and the build-up of $AlCl_3$. Similar processes take place on aluminum in CCl_4 where hexachlorethane has been found as well as $AlCl_3$.[159] Small quantities of water, ultra-violet radiation, or salt impurities (e.g., $AlCl_3$) facilitate the breakdown of CCl_4, with the catalytic activity of the metals falling in the series: low carbon steel–aluminum–copper–zinc.[160] It is known that amines have been successfully used for corrosion protection in such systems. These neutralize HCl and, being adsorbed on metals, block the surface and thereby inhibit the decomposition of the chlorinated hydrocarbons. In the case of an even more corrosive medium, such as, according to Gerasyutina *et al.*,[161] the 8% DMFA + 92% CCl_4 system acting on steel St3[C:0.14–0.22%], soluble $[Fe(DMFA)_4Cl_2][FeCl_4]$ complexes are formed and the use of stronger complex-forming agents is necessary for inhibition. The authors of that work obtained a high coefficient of inhibition, Z, of 90–99.9% by using such well known chelating agents as 8-hydroxyquinoline, oxalic acid, or succinic acid. The success of such an approach was confirmed by the effective inhibition by triazoles of the corrosion of copper in a methylene chloride–ethanol (9:1) system containing 0.1mM HCl, 0.2mM CH_3COOH, and 0.2 M H_2O.[162]

Despite the relatively limited number of studies of corrosion inhibitors in non-aqueous media, it may be suggested that the use of chelating reagents would be extremely promising. For example, one could point out that the type of passivity that would result would be similar to salt passivity, which is frequently seen in many non-aqueous media or mixed solvents that are rich in non-aqueous components.[157]

3

The Effect of Organic Compounds of the Initial Stages of Localized Corrosion

3.1. ADSORPTION ON PASSIVE METALS

3.1.1. The General Situation

Many technically important metals and alloys begin to corrode in neutral solutions from the passive state, or, at least, from a condition in which a metal surface has been partially passivated by an oxide. Hence, in the majority of cases, the initial stages of corrosion are localized at defects in the primary oxide film, which is normally formed on contact of the metal with air. It is then reasonable to expect that the ability of components of a solution to initiate or inhibit these events will depend to a marked extent on the adsorbability of either the actual components, or the products of their interactions with cations of the corroding metals.*

However, it is unclear whether the defects actually "pass through" the passive film (whether they penetrate through to the metal), or whether the surface at such points in the oxide is in a more active condition with respect to adsorption processes. Arriving at a definitive answer to this question would be facilitated by a theoretical examination of the problem, since it is well established that differences in the structures of the electrical double layers on an oxide and on a metal, that are associated with the localization

*The action of special reducing agents that can lower the concentration of dissolved oxygen in the solution is not restricted only to decreasing the rate of the cathodic reaction since, to a certain extent, their adsorption can also affect the nature of the anodic process.

of charges on the surface of the metal, can have a marked effect on the adsorption of molecules or ions from aqueous solutions.[163]

Despite the uncertainties that exist in characterizing the surfaces of oxidized metals, there are some general statements that can be made concerning the theory of adsorption in relation to such surfaces. According to Rehbinder's law of compensating polarities, electrolytes, i.e., various salts, should be capable of being adsorbed from water onto a metal since their polarity lies between that of the solid adsorbent and the solvent. Adsorption on a more polar oxide is more difficult from water than from a non-polar solvent. This law is, however, not fulfilled in the formation of chemical (hydrogen, covalent) bonds which often occur in the adsorption of organic substances on oxides.

Nechaev and Urbakh[164] have treated chemisorption as an exchange reaction resulting from the overlapping of the wave functions of electrons of the adsorbate and the adsorbent. They concluded that instead of an electron level rigidly fixed in an isolated organic molecule and characterized by the first ionization potential J, there will occur, as the molecule approaches the adsorbent, a set of energy levels of width F and a value Λ. The latter is a function of J and has a minimum value depending on the adsorbent (metal or oxide) of $J = J_p$. If the part of Λ which is effectively independent of J is designated as Λ_0 (this would include the energy of interaction of the electron with image forces, both of itself and of an ion, and also the energy of interelectron interactions by the adsorbed particle), then for chemisorption on a metal one can write:

$$J_p = \Lambda_0 + \lambda(X)eE + V/2 \tag{3.1}$$

where $\lambda(X)eE$ is a term that takes into account the effect of the electrode potential E on the position of the molecule of adsorbate. X is the distance from the surface of the adsorbent to the centre of the electron orbital, with X of the order of the thickness of the Helmholtz layer, the coefficient λ varies between 0.5 and 1.0; V is the energy difference between the bottom of the conductivity zone of the metal and the chemical potential of the electrons in the surrounding medium; and e is the electronic charge.

According to these ideas of chemisorption, the energy of the bond (similar to a covalent bond) that is formed, the amount of adsorbed substance, Γ, and the free energy of adsorption, ΔG_A, have maximum values when $J = J_p$ with its width dependent on the width of the electron zone of the adsorbent closest to the level of the molecule. In the case of chemisorption on oxides, it will be necessary to give attention to their significantly lower valence zone widths, which will reduce the maximum of the function $\Delta G_A(J)$.

In the work of Nechaev and Kuprin and their co-workers,[163–165] an experimental study was made of the adsorption from aqueous solutions of a large number of organic compounds on various metals and oxides. They established that J_p is actually an important characteristic of an adsorbent. For example, in the case of powdered iron there is noticeable adsorption of organic compounds having $J \approx 7.3$, 7.9, or 9.4–9.6 eV (Fig. 3.1). The existence of several values of J_p is not surprising, bearing in mind that iron becomes covered with an oxide film even after brief contact with air or with oxygen dissolved in water. Comparing these data with the results of adsorption measurements on powdered αFe_2O_3, one can note that two maxima coincide, with the third being, as the authors propose, J_p of non-oxidized iron. The polyextremal Γ, J dependencies on copper ($J_p = 7.2$), and its oxides ($J_p = 7.9$) have been treated in a similar manner.

The experimental difficulties involved in obtaining a pure, unoxidized iron surface have stimulated the search for a theoretical approach to the prediction of J_p. With certain assumptions the following has been claimed to hold[163,165]:

$$J_p \approx 2\Phi_{Me(H_2O)} = 2(\Phi_{Me} - {}_{Me}\Psi_{H_2O}) \tag{3.2}$$

where $\Phi_{Me(H_2O)}$ and Φ_{Me} are the electronic work functions in an aqueous solution of an indifferent electrolyte and in vacuum, respectively, and ${}_{Me}\Psi_{H_2O}$ is the Volta potential at the interphase boundary (at the zero charge potential $E_q = 0$). A comparison of the calculated and experimental values of J_p, obtained from adsorption measurements with the exclusion of "oxide" J_p values, led the authors to conclude that there was a satisfactory agreement for a series of technically-important metals (Table 3.1). The reasons for discrepancies lay in the overlooking of van der Waals interactions and the fact that measurements of J_p were made at the steady-state potential, E_{st}, and not at $E_{q=0}$.

This approach is undoubtedly of interest in the theory of the inhibition of corrosion. Moreover, an analysis by Nechaev of the ΔG_A dependencies on J for bismuth has shown that even with the same values of J, a substance can be adsorbed to a greater or lesser extent depending on its molecular mass. Evidently, the hydrophobicity of an adsorbate, i.e., the interactions which tend to 'squeeze' it out from the bulk of the solution, can introduce a noticeable contribution to the chemisorption interaction.

Another important factor, which has not been fully taken into account in the above treatment, is the presence of extremely strong electrostatic interactions between the adsorbate and the metal (oxide). We have been interested both in the region of potentials that are usually remote from $E_{q=0}$ where, as a consequence, small inorganic ions will tend to displace

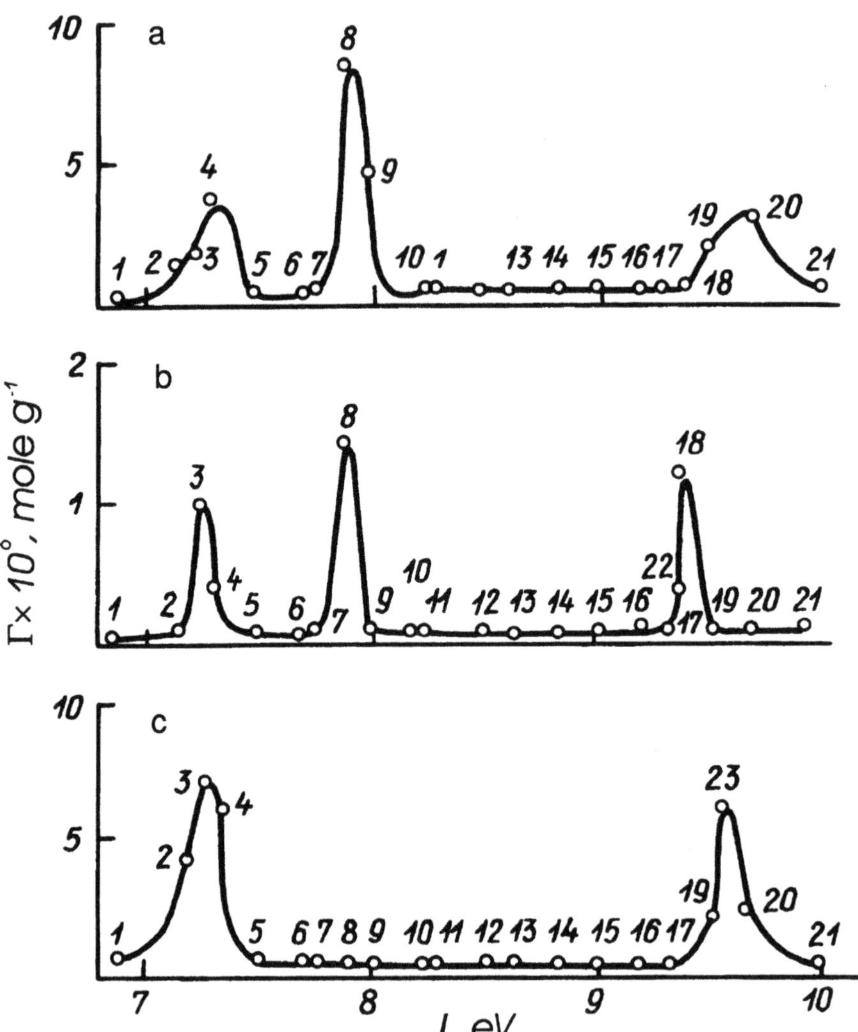

Figure 3.1. The dependence of the adsorption of organic substances in 0.1 M $NaClO_4$ on (a) powdered iron, (b) steel, (c) Fe_2O_3 on the ionization potential.[163] 1, benzidine; 2, N,N dimethylaniline; 3, diphenylamine; 4, naphthylamine; 5, m-toluidine; 6, aniline; 7, α-naphthol; 8, 8-hydroxyquinoline; 9, methylnaphthalene; 10, pyrogallol; 11, m-cresol; 12, resorcinol; 13, phenol; 14, quinoline; 15, 2,6 lutidine; 16, 2-picoline; 17, furfural; 18, pyridine; 19, benzaldehyde; 20, benzonitrile; 21, nitrobenzoic acid; 22 pyrocatechin; 23 p-cresol.

Table 3.1. Resonance Potentials of Ionization of Metals in Aqueous Solutions[165]

Metal	Φ_{Me}(eV)	${}_{Me}\Psi_{H_2O}$(V)	J_p(eV)	
			calc	exp
Bismuth	4.29	0.29	8.00	7.8
Lead	4.06	0.27	7.58	7.8
Tin	4.30	0.35	7.90	7.85
Iron	4.80	0.85	7.90	7.90
Copper	4.55	0.95	7.20	7.20
Chromium	4.45	0.50	7.90	7.90
Nickel	5.04	0.90	8.28	7.90
Aluminum	4.40	0.68	7.24	—

large organic molecules and even anions in the formation of the electrical double layer, as well as in the fact that the charge on the ion apparently has a strong influence on the nature of the adsorption on the oxide. According to Nechaev,[163] the adsorption of univalent anions (excluding F^-) on oxides is of an electrostatic nature and obeys—at least for Al_2O_3, Fe_3O_4, Fe_2O_3, TiO_2, SnO_2, ZrO_2, and PbO—an equation of the Freundlich isotherm type

$$\Gamma = K a_{H^+}^{\gamma} a_{An^-}^{\delta} \tag{3.3}$$

where a_{H^+} and a_{An^-} are the activities of the respective ions and γ and δ are coefficients. The specific adsorption of multicharged anions takes place by another law and leads to overcharging of the surfaces of oxides such that the zeta potential at all pH values is negative and the diffuse part of the double layer is formed by cations. Furthermore, at potentials remote from $E_{q=0}$ the charge on the adsorbent can stimulate the adsorption of surface-active substances of opposite charge, which at concentrations close to the critical micelle concentration will be polymolecular and will therefore act as a diffusion barrier for the chemisorption of species. To be correct, it should be pointed out that Eqs. (3.1) and (3.2) provide the possibility of calculating changes in J_p when the potential of the adsorbent varies. If any potential (especially $E_{q=0}$) is taken as the starting point then on changing its value by ΔE, the following will hold:

$$J_p = J_p^0 + \lambda e \Delta E \tag{3.4}$$

Kuprin and Nechaev[165] describe a case of good agreement of Eq. (3.4) with the experimentally determined dependence of J_p on E, while in the case of adsorption on tin, a second value of J_p was found which was independent of *E*. There is still insufficient basis for explaining the latter, pointing to the need for further studies in this direction. It is also significant that this approach has been developed and demonstrated with the example of the chemisorption of organic compounds when present as molecular additions. However, the concepts of nucleophilic substitution of ligands in the surface complex indicate that the greatest activity would be expected from anions.

Unfortunately, the limited information on J values of anions is a serious impediment for assessing their chemisorption capabilities. Nechaev has used the J values given in Table 3.2 and concluded that on oxide-free cadmium, tin, bismuth, mercury, and iron the surface activity should decrease in the order $I^- > Br^- > Cl^- > F^-$ since the $J - J_p$ difference increases in that order.[163] Following from this suggestion, it is then not difficult to make calculations for those cases where the adsorption occurs on oxides, using values of J_p in this table. Such a calculation shows that on oxides the chemisorption of Cl^- and Br^- is more probable than that of I^-, and that the benzoate anion, $C_6H_5COO^-$, which in neutral solutions is known to inhibit pitting corrosion, can in many cases (Bi_2O_3, SnO_2, Fe_2O_3, CdO) successfully compete with halides in adsorption. Unfortunately, a direct check of the sequence of adsorbability of these ions has not been made, although data obtained by radiometric, ellipsometric, and other methods exists for their adsorption on oxidized metallic electrodes.

Table 3.2. Ionization Potentials of Anions and the Possibility of Their Adsorption on Oxide-covered Metals

Anion	J(eV)	Bi_2O_3	SnO	SnO_2	γAl_2O_3	ZnO	αFe_2O_3	CdO	CuO Cu_2O
		J_p							
		9.59	9.14	9.53	8.00	7.3	9.5	9.52	7.9
		$J-J_p$							
I^-	8.9	−0.69	−0.24	−0.63	0.9	1.6	−0.6	−0.62	−1.0
Br^-	9.1	−0.49	−0.04	−0.43	1.0	1.8	−0.4	−0.42	−1.2
Cl^-	9.2	−0.39	−0.06	−0.33	1.2	1.9	−0.3	−0.32	−1.3
F^-	10.0	0.41	0.86	−0.47	2.0	2.7	0.5	0.48	−2.1
SO_4^{2-}	10.8	1.21	1.66	1.27	2.8	3.5	1.3	1.28	2.9
NO_3^-,ClO_4^-	11.3	1.71	2.16	1.77	3.3	4.0	1.8	1.78	3.4
$C_6H_5COO^-$	9.3	−0.29	0.16	−0.23	1.3	2.0	−0.2	0.22	1.4
CH_3COO^-	9.97	0.38	0.83	0.44	1.97	2.67	0.47	0.45	2.07
CO_3^{2-}	10.2	0.61	1.06	0.67	2.2	2.9	0.7	0.68	2.3

3.1.2. Aluminum and Its Alloys

The adsorption of anions, including those of organic acids, from neutral solutions onto passive metals has probably received the most attention in the case of aluminum, the surface of which, without doubt, is well oxidized. It has been pointed out by Vermilyea *et al.*[166] that those substances which inhibit the solution of Al_2O_3 in water are also effective inhibitors of the corrosion of aluminum. The authors of that work, using perchlorate at pH 5 as the background solution (the pH being controlled by NaOH and $HClO_4$ additions), established that an anodized aluminum foil is easier to stabilize with anions of surfactants. This is associated with the facts that at such a pH, Al_2O_3 has a small but positive charge, and that inhibiting anions hydrophobize to various extents the oxide surface. Some surfactants of the molecular type intensify their protective action with time and it is suggested that this results from their accumulation in pores of the AlOOH layer deposited as a corrosion product on the aluminum surface. Alkylphosphates are found to be more effective inhibitors than sulfates or carboxylates and form quite thick layers (up to 20nm) on the metal surface. Infra-red spectroscopical measurements indicate that the functional group of the inhibitor molecule is directly bound to the oxide and that the lines corresponding to aliphatic C–H groups continue to be visible even after attempts to remove the film by organic solvents. However, holding a test specimen in hot water leads gradually to the dissolution of the inhibiting layer.

Adsorption plays a decisive role in the protection of aluminum or its alloys by sodium oleate, which is one of the most effective inhibitors in neutral media.[74] In this case, the conditions of the formation of the adsorbed layer are found to have a significant influence on the effectiveness of the protection (Fig. 3.2). Thus, when an aluminum electrode was held for 1h in an oleate solution before introduction of chloride ions, the pitting potential, E_{pit}, was 0.7 V more positive than with the simultaneous addition of chloride and the inhibitor. Similarly, the protective effect was not reduced when the oleate was added after the chloride. This points to strong adsorption of the inhibitor at potentials close to the steady-state value, E_{st}. Thus, one can understand the reason for the strong retardation of the cathodic reduction of oxygen on aluminum that we have observed with oleate. This effect points to a polymolecular mechanism for oleate adsorption since the reduction of oxygen, which occurs under diffusion control, is inhibited effectively only when there is a high surface coverage by organic compounds.

Taking advantage of the fact that the protective action was effectively preserved after treatment of aluminum with oleate solution for 5 h, followed by exposure of the metal to a dry atmosphere (5 h), and washing with

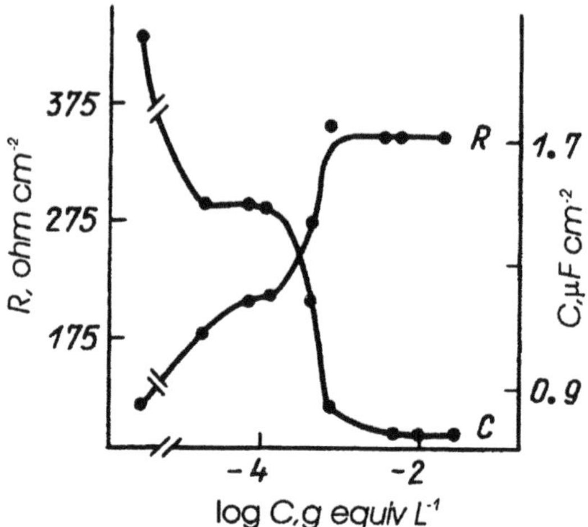

Figure 3.2. Dependence of the capacitance and resistance of an electrode of the aluminum alloy D16 on the concentration of sodium oleate in a borate buffer at $E = 0.3$ V.

water, the adsorption of this inhibitor could be studied by several of the methods that require removal of the electrode from the solution.[74] Piezo-quartz microbalance measurements showed an increase in mass of an oxidized aluminum specimen as a result of adsorption of oleate from water or a borate buffer solution. This increase in mass showed a strong dependence on oleate concentration. Thus, with 3–5 mmol L^{-1} oleate, the mass increase calculated on the geometric surface area was 5–8 $\mu g\ cm^{-2}$, but with 10–100 mmol L^{-1} it was 20–80 $\mu g\ cm^{-2}$. After washing the specimen immediately with water, a layer of mass 0.8–1 $\mu g\ cm^{-2}$ remained on the surface of the specimen. In this case, Auger spectroscopy revealed the absence of Na^+ on the surface, which indicates a predominating adsorption of the oleate anion.

However, the reversibility of the adsorption was strongly dependent on the duration of exposure of the specimen to the dry atmosphere. Thus after 5–10 days exposure, washing with water did not affect the adsorbed layer. Adsorption of sodium oleate from aqueous solution had almost no effect on the electronic work function which, irrespective of the oleate concentration and washing of the specimens, decreased for the aluminum alloy AMG-6 [Al–Mg 6%] by 0.03–0.05 eV. Within the limits of sensitivity of the infra-red method of a single reflection, it was not possible to prove chemisorption of oleate on aluminum, but undoubtedly the adsorption is of a polymolecular character. Furthermore, the quantity of sodium oleate irreversibly bound to the metal, i.e., that which is not removed on washing with water is—allowing for the surface roughness—several monolayers. Upon contact with air, bond rupture occurs, with polymerization of the layer leading to the formation of a 3-dimensional structure.

Polymerization plays an important role in the protective action of oleate. In fact, if the protection were caused only by the blocking of active centres, then carboxylates which form closely-packed adsorption layers would be more effective. Since the presence of the unsaturated bonds in the middle of the chain prevents close packing, then a greater effectiveness would be expected from stearate than from oleate. In fact, an increase in the number of double bonds in the higher fatty acids does improve their protection of aluminum in neutral solutions.[167] It is significant that the hydrophile–lipophile balance as a measure of the surface activity of an additive decreases in the series: linolenate–linoleate–oleate–stearate (sodium salts).

Finally, in the case of unsaturated compounds there is the possibility that the adsorption bonds are strengthened either as a result of high polarity and polarizability, or because of the formation with the positively-charged aluminum oxide surface of complexes with transfer of charge from a conjugated double-bond system. However, a more probable explanation is that physical adsorption of these anions with subsequent polymerization of the adsorbed layer occurs. Support for an adsorption–polymerization mechanism also comes from the results of impedance measurements which we have made on the aluminum alloy D16 (Fig. 3.2).

Even small [10^{-5} M] additions of oleate lower the electrode capacitance by more than a quarter of its initial value, and with further increase in concentration there is an abrupt fall to 0.75 μFcm^{-2}. The dependence obtained is of the same form as that of the isotherms for polymolecular adsorption which have been quite well characterized for the adsorption of anionic surfactants on aluminum oxide.[168] The usual critical micelle concentration, CMC, of oleate in water is 10^{-3} M (depending on the concentration of indifferent electrolyte), but the formation of polymolecular layers on an electrode surface evidently takes place at lower concentrations, that is, in that concentration region where the sudden change in impedance occurred. This was later confirmed in ellipsometric studies of the adsorption of sodium oleate on smooth aluminum.[169] This work showed that with concentrations of oleate above the CMC, two layers are formed: a closely-packed hydrophobic layer with vertically-oriented molecules and, adsorbed on this, a monolayer which is removed by a subsequent washing with water (without the exposure to air which would facilitate polymerization).

It should be pointed out that the adsorption of anions on aluminum having an air-formed oxide film can be quite strong, even in the absence of such stabilizing factors as polymerization. Thus, the chloride ion, despite the large difference in its J value from the J_p of aluminum oxide (Table 3.2), is adsorbed from aqueous solutions at pH 6.3–7.5 sufficiently strongly that, according to radiometric measurements,[170] washing of the specimen with distilled water for 1 h led to no large changes, and it was only after a 3-day holding period in water that the quantity of sorbed Cl^- was reduced by 75%. It was also found that the kinetics of chloride adsorption were non-linear with respect to time and that anions that inhibit pitting formation (NO_3^-, SO_4^{2-}, and OH^-) impeded, but did not prevent, the adsorption of Cl^-. The order of increasing adsorbability of inhibitors does not agree with the order of increasing inhibitor action, which fact the authors associated with the possibility that the number of adsorption centres did not agree with the number of pitting nucleation centres. We have found irreversible adsorption of Cl^-, and also of inhibiting ortho-substituted benzoate anions on aluminum and its alloys by X-ray electron analysis of surface layers.[171,172]

In particular, the adsorption of phenylanthranilate on aluminum takes place over a wide range of potentials, a fact which is supported by the retardation, as with oleate, of both electrochemical reactions. The anodic behavior of aluminum in solutions of both inhibitors is similarly dependent on the method of both their introduction and that of the activator. Apparently, the adsorption of Cl^- ions depends to a large extent on the potential of the electrode such that displacement of the inhibitor by Cl^- ions occurs as the potential is moved in a positive direction. In this case, an important role is played by the ratio of the concentration of the anions, which can be illustrated by the example of the inhibiting action of isodifluorant [sodium N-(m-difluoromethylthiophenyl)-anthranilate].

X-ray electron spectroscopic analysis of an aluminum surface maintained for 15 h in a borate buffer (pH 8.1) containing 0.01 M NaCl has shown the presence of oxygen, chlorine, and carbon, as well as aluminum (Table 3.3). When the solution also contained 0.001 M isodifluorant the quantity of chlorine on the surface was somewhat decreased, while the quantity of carbon increased and nitrogen also appeared in the spectra.

However, the last two elements cannot uniquely indicate the presence of isodifluorant, especially since fluorine and sulfur, which are present in this compound, were not detected after exposure to the 0.001 M solution (the nitrogen could have been adsorbed from the atmosphere). With increase in the inhibitor concentration to 0.01 M, the displacement of chloride from the surface was more effective and was accompanied by the appearance of fluorine and sulfur in the surface layer. The content of these elements in the surface layer increased when the inhibitor concentration was raised to 0.015 M, but showed little further change on increase to 0.05 M; chlorine was not detected in the spectra.

Table 3.3. The Protective Effect of Isodifluorant and the Ratio of the Elements in the Surface Layer of Aluminum from X-ray Electron Spectroscopy

Concentration of isodifluorant (mol L^{-1})	Element							Protective effect ΔE (V) after 15 h in solution
	Al	O	Cl	C	N	F	S	
–	1	2.65	0.14	0.46	N.F.[a]	—	—	
0.001	1	3.40	0.10	1.29	0.015	N.F.	N.F.	0.21
0.010	1	2.64	0.05	11.6	0.45	0.55	0.68	0.25
0.015	1	7	N.F.	56	2.9	3.3	5.6	0.33
0.050	1	6.05	N.F.	76.1	3.04	3.18	4.9	0.70

[a]N.F. = not found.

Since the depth of the layer that could be examined by this technique is not greater than 2.0–2.5 nm and the Al peak can still be seen even with maximum surface coverage, θ_{inh}, by the inhibitor, it may be assumed that the adsorbed layer of isodifluorant has a thickness of about 1.5 nm. This is supported by the fact that a first etching of the specimen by an argon beam removes the inhibitor from the surface. A calculation made using Stuart molecular models based on the values of internuclear distances and bond angles showed that a projection of the isodifluorant on to the normal to the metal surface was equal to 1.0–1.3 nm. Consequently, with an isodifluorant concentration greater than 0.015 M the chloride ion at $E \approx E_{st}$ is completely displaced from the surface of oxidized aluminum and the inhibitor forms a coating close to a monolayer in thickness. With further anodic polarization of the electrode, the concentration of the inhibitor is found to have a significant effect on the value of $\Delta E = E_{pit}^{inh} - E_{pit}^{background}$ (where the superscripts refer to values in solutions with and without inhibitor present). However, this dependence does not agree with that for the effect of θ_{inh} on ΔE, for which a stepwise growth is observed with practically no change in θ_{inh}.

This last observation can be explained, either by the fact that the value of θ_{inh} measured after removal of the electrode from solution does not correspond to the true value—for example, because of a disruption of the polymolecular coating which has its inner layer firmly bound to the surface—or, because of a change in the mechanism of pitting formation at larger values of θ_{inh}. A further reason could be the non-equilibrium adsorption of anions as indicated by Nguen and Foley[134] in studies with aluminum oxide and specially-purified aluminum powder. As will be shown below, the last two of these reasons are the more probable. However, it is equally important to note the strength of the bond between various anions and the oxide film on aluminum, and the varying dependence of this on the potential of the electrode. This signifies an important contribution of the electrostatic interaction in the adsorption and despite any analogy, the absence of similar adsorptive properties of the passive metal and its insulating oxide. The metallic substrate of the oxide is capable of influencing the structure and defect nature of the oxide, thus changing its solubility and energetic condition and consequently, the adsorption on the oxide of various components of the solution. Therefore, the direct measurement of adsorption *in situ* onto the surface of passive electrodes is of principal importance in the theory and practice of inhibiting the corrosion of metals.

3.1.3. Oxidized Iron

Differing from aluminum, the surface of iron in near-neutral media exists in various states of oxidation and consequently, has various adsorption activities. It has already been pointed out in Section 1.4 that benzoate stimulates the dissolution of iron in the active state but inhibits corrosion when adsorbed on an oxidized surface, with the quantity of benzoate adsorbed (as measured using ^{14}C-labelled benzoate) not exceeding 0.1 of a molecular layer.[173]

Oxidized iron is characterized by the presence of not one but several minimum values, J_p, of the first ionization potential, which partly explains the chemisorptive activity on such surfaces of many compounds, particularly benzoate and substituted benzoates. Among the latter is phenylanthranilate, which is one of the more effective corrosion inhibitors and which is adsorbed over a wide range of potentials on iron electrodes. Ellipsometric studies of its adsorption on iron have been conducted at E = 0.2–0.4 V, that is, in the passive region of potentials, both with and without the prior formation of an oxide film on the electrode surface.[73] To obtain the oxidized surface, the specimen, after cathodic reduction of the air-formed oxide film, was held for 8 h in a borate buffer at E = 0.2 V. Over this period, the values of the phase angle shift, Δ, and polarization restoration angle, ψ, reached steady-state values—changes in these values then being within the limits of experimental error—and the oxide film thickness reached 10 nm. Ellipsometric measurements of adsorption were then recorded automatically over a period of 1.5–2.0 h for each of several concentrations of phenylanthranilate introduced into the borate buffer from appropriate volumes of a concentrated phenylanthranilate (0.1 M) solution. The introduction of the inhibitor immediately led to a deviation of

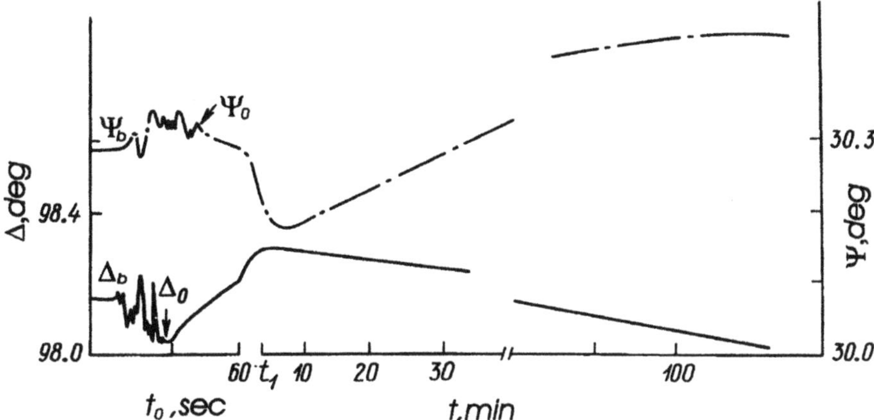

Figure 3.3. Changes in the ellipsometric parameters Δ and ψ in the process of adsorption of phenylanthranilate from a borate buffer solution on to oxidized iron at $E = -0.2$ V.

the ellipsometric parameters from the Δ_b and ψ_b values established in the pure borate buffer (Fig. 3.3). In the period from 0 to t_0 the deviation could be associated with a change in the refractive index of the solution. From the time t_0, corresponding to the onset of adsorption, the stepwise changes in Δ, ψ, and in the light intensity associated with movement of the solution stopped and the ellipsometric parameters began to change monotonously.*

Generally, the actual adsorption process consisted of two parts: (i) the period from t_0 to t_1, in which Δ increased and which in principle corresponds to a "removal" of the film from the surface; (ii) the period with $t > t_1$ in which Δ decreased, reflecting film growth associated with adsorption of inhibitor. In the first period, the displacement from the surface of part of the supporting electrolyte is simultaneously accompanied by covering of the surface by the adsorbate, but this could not be monitored from changes in Δ and ψ, which thus appears as a deficiency of the method. Changes in values of Δ and ψ during the adsorption process were therefore recorded in relation to their values at time t_1 and then examined as a function of phenylanthranilate concentration.

It was found that the limiting values of Δ and ψ and consequently, the adsorption, were reached at an inhibitor concentration of about 0.05 M for the non-oxidized surface but at 0.02 M for the oxidized surface. The amount of inhibitor on the oxidized surface at $E = 0.2$ V was equivalent

*The time scale in Fig. 3.3 is not uniform. The section from 0 to t_1 occupied 2–3 min and therefore could not be recorded on a manual ellipsometer.

to a monolayer (1.3nm), whereas on the non-oxidized electrode in the borate solution of 0.05 M phenylanthranilate the thickness, at this potential, reached 4–5 monolayers (see pp. 43–45). Evidently, in the first case a large part of the potential drop is localized in the oxide and there is a smaller contribution of the electrostatic component to the energy of adsorption. In the absence of the primary oxide, the adsorption of the inhibitor increases as a result of the migration of phenylanthranilate anions under the action of the electrical field until it is compensated by the potential drop in the adsorbed layer. However, cases of monomolecular adsorption (at cathodic potentials or on an oxidized surface) are formally described by Frumkin's equation

$$BC = -\frac{\theta}{1-\theta}\exp(-2a\theta) \tag{3.5}$$

in which the interaction constant, $a > 0$, indicates an attractive interaction between the adsorbed particles, and B is an adsorption equilibrium constant associated with the standard free energy of adsorption, ΔG°_{A}, by the following relation:

$$B = -\frac{1}{55.5}\exp\left(-\frac{\Delta G^{\circ}_{A}}{RT}\right) \tag{3.6}$$

The calculated adsorption parameters are given in Table 3.4. Formally, ΔG°_{A} reflects only the electrostatic interaction characteristic of an interphase surface at very small θ[174] and is only part of the free energy of adsorption.

Electrostatic interaction facilitates the adsorption of an anion on a positively-charged surface at very low values of θ. For the case of a negatively-charged surface and anions, the electrostatic interaction should prevent adsorption. Therefore at $E = 0.2$ V, and even more at $E = 0.4$ V, the

Table 3.4. Adsorption Isotherm Parameters for Phenylanthranilate on an Iron Electrode

Potential (V)	a	B(L mol^{-1})	$-\Delta G_A{}^{\circ}$(kJmol^{-1})	Method of measuring adsorbtion
−0.65	1.62±0.15	10.77±0.66	13.66±0.17	ellipsometric
0.00	1.40±0.05	53.3±7.1	19.72±0.29	impedance
0.20	2.11±0.34	18±2.4	16.80±0.33	ellipsometric
0.40	1.5±0.2	40±2.0	18.72±0.33	ellipsometric

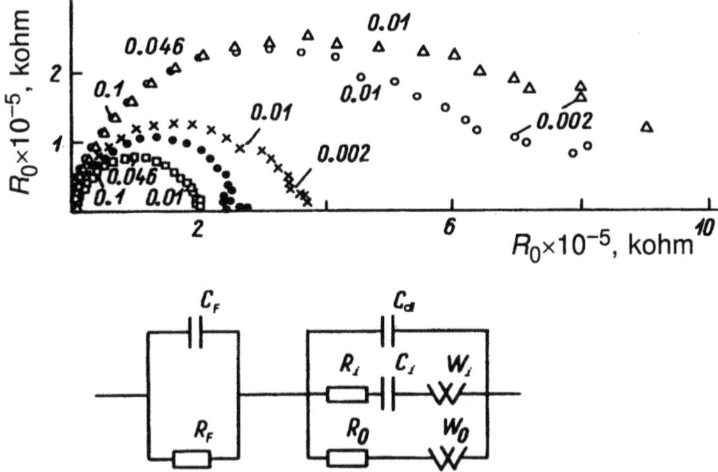

Figure 3.4. Equivalent circuit (bottom) and Nyquist (top) plots for passive iron electrode in a borate buffer without (□) and with additions of phenylanthranilate in (mM): 0.4 (●); 10 (X); 20 (O); and 50 (Δ). The ciphers indicate frequencies in Hz.

value of ΔG°_{A} increases as a result of the electrostatic forces. The isotherms obtained are in agreement with the results of impedance studies of adsorption.[175] In these, the equivalent circuit of a passive oxidized electrode is represented as a combination of the impedance of the adsorption of an electrochemically-indifferent substance (the Frumkin, Melik–Gaikazyan impedances* C_d, R_i, W_i, C_i) with elements that characterize the oxide film, C_F, R_F, the electrolyte resistance R_e, and the impedance of the electrochemical reaction R_o, W_o (Fig. 3.4). The Warburg impedance W, which describes the occurrence of diffusion processes in this circuit, is caused either by the transport of adsorbate to the electrode surface, W_i, or of charge carriers in the oxide, W_o. Values of θ were calculated from the formula $\theta = (C_o - C)/(C_o - C^1)$ where C_o and C^1 are the values of C_d in the supporting electrolyte and with $\theta = 1.0$. An isotherm constructed in this way is also described by the Frumkin equation (Table 3.4) and its parameters are close to those for the ellipsometric isotherms for phenylanthranilate adsorption on the same electrode.

Attention should also be given to the fact that the value of the adsorption free energy of phenylanthranilate is commensurate with that for the

*Here C_d is the capacitance of the electrical double layer, R_i is the resistance of the adsorption process, and C_i is a capacitance characterizing the change in the volume charge of the oxide.

free energy of formation of the water-insoluble complexes of phenylanthranilic acid with Fe^{3+}, as calculated from the corresponding stability constant (K_s) determined by pH–potentiometric titration in an aqueous–dimethylformamide solution.[176]

$$\Delta G^o = -RT\ln K_s = -30.5 \text{ kJ mol}^{-1} \tag{3.7}$$

The standard potential, $E^o = -0.143$ V, is noticeably more negative than the potential of an electrode with an oxidized surface and therefore, from the thermodynamic point of view, it is quite acceptable to propose that the adsorption of phenylanthranilate is accompanied by formation of a complex with Fe^{3+}. On the other hand, it is known that chemisorption from solutions is characterized by very much smaller values of the free energy of adsorption than chemisorption from the gas phase. Evidently, in these cases the measurements are not of a true value of ΔG^o_A but of an apparent value, since the adsorption occurs with the displacement (substitution) of solvent molecules requiring an additional expenditure of energy. Consequently, it can be suggested that the true value of ΔG^o_A for phenylanthranilate is larger than that measured by at least the value of the energy of desorption of the solvent molecules.

This situation is also indicated by the results of studies of the adsorption of some sulfur-containing compounds which can often be chemisorbed on an iron electrode.[6] In the work of Bockris, *et al.*,[177,178] a radiometric method was used to show that in a borate buffer at pH 8.4 the adsorption of thiocyanate and thiourea on oxidized (50 min at $E = 0.3$ V) iron occurs over a wide range of potentials with the displacement of water molecules (the Bockris–Swinkels isotherm). Adsorption increased with increase in potential reaching, in the case of thiourea, a maximum at $E = 0.9$ V ($\Delta G^o_A = -20.25$ kJ mol^{-1}). It is of interest that SCN^-, which is known for its high complex-forming capabilities with Fe^{3+}, is characterized by somewhat lower values of ΔG^o_A even at anodic potentials.

The possibility of the formation of polymeric adsorbed films on an iron oxide surface should also be considered. According to Goledzinowski, *et al.*,[179] this was the case when 2-pentylaminobenzimidazole (PABI) was introduced into a borate buffer after 50 min of oxidation of iron. Blocking of the oxide surface prevented adsorption of OH^- and lowered the double-layer capacitance. The X-ray photoelectron spectra of such a film included a weak signal from nitrogen which disappeared after 10 sec of argon-ion etching, i.e., the PABI was covering only the external surface of the oxide. When the oxide film was formed in the presence of the PABI, the concentration was constant throughout the thickness. In this case, the nitrogen signal was very intense and remained practically unchanged after 10 sec of argon-ion etching. Coulometric measurements showed that the film thickness of PABI decreased somewhat (from 5 to 4 nm). The adsorbed film that is formed on the oxide at potentials below that for oxygen evolution, E_{O_2}, is almost completely removed on washing with water. Irreversible adsorption is observed only with $E > E_{O_2}$. It is suggested that PABI is adsorbed in the polymeric as well as the monomeric form of a complex with iron ions.

A similar type of interaction with oxidized electrodes often occurs in the protection of copper and its alloys by some chelating reagents and we have examined this in detail in a review.[119] However, their action is not restricted to adsorption of an organic ligand and they are not always effective in combating the local depassivation of metals. Furthermore, it should be noted that the concept of a clear distinction between adsorption and complex formation in the development of a film of difficultly-soluble complexes on the surface of a protected metal often leads to considerable difficulties. It is therefore important to evaluate the kinetic aspects of the depassivation of metals taking into account all those factors that influence this process so that the most rational method of protecting a metal from pitting corrosion can be devised.

3.2. THE INDUCTION PERIOD AND EFFECTIVE ACTIVATION ENERGY FOR PIT FORMATION

The concept of the critical pitting potential, E_{pit}, as an energetic characteristic of the depassivation of a metal and the impossibility of predicting its value solely from a thermodynamic analysis make the assessment of a free energy of activation, $\Delta G^{\neq}$, for pitting initiation particularly important. As pointed out earlier, since the pitting process is multistage in nature, the determination of a true value of $\Delta G^{\neq}$ is hardly possible. Furthermore, even for the determination of an effective value, $\Delta G^{\neq}_{eff}$, it will be necessary to have reliable kinetic criteria of the process.

As such a criterion, the induction period, τ, for the formation of the first pit had already begun to be used in 1959.[180] To determine τ, the potential of an electrode was first kept at a value more positive than the expected E_{pit}, an activator was then added to the solution, and the time recorded to the first increase in current resulting from the local depassivation of the surface. In later work[181] using a rotating ring-disc electrode, it was shown that in a borate buffer, iron goes into solution in the form of Fe^{3+} ions from the passive state following the introduction of chloride ions into the solution. However, at the end of τ, the current increased with the appearance of Fe^{2+} and the first stage of the process was completed with the formation of the first pit. In subsequent stages, a large number of pits will often occur simultaneously. Since the first pit appears after a time τ, the mean rate of its initiation will be proportional to $1/\tau$. However, it does not follow that this can be correlated with the rate of pitting occurrence, which is actually determined by the number of points of damage formed on the surface per unit time.

An assessment of $\Delta G^{\neq}_{eff}$ from the temperature dependence* of $1/\tau$ was first made by Hoar and Jacob[182] for the case of the chloride depassivation of a chromium–nickel steel but the value obtained (240 kJ mol^{-1}) was unreliable because of the narrow temperature range used. Foley,[133] in a study of the depassivation of aluminum alloys, found that the nature of the activating anions had a marked effect on $\Delta G^{\neq}_{eff}$. This can be explained on the basis of the concept of formation of complexes using the idea of local depassivation as a consequence of the occurrence of a nucleophilic substitution of ligands in a surface complex.[183]

In this case, as we have shown,[184] the theory of absolute reaction rates gives the equation

$$V = \left(\frac{kT}{h}\right)\kappa \frac{a_{An}^{m(s)}}{\gamma^{\neq}} \quad \exp(-\Delta G^{\neq}/RT) \tag{3.8a}$$

and taking into account that $1/\tau = KV$:

$$\frac{1}{\tau} = K\left(\frac{kT}{h}\right)\kappa \frac{a_{An}^{m(s)}}{\gamma^{\neq}} \quad \exp(-\Delta G^{\neq}/RT) \tag{3.8b}$$

where V is the reaction rate; k is the Boltzman constant; h is Planck's constant; κ is the transmission coefficient; $m(s)$ is the order of reaction with respect to the anion taking into account its adsorbability; and $\gamma^{\neq}$ is the activity coefficient.

Evidently, Eq. 3.8b is true only for steady-state conditions—at least as far as the potential of the electrode is concerned.† Furthermore, at

*According to the theory of absolute reaction rates, the temperature dependence of the constant rate provides a value $\Delta G^{\neq}$, but not $\Delta G^{\neq}$. However, to a first approximation for a not too wide temperature range when the entropy of activation $\Delta S^{\neq}$ is rather small, we can assume that $\Delta H^{\neq} \approx \Delta G^{\neq}$, especially when we are considering only some effective values.

†Strictly speaking, one should use the overpotential of the anodic reaction $\eta = E - E_{rev}$ but because of a lack of information on the temperature dependence of E_{rev} the condition that E = constant has to be used.

sufficiently low concentrations of an activator, with pH and ionic strength kept constant, the dependence

$$\frac{1}{\tau} = K_o \exp(-\Delta G^{\neq}_{eff}/RT) \tag{3.8c}$$

will hold, and this allows an assessment to be made of the kinetic limitation of the reaction.

For iron in a borate buffer solution (pH 7.4) at $E = 0.2$ V and $a_{An} =$ 3.2 mM, the dependence of the rate of pitting initiation on temperature (283–353 K) has been described by Eq. 3.8c but the calculated values of $\Delta G^{\neq}_{eff}$ varied with the nature of the activator:

Anion	$\Delta G^{\neq}_{eff}$ (kJ mol^{-1})
Cl^-	37 ± 6
Br^-	41 ± 6
I^-	44 ± 7
F^-	86 ± 11
NO_3^-	81 ± 10
N_3^-	75 ± 11
SCN^-	76 ± 10
$HCOO^-$	118 ± 18
CH_3COO^-	111 ± 18
$C_2H_5COO^-$	90 ± 17
$C_3H_7COO^-$	78 ± 12

It is interesting that for the same composition of the complex $(FeL)^{2+}$, the values of the stability constants, K_s, increase in the order $Br^- < Cl^- < I^- < SCN^- < F^-$[75] which effectively agrees with the order of increase in $\Delta G^{\neq}_{eff}$. However, the description of $\Delta G^{\neq}_{eff}$ cannot be made only on the basis of the K_s values of the complexes. For example, the stability of $(FeOOCCH_3)^{2+}$ is little different from that of $(FeSCN)^{2+}$, but $\Delta G^{\neq}_{eff}$ for acetate is noticeably higher, not only than that for SCN^- but also that for F^-, which form the most stable complexes. The decrease in $\Delta G^{\neq}_{eff}$ with increase in alkyl chain length in the carboxylates that are studied is also important. This can be associated with an increase in polarizability and hydrophobicity of the anion. By using the Hansch π constants of hydrophobicity and the logarithms of the molar refractivities of the atoms or their groups, log*MR*, which are proportional to the polarizability of the anion, $\Delta G^{\neq}_{eff}$ can be successfully described by the equation:

$$\Delta G^{\neq}_{\text{eff}} = 79.1 - 39.8\,\pi + 6.2\,\log MR \tag{3.9}$$

i.e., using those characteristics which were used to describe E_{pit} of iron [see Eq. (2.29)]. An analysis of Eq. 3.9 shows that the greatest contribution to the value of $\Delta G^{\neq}_{\text{eff}}$ is made by solvation effects. Thus, the greater the hydrophobicity of the activator the more readily it, by being desolvated, will displace OH^- from the surface complex into solution. This conclusion has been confirmed experimentally by a negative first order with respect to OH^-, independent of electrode potential, and by a positive order with respect to Cl^- (Fig. 3.5). The latter, however, is fractional and dependent on temperature. By taking into account the fact that Cl^- is adsorbed by a Freundlich isotherm on iron oxide, we have put forward[184] an equation based on the assumption of a quadratic dependence of the equilibrium adsorption constant on potential, that is, by analogy with adsorption on mercury:

$$p = mp' = \frac{A}{\ln a_{Cl^-}} - \frac{\delta FE}{RT \ln a_{Cl^-}} + \frac{\gamma p'(E - E_m)^2}{\ln a_{Cl^-}} \tag{3.10}$$

where m is the power index in the Freundlich equation, p and p' are the conditional and true orders of reaction with respect to Cl^-, E_m is the point on the abscissa where the lines intersect on the graph of electrode charge

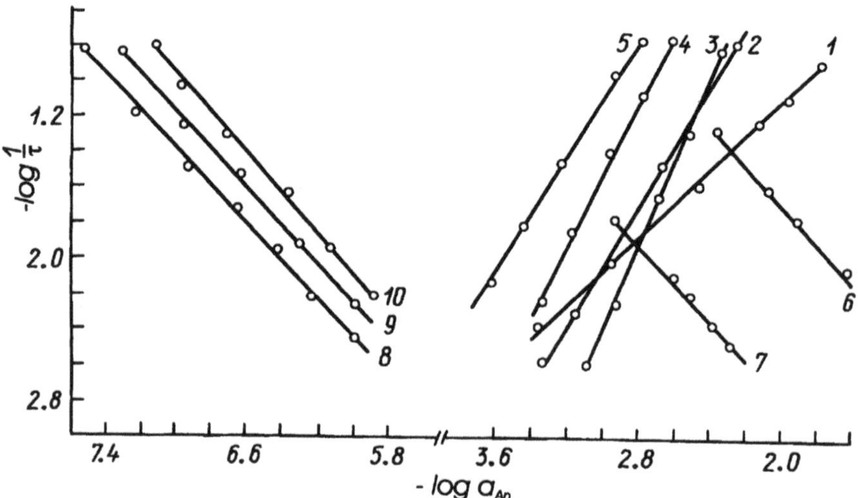

Figure 3.5. The effect of the activity of ions: Cl^- (1–5) and OH^- (6–10, $C_{Cl} = 2$ mM) on the rate of pitting initiation on iron at various potentials (V): 1,6,8–0.2; 2–0.4; 3,7,9–0.6; 4–0.8; 5–1.0.

$g = f(E)$, A and δ are constants, and γ is a constant related to the adsorption maximum. From this it can be seen that the experimentally determined order of reaction for Cl^- will, in general, be fractional and vary parabolically with potential.

It was also noticed that the orders of reaction with respect to not only OH^- but also phenylanthranilate were close to -1.0, which indicates a significant difference in the mechanism of their interaction with the electrode, compared with chloride.

If in a non-inhibiting solution or at low θ_{inh} the "departing groups" are solvent molecules, then in relatively concentrated phenylanthranilate solutions, such groups will be the anions of the inhibitor. In the latter case, the inhibitor is capable of significantly increasing the potential barrier for the pitting process and its negative reaction order will be in accordance with the assumption of a nucleophilic substitution of the inhibitor.

Upon varying the chemical structure of the inhibitor, changes in $\Delta G^{\neq}_{eff}$ could be computed using the LFER principle and the σ constants of substituents. This was shown for the initiation of pitting on iron in a borate buffer with the examples of meta- and para-substituted phenylanthranilate containing electron-donating groups. Values of $\Delta G^{\neq}_{eff}$ calculated for the case of $C_{inh} = 2.5$ mM and $C_{NaCl} = 3.2$ mM are given in Fig. 3.6 as a function of the normal Wepster σ^n constant which, unlike the Hammett σ constant, is free from the effect of a direct polar conjugation.[185] The latter does not occur because of the presence of the same type (electron-donating) of reaction centre (COO^-) and substituents. The linear dependence shown by $\Delta G^{\neq}_{eff} = f(\sigma^n)$ confirms the validity of the use of the LFER principle for describing the effects of organic anions.

The value of $\Delta G^{\neq}_{eff}$ is greatest for sodium mephenamate (N-2,3-xylyl-anthranilate) (MEP) which supports our proposed mechanism for the action of substituted phenylanthranilates. With $C > 1.8$ mM, the increase in the energy barrier of the pit initiation process is caused by the strengthening of the chemisorption bond between the metal and the inhibitor, brought about by the increase by the substituent of the electron density at the reaction centre. Consequently, the electronic effect of the substituent amounts to a change by the inhibitor of the activation energy of the pitting initiation. A value of $\Delta G^{\neq}_{eff} = 21$ kJ mol^{-1} is quite low for the background solution (without inhibitor) since its value will decrease as the $E - E_{pit}$ difference is increased (with reduction in the height of the transition-state barrier). The addition of a substituted phenylanthranilate significantly increases the value; thus, in the presence of mephenamate, $\Delta G^{\neq}_{eff}$ reaches 91 ± 7 kJ mol^{-1} which is commensurate with $\Delta G^{\neq}$ of homogeneous reactions of nucleophilic substitution.[127] Consequently, an increase in the electron-donating properties of the inhibitor can impede its displacement from the surface by the activator and increase $\Delta G^{\neq}_{eff}$. It is therefore logical to expect that a change in the energy of the reaction of pitting nucleation will also affect the value of E_{pit}.

In fact, the similar dependence of E_{pit} on the σ^n constant of the substituent confirms this (Fig. 3.6) and indicates the correctness of the assumption that a change in E_{pit} caused by a change in the composition of the solution is, to a first approximation, proportional to the corresponding change in the effective activation energy of pitting nucleation. In other words, the higher the activation energy of this process, the more positive is the potential for pitting initiation. This proposition, which we have clearly formulated[183] was put forward on the basis of a study of regularities in the effect of solution composition on E_{pit}.[139,148,186–189] For example, in later sections of this book we examine in more detail the inhibiting effect $\Delta E = E^{inh}_{pit} - E^{background}_{pit}$ as essentially proportional to $\Delta(\Delta G^{\neq}_{eff})$ and apply the LFER principle with the appropriate correlation equations. A correspondence has also been observed between changes in $\Delta G^{\neq}_{eff}$ and E_{pit} with increase in the concentration of the activator. It is of interest that this takes place not only in chloride solutions, in which both values usually decrease exponentially with increase in a_{Cl^-}, but also in those cases where similar dependencies have a more complex character.[183] Finally, it can be pointed out that the alloying of iron with chromium (up to 5%) also leads to a symbiotic growth of $\Delta G^{\neq}_{eff}$ and E_{pit}.[184]

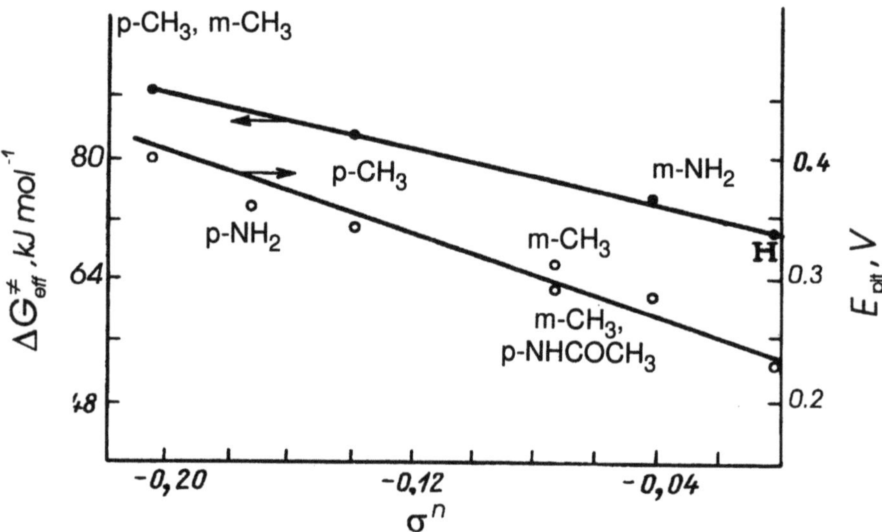

Figure 3.6. The effect of Wepster constants σ^n of substituents in the phenylanthranilate molecule on $\Delta G^{\neq}_{eff}$ for the pitting of iron in a borate buffer at $E = 0.2$ V, $C_{inh} = 2.5$ mM, $C_{NaCl} = 3.2$ mM and on E_{pit} at $C_{inh} = 2$ mM, $C_{NaCl} = 5$ mM.

An examination of $\Delta G^{\neq}_{\text{eff}}$ as a characteristic of the height of the potential barrier of the transition state for the reaction of complex formation also allows an understanding to be reached of some aspects of the effect of potential on pitting initiation. Apart from the changes in the formal order of reaction on iron that have already been considered with respect to Cl^-, an empirical equation also merits attention; this relates τ and E_{pit} by:

$$\log \tau/\tau_0 = E_0/(E - E_{\text{pit}}) \tag{3.11}$$

in which τ_0 and E_0 are constants. Heusler and Fischer[181] showed that the constant $\tau_0 = 2.5 \pm 0.5$ s and that it is independent of pH and Cl^- concentration for the case of a deaerated borate buffer and electrolytically-pure iron. Our investigations of the anodic behavior of Armco iron in an aerated borate buffer solution have indicated that, with the same error, $\tau_0 = 2.8$ s and that it is independent of the nature of the halide ion (Fig. 3.7). It is evident that τ_0 will represent a minimum induction period as $E \to \infty$. This allows τ_0 to be viewed as a characteristic of the stability of a passivated surface complex towards attack by an aggressive anion. In the light of such

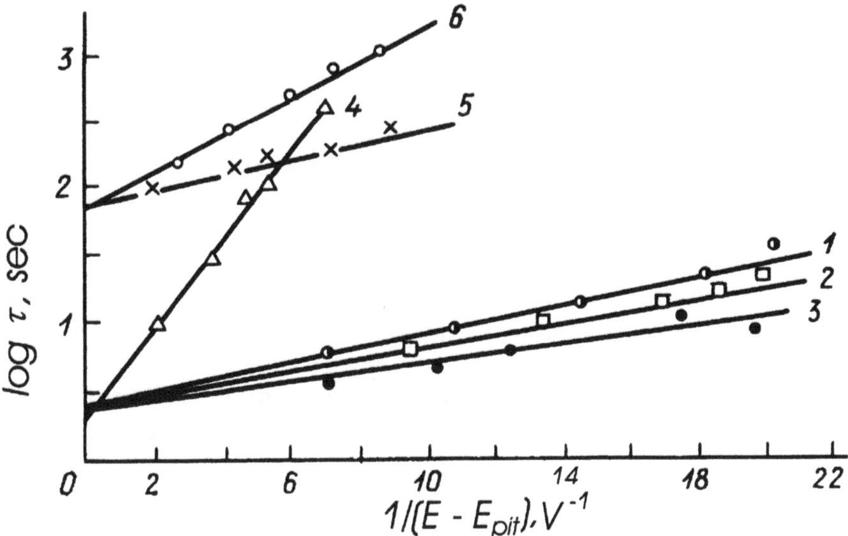

Figure 3.7. The connection between the induction period and E_{pit} of iron in a borate buffer (pH 7.4) with additions of the salts: 1, 3.3 mM NaI; 2, 3.3 mM NaBr; 3, 3.3 mM NaCl + 0.5 mM phenylanthranilate; 4, 1.3 mM NaCl; 5, 1.8 mM phenylanthranilate + 4 mM NaCl; 6–0.1 M phenylanthranilate + 0.38 M NaCl.

considerations we can examine the initiation of pitting in the presence of phenylanthranilate in buffered borate solutions of NaCl. With small, i.e., < 0.5 mM, additions of inhibitor, τ_0 remains practically unchanged at 3.2 $\pm$ 0.5 s but on increasing C_{inh} to 1.8 mM it reached a value of 65 $\pm$ 10 s. A similar value of $\tau_0 = 70 \pm 10$ s was obtained for the nucleation of pitting on iron which had been previously passivated in 0.1 M phenylanthranilate when its surface was free from an oxide film. These results can be explained by the fact that in both cases the necessary condition for the nucleation of pitting is the replacement by the activator of the passivating species in the adsorbed complex at the most advantageous sites for this process on the surface: in one case, for small C_{inh}, the passivating species is OH^- and $\tau_0 = 3.2 \pm 0.2$ s, in the other case, for larger C_{inh}, the passivating species is the inhibitor anion and $\tau_0 = 6.5$–70 ± 10 s.

It should be pointed out that when $1/(E-E_{pit}) \rightarrow 0$, i.e., in conditions where the electrostatic factor is constant, pitting formation can hypothetically be represented as a chemical reaction and τ_0 as a value which is inversely proportional to its rate, V_{chem},[190] for such a case $\Delta G^{\neq}$ in Eq. (3.8) does not contain an electrostatic component. In chloride solutions, the order of reaction with respect to Cl^- becomes constant, but τ_0 will depend on a_{Cl^-}, as observed in refs.[181,184], only in conditions such that the activity of the activator in the defects in the passivating layer $a_{Cl^-,s} = 1$.* Since τ_0 is determined by extrapolation to the limit of a large $E-E_{pit}$ difference, it can be concluded that the basic driving force for Cl^- adsorption on passive iron will be its electrostatic attraction to the positively-charged surface of an electrode. This is in agreement with the data referred to above[163] concerning the adsorption of Cl^- on oxides, which is of an electrostatic nature. The fact that $\tau_0 \neq 0$ confirms that adsorption is only the initial stage of the nucleation of pitting, and the non-random increase in τ_0, with the consequent retardation in V_{chem} in a chloride solution, occurs at significant values of θ_{inh} (in the case where the "departing group" is the inhibitor itself).

The adsorption of carboxylates is apparently significantly less dependent on the electrostatic factor. This can be seen not only from the results of studies with phenylanthranilate but also with simpler anions. For example, the kinetics of the initiation of pitting by butyrate ($C_3H_7COO^-$) on nickel and iron can be distinguished from chloride depassivation. Thus, over a wide range of concentrations, τ and $\Delta G^{\neq}_{eff}$ are independent of poten-

*Since for a chemical reaction $V = k[a_{Cl^-}]^n$, i.e., $\frac{1}{\tau_0} = K[a_{Cl^-}]_s^n$ it is clear that for $\frac{1}{\tau} = \frac{1}{\tau_0} =$ constant provided that $[a_{Cl^-}]_s = 1.0$. A second condition for the independence of τ_0 on a_{Cl^-} can be a zero true reaction order with respect to Cl^-, but it follows from Eq. (3.10) that $p' = 0$ since $p = mp' = 0$.

tial, and the latter—as well as E_{pit}—are independent of the activity of the butyrate.[138,190] It follows that $\tau = \tau_0$ and for iron with a_b = 5–60 mM butyrate

$$\log 1/\tau_0 = -0.61 + 0.96 \log a_b \quad (3.12)$$

The order of reaction with respect to butyrate, $p = 0.96 \pm 0.1$, is close to 1.0 and independent of potential, which points to only a small contribution of the electrostatic factor to the interaction of this anion with the electrode. Furthermore, the value of $\Delta G^{\neq}_{eff}$ introduced above indicates significant kinetic limitations to the process of pitting initiation. Unfortunately, the absence of information about the adsorbability of butyrate on iron oxide does not allow τ_0 to be calculated for the $a_{b(s)} = 1.0$ condition. However, even with $a_b = 1.0$, τ_0 can be assessed as 4.1 ± 0.6 s. This value is somewhat above that of $\tau_0 = 3.1 \pm 0.6$ s found for the initiation of pitting of iron by halide additions (I^-, Br^-, Cl^-) electrostatically adsorbed on its oxide. It is probable that a further study of τ_0, for example, of its dependence on temperature of the solution or on alloy composition, in combination with results of adsorption measurements on oxides, will lead to a refinement of the energetics of the initial stages of pitting and a deeper understanding of the mechanism.

3.3. THE DEPENDENCE OF THE EFFECTIVENESS OF INHIBITORS ON THE NATURE OF THEIR FUNCTIONAL GROUPS

It follows from the mechanism of the local corrosion of metals that the ability of an organic compound to be adsorbed over a wide range of potentials will facilitate the development of its protective properties. The actual adsorption of a compound will depend on its charge and the nature of any functional group taking part in the formation of a bond with the metal surface. Hence, surface-active anions would be expected to be highly effective for a positively-charged surface, such as iron or aluminum, in neutral media.

In fact, sodium oleate not only ennobles E_{pit} of aluminum but also effectively suppresses the cathodic process (Fig. 3.8) and an electrode in sodium oleate solutions remains passive over a wide range of potentials. On the other hand, surfactants of the cationic type, e.g., alkylbenzyldimethylammonium chloride (ABDM) inhibit only the cathodic reaction—displacing E_{pit} by only a small amount, ~50 mV. This is explained by the fact that at a sufficiently negative potential the adsorption of organic cations is

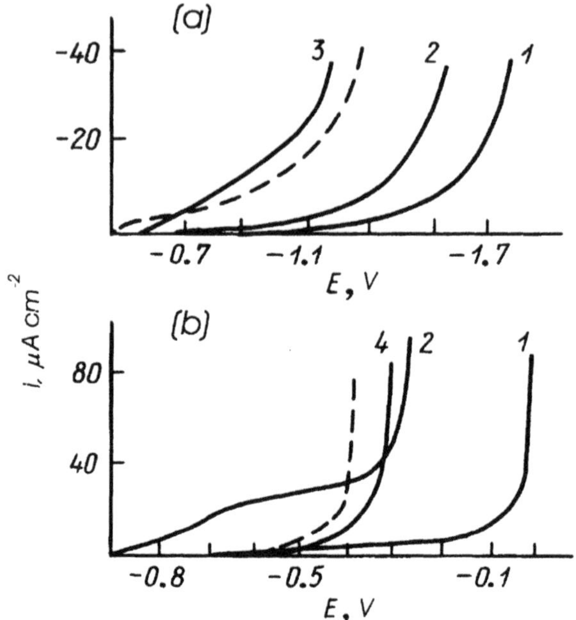

Figure 3.8. Polarization curves for aluminium in a borate buffer containing NaCl 1.5 + Na_2SO_4 3.5 g L^{-1} (dashed line) and 0.01 M (a) or 0.02 M (b) sodium oleate (1), ABDM-chloride (2); OP-7(3); and Sintamide-5 (4) (see Table 3.5 for description of abbreviations).

encouraged, but upon displacement of the potential in a positive direction they are displaced by chloride ions. The importance of the value of the charge of the adsorbate is also seen in the case of nonionic surfactants, e.g. OP7 and sintamide 5, (see Table 3.5) which despite their high surface-active properties, do not prevent pit formation.

Analogous data have been obtained in studies of other surfactants the most typical of which are listed in Table 3.5.[167] The best inhibitors were the anionic surfactants with unsaturated bonds, particularly oleate and olefinic sulfonates having C_{18}–C_{22} chains. Sodium volgonate, which is close in composition to olefinic sulfonates but which does not contain double bonds, was less effective. A carboxylate will ennoble E_{pit} more than a sulfonate or sulfate having similar structures of the hydrophobic radical. The role of the functional group has been investigated on many occasions but its understanding requires consideration of the mechanism of protection by inhibitors.

According to the ideas that have been developed, the inhibiting effect ΔE, is caused, to a first approximation, by the difference in $\Delta G^{\neq}$ of the

Table 3.5. The Protective Action of Surface Active Substances in the Depassivation of Aluminum by Chlorides in a Borate Buffer (pH 8.08) Containing NaCl 1.5 + Na_2SO_4 3.5 gL^{-1}

Surfactant	Mean molecular mass	Concentration of surfactant (mol L^{-1})	Quasi-steady-state potential E_{st}(V)	E_{pit}(V)
Background solution	—	—	−0.574	−0.375
		Surfactants of the anionic type		
Sodium oleate	304	0.02	−0.607	−0.030
Olefin sulfonate	405	0.02	−0.888	−0.250
Dispersant NF[a]	520	0.02	−0.710	−0.270
Wetting agent NB[b]	310	0.02	−0.556	−0.284
Sulfonol[c]	386	0.02	−0.638	−0.300
Sodium Volgonate[d]	322	0.02	−0.538	−0.313
Sodium lauryl Sulfate	256	0.02	−0.580	−0.315
		Surfactants of the cationic type		
Alkylbenzyldi-Methylammonium chloride	360	0.02	−0.900	−0.325
Alkamon OS-2[e]	601	0.02	−0.910	−0.332
		nonionic surfactants		
Sintamide-5[f]	383	0.02	−0.573	−0.353
OP-7[g]	542	0.02	−0.560	−0.371
Proksanol[h]	11000	5.10	−0.575	−0.385

[a]disodium methylene-bis-(naphthalenesulfonate), $NaO_3SC_{10}H_6CH_2C_{10}H_6SO_3Na$
[b]sodium butylnaphthalene sulfonate, $C_4H_9C_{10}H_6SO_3Na$
[c]sodium alkylbenzenesulfonate based on kerosene, $C_nH_{2n+1}HC_6H_4SO_3Na$
[d]a sodium alkyl sulfonate [$C_nH_{2_{n+1}}$, $C_mH_{2_{m+1}}$) $CHSO_3Na$ where n + m = 11-17
[e](alkyldihydroxyethylene)methyldiethyl ammonium benzene sulfonate [$C_nH_{2n+1}O(C_2H_4O)_2CH_2N^+(CH_3)$ $(C_2H_5)_2$ $C_6H_5SO_3$ where n = 16-18
[f]polyethylene glycol esters of monoethanolamine synthetic fatty acids N-mono-(2-polyethylene glycol) amides of synthetic fatty acids C_nH_{2n+1} $CONHCH_2CH_2O(C_2H_4O)_mH$ where n = 10-16, m = 5-6
[g]monoalkyl phenolic esters of polyethylene glycols $C_nH_{2n+1}HC_6H_4O(C_2H_4O)_mH$ where n = 10-16, m = 5-6
[h]block copolymer of ethylene oxides (n) and propylene (m) where n = 48-61 and m = 24-31 hydrophobic radial.

two surface reactions. Consequently, the inhibiting properties of organic compounds should depend on their electronic structure and hydrophobicity. However, as already noted when considering the induction period, different mechanisms for the initiation of pitting are possible in the presence of an adsorbing inhibitor. The adsorption of an inhibitor on the most active centres of a passive electrode forces an activator to interact with the surface at less energetically favorable parts; therefore, a greater $\Delta G^{\neq}$ is required. With small and medium values of surface coverage by the inhibitor, θ_{inh}, the mechanism of pitting initiation will remain unchanged, taking place as shown in Eq. (2.26) with nucleophilic substitution of solvent molecules. Higher values of θ_{inh} essentially mean that Eq. (2.26) takes place with the formation of an inhibitor complex, i.e., $An^- \equiv In^-$:

$$[Me(OH)_k S_l]_{ad}^{z-k} + m In^- \rightarrow [Me(OH)_{k-m} S_{l-n} In_m]_{ad}^{z_1-k} + m OH^- + nS + (z_1 - z)e \quad (3.13)$$

Then the subsequent nucleophilic attack by the activator is represented by:

$$[Me(OH)_{k-m} S_{l-n} In_m]_{ad}^{z_1-k} + q An^- \rightarrow [Me(OH)_{k-m_1} S_{l-n_1} In_{m-y} An_q]_{ad}^{z_2-k-q+m_1+y} + (m_1 - m)OH^- + (n_1 - n)S + y In^- + (z_2 - z_1)e \quad (3.14)$$

It is clear that Eq. (3.14) is a more general scheme which includes, for example, the case where $m = y = 0$ as in Eq. (2.26), as well as with the appropriate combination of coefficients, mechanisms of inhibition.

The theory and practice of correlation analysis in chemistry shows that taking account of the reactivity of In^- even in the most simple case when it fulfills the role of a reagent, [as in Eq. (3.13)], and not a "departing" group [as in Eq. (3.14)], can become quite complex if the nature of the reaction centre is changed. Because of this, aluminum in chloride solution containing various substituted benzenes C_6H_5R (Table 3.6)[191] is particularly suitable for study.

On varying the substituent in these compounds, the value of ΔE for aluminum resulting from the introduction into the solution of 0.1 M of the sodium salts is described by the equation

$$\Delta E = 0.489 - 0.038\, pK_a + 0.062\pi \quad (3.15a)$$

where π is the Hansch constant related to hydrophobicity (section 2.2). If a beneficial effect of the hydrophobicity of the reaction centre on the effectiveness of protection [ΔE increases with increase in the π-constant]

Table 3.6. The Dependence of ΔE on Aluminum in 0.1 M Solution of Monosubstituted Benzenes on the Basicity and Hydrophobicity of the Substituents[a]

Substituent	COOH	CH_2COOH	CH_2CH_2COOH	SH	$B(OH)_2$	OH	CON HOH	SO_3H	NH_3^+
π	−0.86	−0.67	−0.10	0.39	−0.55	−0.67	−1.87	—	−4.19
pK_a	4.12	4.31	4.66	9.43	8.86	9.98	8.89	0.69	4.58
$\Delta E(V)$	0.31	0.245	0.325	0.14	0.17	0.03	0.05	0.04	0.01

[a]Background solution—borate buffer, pH 7.4, containing 0.01 M NaCl.

is in agreement with the proposed mechanism of inhibition, then a fall in ΔE with increase in the basicity of the inhibitor would appear to be unexpected. However, one must take into account that Cl^- as a ligand in nucleophilic substitution is in competition not with the whole of the concentration of the organic additive but only with the anions that are formed on its dissociation in the solution. Since the degree of dissociation increases with decrease in pK_a, the value of the latter will have a negative sign in Eq. (3.15a). On recalculating for $C = 0.1$ g ion L^{-1}, the dependence of ΔE on the anion parameters acquires the form

$$\Delta E = 0.348 + 0.269\pi + 0.014pK_a \tag{3.15b}$$

In agreement with the theory, the effect of pK_a is minimal and the protective effect can be described solely by the hydrophobicity of the functional group

$$\Delta E = 0.448 + 0.22\pi \tag{3.15c}$$

Unfortunately, since it is impossible to establish the required concentrations of anions in solutions of weak acids, the values of ΔE have to be determined by extrapolation of the concentration dependence, and this lowers the precision of Eqs. (3.15 b and c). Furthermore, this relationship is fulfilled only for $C > C_{cr}$ from which the linearity of the $E - \log C$ function begins (Fig. 3.9). At lower concentrations, the organic additives have no effect on the local dissolution of aluminum, due to their inability to establish a competition with chloride in complex formation on the electrode surface.

It is a characteristic of the situation that the sensitivity of the protective effect to a change in the concentration of inhibitor, as represented by the tangent of the angle of the dependence of ΔE on $\log C$, is also determined by pK_a and π

$$\tan\alpha = 1.418 - 0.127pK_a + 0.158\pi \tag{3.16}$$

This relationship indicates a possible quantitative prediction of the inhibiting effects of anions of aromatic acids as a function of their nature and concentration. In fact, the following relationship has been found to hold for the series of aromatic monobasic acids in Fig. 3.9a:

$$\begin{aligned}\Delta E = 2.35 - 0.191pK_a + 0.276\pi + 1.6891\log C \\ - 0.157pK_a \log C + 0.0207\pi \log C\end{aligned} \tag{3.17}$$

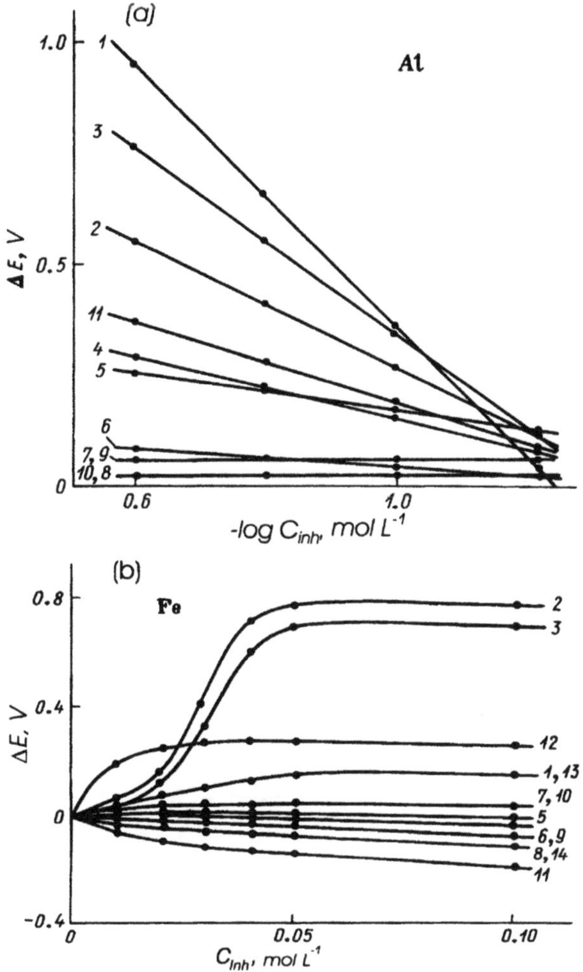

Figure 3.9. Dependence on concentration of the effectiveness of monosubstituted benzenes C_6H_5Y (where Y is a reaction center) in the protection of Al (a) and Fe (b) from pit formation in 0.01 M buffered NaCl (pH 7.4). Y = COOH (1); CH_2COOH (2); CH_2CH_2COOH(3); SH (4); $B(OH)_2$ (5); OH (6); CONH(OH) (7); $C_6H_5CONH(OH)$ (8); AsO_3H (9); NH_2 (10); $PO(OH)_2$ (11); C_2H_2COOH (12); $CONHCH_2COOH$ (13); CH(OH)COOH (14).

The effectiveness of phenylphosphonic acid [$\pi = -1.59$, $pK_{a_1} = 1.83$, $pK_{a_2} = 7.07$] cannot be predicted by this equation in view of the fact that by pH 7.4 it has already dissociated twice (> 50%). An increase in electrostatic interaction of an organic additive with aluminum can apparently compensate for a lack of hydrophobic properties. Thus, sodium phenylphosphonate (E = 0.205 V) is superior in its protective properties to all the monosubstituted benzenes that have been studied, apart from the carboxylates. The latter are commercially available and are of low toxicity. Various substituted benzoic acids—the arylcarboxylates—are, in fact, widely used as corrosion inhibitors in neutral media.

It is significant that arylcarboxylates are effective in protection against pitting of not only aluminum but also other metals, including iron (Fig. 3.9b). Whereas benzene sulfonic acid can initiate the local dissolution of iron, and some other derivatives of benzene can either facilitate, or fail to prevent, pitting in chloride solution, benzoate is one of the better inhibitors. However, it is inferior in effectiveness to certain other aromatic carboxylic acids, with the exception of hippuric acid (R = $CONHCH_2COOH$), which is hydrolyzed in water to benzoic acid. The inhibitive effects caused by additions of benzoic, phenylacetic, phenylpropionic, cinnamic, benzhydroxamic, phenol, and aniline were described by the equations[191]:

$$\text{at } C_{inh} = 0.01\text{ M: } \Delta E = 0.010 + 0.103 \log MR + 0.042\pi \quad (3.18a)$$

$$\text{at } C_{inh} = 0.02\text{ M: } \Delta E = 0.013 + 0.170 \log MR + 0.066\pi \quad (3.18b)$$

$$\text{at } C_{inh} = 0.03\text{ M: } \Delta E = 0.030 + 0.318 \log MR + 0.122\pi \quad (3.18c)$$

where MR is the molar refraction.

Here, the values of π and $\log MR$ used, as in Eq. (3.15) to characterize the reaction center are not those of the nucleophile itself (no literature data is available for the majority of the compounds) but those of the conjugate Brønsted acids (Table 3.6).

An analysis of Eq. (3.18) shows that the protective action of the inhibitor increases with increase in its polarizability and hydrophobicity. The positive effect of the first factor is associated with intensification of the covalent interaction of the metal with the nucleophile and with the facilitation of the formation or strengthening of the inhibitor complex. Simultaneous contributions of hard interactions and of non-specific solvation effects to the value of the π constant make it difficult to understand the effect of this characteristic on the inhibiting action of the nucleophiles. However, to a first approximation it can be considered to be a contribution of the chemical nature of the nucleophile to its affinity to water. It then becomes clear why the increase in π constant increases ΔE, since an increase in hydrophobicity of the nucleophile intensifies its "ejection" from the bulk of the electrolyte onto the surface, i.e. its adsorption.

In this region of C_{inh}, the approximately constant contributions of the π and $\log MR$ factors to the protective effect confirms the conclusion that the mechanism of depassivation remains constant. A further increase in C_{inh} should, as already observed, change the mechanism of the nucleophilic substitution with the departing group becoming the inhibitor. Therefore, with $C > 0.03$ M, when for various compounds different mechanisms of depassivation exist, a relationship of the type in Eq. (3.18) will not be fulfilled. The reasons for this are the lack of any constancy in the mechanism of the process and the fact that it has still not been possible to overcome the problem of finding chemical methods for the quantitative identification of the departing groups. Despite this, it is already possible to use the criteria explored above for assessing the role of the reaction centre of an inhibitor in the selection of the most promising classes of organic corrosion inhibitors.

3.4. THE EFFECT OF THE CHEMICAL STRUCTURE OF CARBOXYLATES ON THEIR PROTECTIVE PROPERTIES

3.4.1. Arylcarboxylates

The role of the chemical structure of inhibitors of localized corrosion was first examined[148] using the linear free energy relationship principle with anions of those substituted benzoic acids which had earlier been considered by Hammett in some detail in his study of composition—property relationships. Initially, the most detailed study was of the inhibiting action of ortho-arylcarboxylates, for which a V-shaped dependence of $\Delta E = E_{pit}^{inh} - E_{pit}^{background}$ or of $\Delta E_R = E_{pit}^{R} - E_{pit}^{H}$ where R = substituent, and H = hydrogen, i.e. unsubstituted carboxylate, reflected the relative dependence of the protective effects brought about by the introduction of a substituent on the σ-constant of the substituent. This characteristic was shown, for example, in the protection of duralumin by some substituted benzoates in solutions of a number of activators at various pH and temperatures (Fig. 3.10).[192] It is of interest that unlike iron, for which ΔE correlates with the Taft σ_0^* constants which take into account the total induction and mesomeric effects of the substituent; the correlation in this case is only with the induction constants, σ_I.

Evidently, during adsorption of ortho-arylcarboxylates on an aluminum alloy the inhibitor is subject to deformation, which impedes effective conjugation of the substituent with the reaction centre. As a result of this,

$-NH_2$ or $-OCH_3$ groups do not exhibit electron-donating properties and become acceptors. The introduction of acceptor groups is less effective, as can be seen from the less steep course of the corresponding lines in Fig. 3.10. Compounds containing an electron-donating substituent possess increased electron density at the reaction centre and form stronger bonds with the surface. On the basis of the concepts of chemisorption that have been considered above, the precursor of the reaction series, i.e., benzoate, should not have a high reactivity, which means that reaction 3.13 is significantly impeded. This agrees with the fact that it is necessary to add significant quantities of arylcarboxylates to obtain inhibiting effects. The probability of the chemisorption of the activators (on the basis of Table 3.2) increases in the series $F^- < Cl^- < Br^- < I^-$, which is in the same order as their effects in decreasing the protection given by arylcarboxylates, the increase in their basicity, and their reactivity in nucleophilic substitution in protonated solvents. Consequently, it may be suggested that in the initiation of pitting, the scheme described by Eq. (3.14) is operative, i.e., in which the surface complex with the inhibitor is the substrate, and the nucleophile released by the activator An^- can be OH^- or the inhibitor itself. Here, as in the simpler homogeneous reaction of S_N2 nucleophilic substitution, importance is attached to the change in electron density at the reaction centre during the transition from the initial conditions to the activated complex. The electronic effects of the substituents will only be unimportant for the kinetics of the process if matching substitution occurs, which is usually not the case. The more frequent case is the rupture of the old bond following the formation of the new bond. In the transition to such a "compressed" transition state, the positive charge on the complex former will be decreased and the reaction should be accelerated, with the inhibition effect decreased on lowering the electron-donating properties of the substituent. Then the dependence of ΔE_R on the σ constant should be characterized by a slope $\rho < 0$, as is observed for electron-donating substituents (slope ρ_+).

However, if the transition from the initial to the activated state increases the positive charge, the opposite effects should be observed at the reaction centre. In normal homogeneous reactions, this will occur in those cases where the rupture of the old bond overtakes the formation of the new (a "loose" transition state). It is significant that within one reaction series the "loose-compressed" transition state can change and an inversion of the sign of ρ can occur. Undoubtedly, the depassivation of a metal is a more complex process and additional causes for the increase in positive charge on the complex former in the transition or final state can arise. These could include the ionization of the metal, e.g., $[z_1 - z \geqslant 1]$ in reaction (3.13) but also a prior dehydroxylation of the surface, e.g., $[m_1 - m \neq 0]$

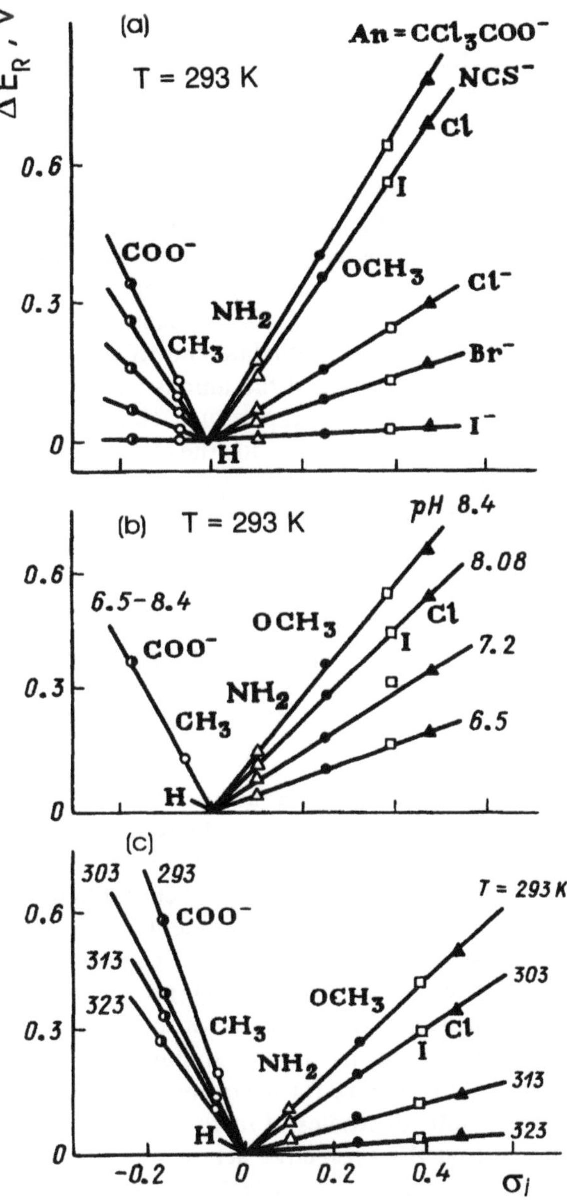

Figure 3.10. Dependence of protection of D16T aluminium alloy by 0.08M ortho-arylcarboxylates in a borate buffer at pH 8.1 (a and c) or at 6.5-8.4 (b) containing 5(a) and 2mM (b,c) NaCl on the σ_i constant of *R* (aggressive anions are identified in a, pH in b, and temperature in c).

in reaction (3.14). In the latter case, a decrease in the rate of reaction (3.14) would be expected, that is, an increase in ΔE upon increasing the pH of the solution, which is in fact observed in the protection of D16 (duralumin) alloy by arylcarboxylates containing electron-accepting substituents (Fig. 3.10b). In support of the proposed mechanism is the fact that such a change in pH has no noticeable effect on the value of ρ_+. This is one more advantage of arylcarboxylates containing an electron-donating substituent since even at pH 8.4, $|\rho_+| = 2.8$ remains above $|\rho_-| = 1.6$. Increase in temperature lowers ΔE_R of both types of arylcarboxylates, but even in this case, $|\rho_+|$ is decreased to a lesser extent than is $|\rho_-|$. For example, it will lead to some reduction in the effectiveness of phthalate and toluate (Fig. 3.10c) and the beneficial effect of substituents and the anions of substituents will be retained even at $T = 373$K. At this temperature, electron-accepting substituents are almost without effect in increasing the effectiveness of the inhibitor, and at $T > 323$K, arylcarboxylates of this type will be no better than sodium benzoate. Such a critical temperature, which can be predicted from the Leffler relation[120] for the reaction series being considered, can be described in the form

$$|\rho_+| = -4.67\,(1 - 467/T) \tag{3.19a}$$

$$|\rho_-| = -11.7\,(1 - 323/T) \tag{3.19b}$$

The value of the isokinetic temperature in the case of arylcarboxylates with electron-donating substituents ($T_0^k = 467$K) is significantly above the T_0^k that is observed in the inhibition of acid corrosion by various organic compounds.[33] This can be explained by the fact that in the system we have studied, the chemical interaction of components of the solution with the metal surface is more important than in acid corrosion. Despite the fact that arylcarboxylates with electron-accepting substituents are characterized by lower T_0^k, they will, according to the proposed mechanism, also be capable of chemisorption on the surface of aluminum and its alloys.

This is shown both by data for adsorption of ortho-iodobenzoate[171] and also from the results of studies of meta and para arylcarboxylates.[193] These have shown that the protective action of quite a number of compounds (20 anions) cannot be described without simultaneously taking into account the electronic, σ; and hydrophobicity, π, constants of the substituents. It turns out that independent of the nature of the latter, the inhibitive effect can be described by the equation:

$$\Delta E = a_2 + b_2\pi + \rho\sigma \tag{3.20}$$

An important finding is that there is a change in the ratio of solvation and electron contributions of the substituent to ΔE on changing the inhibitor concentration, C_{inh}, and pH of the solution (Table 3.7).

A comparative analysis of the values of the coefficients in Eq. (3.20) shows that ΔE is lower for aluminum than for iron, for which, despite the $J - J_p$ difference [where J is the ionization potential of the adsorbed molecule and J_p is the resonance ionization potential of the electrode, i.e., metal, metal oxide, etc. (see section 3.1.1.)], one would have expected chemisorption of benzoate to take place to a greater extent (Table 3.2). The fact that ΔE for aluminum is determined more by the greater hydrophobicity of the substituents in the arylcarboxylates than by their electron-donating properties, also agrees with ideas about the poor ability of arylcarboxylates to chemisorb on aluminum. For iron, the importance of both factors is similar at C_{inh} from 0.01–0.05 M, but with increase in pH, the contribution of the electron-donating capabilities of the substituent to ΔE is reduced, since at pH 9.2 only the π constant of the substituent should be taken into account. Evidently, with increase in a_{OH^-} there will be an impediment to the release of OH^- from the surface by the arylcarboxylate anion. This is also supported by the fact that after holding an electrode in a neutral solution of fluorobenzoate, and despite subsequent washing in water, X-ray electron spectroscopy clearly shows inhibitor firmly adsorbed on the iron surface. When an analogous test was conducted in a solution at pH 9.2, the fluorobenzoate was fully desorbed from the electrode on rinsing.

Fluorinated substituents in arylcarboxylates (Table 3.8) are of special interest because at relatively small size they possess a high hydrophobicity. Thus, for $-OCF_3$, $\pi = 1.04$, whereas the substitution of hydrogen by $-OCH_3$ has practically no effect on the energy of interaction of an aromatic compound with water ($\pi = -0.02$). Fluorinated hydrocarbon substituents have a weaker effect in increasing the protective properties of arylcarboxylates

Table 3.7. The Effect of Inhibitor Concentration C_{inh}, and pH of a Borate Solution Containing 5 mM NaCl, on the Coefficients in Eq. (3.20)

Electrode	C_{inh}(M)	pH	a_2	b_2	ρ
Aluminum	0.01	7.36	0.071	0.053	−0.034
Armco-iron	0.01	7.36	0.240	0.156	−0.151
	0.02	7.36	0.360	0.240	−0.213
	0.05	7.36	0.670	0.398	−0.472
	0.02	6.05	0.380	0.165	−0.0195
	0.02	8.04	0.252	0.237	−0.0112
	0.02	9.20	0.270	0.136	—

Table 3.8. The Protective Effect of Arylcarboxylates (0.01 M) in the Pitting Initiation of Metals in a Borate Buffer (pH 7.4) containing 0.005 M NaCl

Substituent in benzoic acid	ΔE(V)		Substituent in benzoic acid	ΔE(V)	
	Iron	Aluminum		Iron	Aluminum
H	0.095	0.070	p-OC_5F_{11}	—	0.33
o-$OCHF_2$	0.063	0.015	p-$OC(O)CFHCF_3$	0.365	0.170
m-$OCHF_2$	0.348	0.175	o-CF_3	—	0.020
p-$OCHF_2$	0.275	0.109	m-CF_3	0.35	0.12
p-OCF_3	0.330	0.125	p-C_5F_{11}	—	0.21
p-$OCH_2C_2F_4H$	0.287	0.185	p-$CF(CF_3)_2$	—	0.14
p-OCF_2CFHCF_3	—	0.110	p-$C(OH)(CF_3)_2$	0.095	0.070
p-OC_4F_8H	0.340	0.330	p-OC_3H_7	0.370	0.265
p-$OC_5F_{10}H$	—	0.23	p-OC_6H_{13}	0.520	0.320

than oxyfluoralkyl substituents, an effect which is explained not only by the greater hydrophobicity of the former but also by the larger electron-accepting influence (the introduction of a CF_3 group in the para position is characterized by $\sigma = 0.54$, but $\sigma = 0.35$ for an OCF_3 group). Since an increase in length of the fluoralkyl chain has little influence on the electronic effect of the substituent [for p$CF(CF_3)_2$, $\sigma = 0.53$, but for $CF_2(CF_2)_2CF_3$, $\sigma = 0.52$], the higher arylcarboxylates with hydroxyfluoralkyl chains are the more effective. However, even in these cases, in order to guarantee a strong shift in E_{pit} towards positive values the introduction of a very hydrophobic substituent is required and this significantly limits the solubility of the inhibitor (for arylcarboxylates with substituted $-OC_5F_{10}H$ it is even somewhat below 0.01 M).

A still greater constraint is the introduction of an electron-accepting substituent in the position ortho to the carboxylic group at which, in view of the closeness to the reaction centre, the electronic effect of the substituent is realized more strongly. This is consistent with the fact that ortho-arylcarboxylates with fluoride-containing substituents are inferior in protective properties even to benzoate or, as in the case of pentafluorobenzoate they may lose these properties completely. Consequently, in order to increase the effectiveness of inhibition of pit formation by arylcarboxylates, their chemical structure should be such as to both provide some hydrophobicity from the substituent as well as sufficient electron density at the reaction centre for chemisorption of their anions on the surface.

In fact, sodium para-propoxybenzoate is one of the best inhibitors of pitting of metals among the meta and para arylcarboxylates (Table 3.8). The substituent in its molecule possesses the same hydrophobicity as OCF_3, but is distinguished by the presence of electron-donating properties ($\sigma = -0.25$). The introduction into benzoate of OC_6H_{13} is still more effective, but further increases in the length of the hydroxyalkyl chain not only lower the solubility of the arylcarboxylate but also hinders its use in neutral solutions. It is therefore useful to examine the possibility of increasing the protective action of arylcarboxylates containing hydrophilic substituents in the position ortho to the carboxylic group by changing the side chains in the molecule.

3.4.2. N-substituted Sodium Anthranilates

The introduction of an electron-donating substituent into the ortho position relative to the carboxyl group in benzoate can significantly increase the protective properties of this arylcarboxylate. Of the compounds that have been studied, sodium anthranilate has been found to be one of the best for inhibiting the pitting corrosion of iron. Bearing in mind that the NH_2- group is hydrophilic, its favorable effect on the action of an inhibitor must be associated with an increase in electron density at the reaction centre, which will in turn lead to a strengthening of the chemisorption bond with the surface. A further increase in protection can be achieved by the introduction of a substituent either into the benzene nucleus or into the amino group. In this case, the effect of changing the chemical structure could be assessed from the σ_0^* constants of the respective groups (NH_2, $NHCH_3$, NHC_2H_5, etc.) being introduced into the ortho position, relative to the reaction centre. Unfortunately, only a limited collection of σ_0^* constants is available, which means that this approach cannot be used. However, substituted sodium anthranilates can also be viewed as compounds in which the electronic influence of the substituent on the reaction centre ($-COO^-$) appears not through the aromatic nucleus but through the $-C_6H_4NH$ group. We have previously described a V-shaped dependence of ΔE_R for iron on the induction constant of the substituent.[189] As a rule, the protection of iron is characterized by a high sensitivity to the introduction of an electron-donating or electron-accepting substituent into anthranilate, although the differences between the values of ρ_+ and ρ_- are not so significant as in other arylcarboxylate solutions.

The effects of the aromatic substituent, as expected, are not described by their σ_0^* constants, which take into account induction and mesomeric effects. Thus, the specificity of phenyl is, to a significant extent, the realization of a (+)C effect which leads to an increase of electron density at the reaction centre of the organic compound despite the (−)I effect of the phenyl substituent.† The beneficial effect of introducing phenyl at the ni-

†Electron-donating effects according to the mechanism of conjugation are denoted by (+)C while electron-accepting effects are (−)C. Analogously, electron-donating effects according to the mechanism of induction are (+)I and acceptor effects (−)I.

trogen atom is also found in the protection of aluminum. It is because of these effects that phenylanthranilate and substituted phenylanthranilates are of such interest.

As a result of the features of its chemical structure and its high hydrophobicity ($\pi = 2.86$), phenylanthranilate is capable of being strongly adsorbed on metals over a wide range of temperatures. Thus, unlike in the protection of aluminum by benzoate, ΔE increases in phenylanthranilate solutions with an increase in temperature. This is particularly noticeable with increase in C_{inh}.[107] The increase in hydrophobicity upon replacing phenyl by naphthyl raises the effectiveness of the inhibitor which in 5 mM NaCl, at $C_{inh} = 2.3$ mM, ennobles E_{pit} of iron by 0.35 V (phenylanthranilate by 0.16 V). The displacement of such hydrophobic inhibitors from the surface complex requires an increased activation energy, the value of which will depend on the nature of the anion-activator.

As already pointed out, (see section 3.4.1.), and to a first approximation, ΔE of an inhibiting anion in an aqueous solution, i.e., a protonic solvent, decreases with increase in basicity, $H^{\dagger}$, of the activating halide (from F^- to I^-). In the case of phenylanthranilate solutions, this effect is observed on various steels and other metals (Fig. 3.11). Moreover, this dependence also extends to other activating anions while retaining the same sequence of changes in hydrophobicity of the atoms or groups making up the nucleophile:

$$\Delta E_{C,\sigma} = \Delta E_{H_o, C,\sigma} + b_{C,\sigma}\,(H - H_o) \qquad (3.21a)$$

where $b_{C,\sigma}$ is a constant reflecting the sensitivity of ΔE to changes in H of the nucleophile for a fixed concentration of solution (C) and chemical structure of the inhibitor (in a series of inhibitors of one type—the σ-constant of the substituent). The subscript "o" here, and later, relates to a standard arbitrarily selected value of the parameter, for example, $H_o = -3.0$ for Cl^-. This dependence is retained on varying the concentration of the anions although the sensitivity of ΔE to H changes

$$b_\sigma = b_{\sigma,C} + a'_\sigma \log(C/C_o) \qquad (3.21b)$$

The single parametric linear dependencies of (3.21), although obtained using certain assumptions, are convenient to use for the quantitative calculation and prediction of the effect of the composition of a corrosive medium on the effectiveness both of individual inhibitors, and of the reaction series

$^{\dagger}H$ is relative basicity of nucleophile, $H = pK_a + 1.74$ where 1.74 is the correction for the pK_a of H_3O^+.[122]

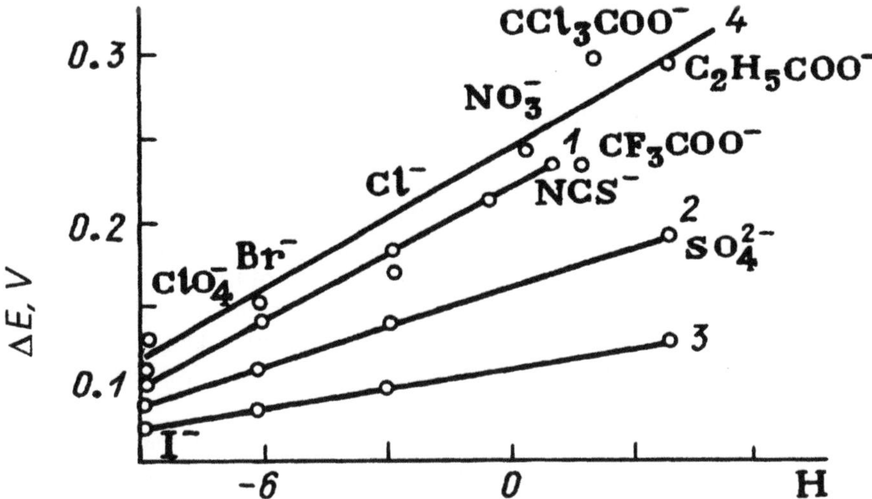

Figure 3.11. Dependence of protective effect of 0.01 M phenylanthranilate on the steels Kh18N10T (1); 30KhGSA (2); Armco Fe (3); and aluminium (4) in a borate buffer containing 0.5 M of various activators on the relative basicity H of the nucleophile (nominal compositions, %, Kh18N10T = = 18:10 Cr–Ni, Ti stabilized; 30KhGSA = C 0.3, 1:1:1 Cr–Mn–Si).

as a whole. However, in order to do this it is necessary to take into account the dependence of ΔE on the concentrations of both types of nucleophiles, the σ constants of the substituent, etc. Such polylinear functions, introduced into chemistry by Pal'm,[123] can be applied to describing the inhibition of acid corrosion and hydrogen charging of metals.[157] We have examined an example of this in the protection against pitting corrosion for solutions of arylcarboxylate and N-substituted sodium anthranilates.[189,192] Furthermore, the examination of a wide circle of aggressive anions using a more rigorous approach requires, as we have seen in section 2.3., taking into account the hydrophobicity and polarizability (measured by the molar refractivity, MR). In a solution not containing an inhibitor, the E_{pit} of metals, including aluminum, is described by the relation in Eq. (2.29). The introduction into such a solution of 10 mM phenylanthranilate changes only the coefficients in Eq. (2.29), so that a simple subtraction of these equations gives

$$\Delta E = 0.66 - 0.07\pi - 0.12 \log MR \quad (3.22)$$

In this equation, the low value of the coefficient before the π constant should be noted. It can be explained by the similarity of the solvation effects in the nucleophilic substitutions of the ligand in the background (uninhibited) and the inhibiting solutions. It follows that this would be expected particularly with low θ_{inh} when the nature of the "departing" groups is unchanged. Obviously in such cases the substitution of Eq. (3.22) by the simple, although less accurate, relationship of the Eq. (3.21a) type would be more appropriate.

An increase in protective properties of organic anions can be achieved by introduction of substituents into phenylanthranilate, as in the case of benzoate and anthranilate. If the carboxyl group and the substituent, *R*, are in different rings then, formally, that fragment of the molecule having both the amino and the carboxyl groups can be treated as the reaction centre, and the effect of the substituent can then be taken into account by the Hammett or Wepster σ^n-constants, as already discussed in section 3.2.

Substituted sodium anthranilates containing electron-donating substituents become, as in the case of other reaction series, more effective than the unsubstituted compounds even with small electronic effects of the substituent (i.e., σ constant values). In the protection of aluminum and its alloys, the dependence of ΔE on the σ^n constant is also linear and the beneficial effect of the substituent is particularly noticeable in dilute chloride solutions[194] and, unlike in arylcarboxylates, it is retained for substituted sodium anthranilates even with C_{NaCl}/C_{inh} = 1–2. In the protection of iron, the effect of the substituent appears still more strongly. Even at C_{NaCl}/C_{inh} = 5 the sodium salt of N-(2,3-xylylanthranilic) acid remains more effective than phenylanthranilate. Unfortunately, the number of electron-donating substituents is not large and acceptor groups raise ΔE of N-substituted sodium phenyl anthranilates only up to σ = 0.45. With larger σ, the protection is lowered and there is a discontinuity in the electron-acceptor branch of the correlation dependence of E on σ (Fig. 3.12). The occurrence of steric hindrances in the adsorption of N-substituted sodium phenylanthranilates is unlikely, even for such bulky substituents as NO_2 or SO_2CHF_2, since the substituent is some distance from the carboxyl group. This can be confirmed from examination of the appropriate Stuart molecular models.

The reduction in hydrophobicity of the substituent can be another reason for the reduction in effectiveness of N-substituted sodium phenylanthranilates, but consideration of this is difficult because of the absence of π constants for SO_2CHF_2, $SCHF_2$ and $OCHF_2$ groups. However, the introduction of two hydrophobic CF_3-groups ($\pi = 1.76$; $\Sigma\sigma = 0.86$) decreases not only the solubility but also the effectiveness of the inhibitor; sodium N-[3,5-bis(trifluoromethyl)]phenylanthranilate at C = 0.001 M is inferior in effectiveness even to phenylanthranilate. Consequently, corrections to the hydrophobicity of the substituent do not alter the conclusions about the negative effect of strong electron acceptors on the effectiveness of inhibitors.

Since the surface activity of the adsorbate depends on its polarity, it was of interest to assess the effect of the substituent on the values of the dipole moments, μ, of the respective acids—the measurements being made in dioxan solutions. This established that the dipole moments increased with increase in the electron-accepting ability of the substituent according to the sequence: H $<$ m-$OCHF_2$ $<$ p-$OCHF_2$ $<$ p-$SCHF_2$ $<$ m-CF_3 $<$ 3,5-bis-CF_3 $\approx$ NO_2 and in agreement with the equation

$$\mu = 1.26 + 9.12\sigma \tag{3.23}$$

with the points which characterize compounds with $\sigma > 0.45$ described satisfactorily by a linear correlation. This indicates that these substituents even increase the polarity of the organic compound. Characteristically, the polarizability of the slightly effective sodium N-(4-difluorosulphomethylphenyl)-anthranilate is practically indistinguishable from that of the better inhibitor, N-(3-difluoromethylthiophenyl)-anthranilate, the log *MR* values of the substituents in these compounds are 1.12 and 1.14, respectively. Thus, the discontinuity in the electron acceptor branches of the ΔE-σ dependencies cannot be explained by a reduction in the polar influence of the substituent. It is more logical to propose that the strong reduction in electron density at the carboxyl group weakens the chemisorption of the inhibitor and facilitates its displacement by the aggressive anion from the surface of the metal or alloy.

An analogous picture is also observed in the protection of other metals (Fig. 3.12). Thus, in the case of zinc, nitrophenylanthranilate is a weaker inhibitor than phenylanthranilate.

To improve the effectiveness of such compounds it is necessary to weaken the acceptor influence of the substituent by the introduction of electron-donating groups. Furthermore, the presence of two different types of substituents, as we first showed in studies of arylcarboxylates,[186] can enhance the protective action of an organic anion by increasing its polarizability. In fact, N-(3-nitro-4-methylphenyl)-anthranilate is a better inhibitor of the corrosion of aluminum than sodium N-(3-nitro)- or (4-methylphenyl)-anthranilate (ΔE = 0.30, 0.28, and 0.25V, respectively). This effect is evidently retained even when the substituents are introduced into the different aromatic rings of phenylanthranilate. Thus, E_{pit} is ennobled by 0.23 V in 0.01 M NaCl in the presence of sodium N-(4-methoxyphenyl)-anthranilate and by up to 0.34 V when a nitro group is introduced in the meta position relative to the carboxylate group. This example points to the fact that both

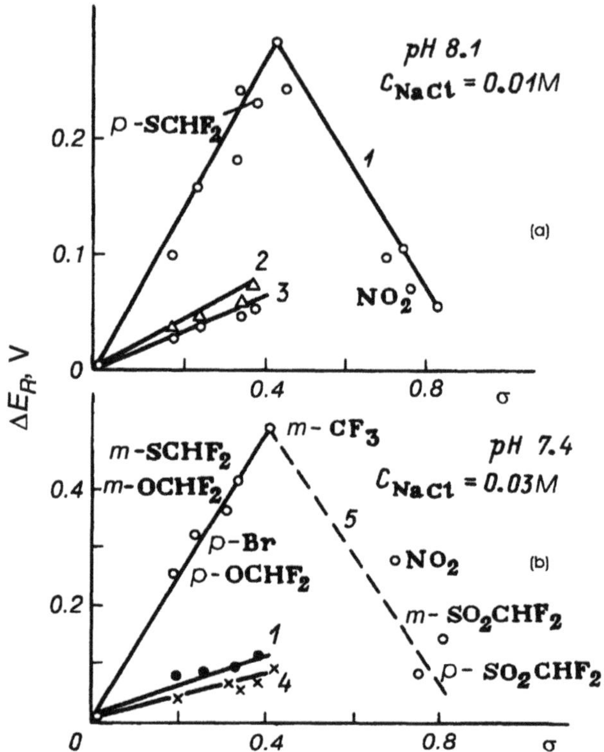

Figure 3.12. The effect of the polarity of electron-accepting substituents in 0.01 M phenylanthranilate on the protection of aluminium (1); the aluminium alloys AMg6 (2); and D16 (3); Zn (4); and Fe (5), in a borate buffer containing 0.01 M (a) or 0.03 M (b) NaCl. (AMg6 = A1–6% Mg; D16 = A1–Cu 3.4–4.9, Mn 0.3–0.9, Mg 1.2–1.8%).

substituents are hydrophilic, so their beneficial effect cannot be explained by an increase in the hydrophobicity of the anion. Some improvement in the inhibition of pit formation in aluminum by introducing the hydrophilic $NHCOCH_3$ group into N-(3-methylphenyl)-anthranilate can be explained only by an increase in polarizability of the anion since in the para position its $\sigma = 0.0$ ($\Delta E = 0.19$ and 0.24 V, respectively).

Corrosion testing confirms the strong influence of the chemical structure of N-substituted sodium anthranilates on their protective properties. Thus, in 0.01 M NaCl, pitting of aluminum occurs within a day, but the addition of 0.01 mol L^{-1} phenylanthranilate extends the induction period for pit initiation to three days, and isodifluorant to 20–22 days. The latter is, of the available substituted sodium anthranilates, the best inhibitor of the corrosion of aluminum alloys, although its meta-isomer is somewhat less effective. However, both compounds are strongly adsorbed on aluminum over a wide range of potentials as confirmed by the x-ray electron studies described in section 3.1.2. Many arylcarboxylates and some N-substituted sodium anthranilates, particularly phenylanthranilate and N-2,3-xylylanthranilate, possess not only high protective properties but also have technological and ecological properties that are in a number of cases extremely valuable in the practice of corrosion protection, as will be described in subsequent chapters.

3.4.3. Alkyl-, Olefin- and Other Carboxylates

Formate and certain lower alkylcarboxylates can cause pitting on iron and aluminum. According to Eq. (2.29), E_{pit} in such solutions can be estimated using the polarizability and π constants of the RCOO groups. On increasing the hydrophobicity of the anion it should lose its aggressive properties, and in fact such compounds can provide a significant inhibiting effect in the pitting caused, for example, by chlorides.

Thus, on introducing RCOONa salts (at either 0.01 or 0.02 M) into a 0.01 M buffer solution of NaCl at pH 8.1, the value (in V) of ΔE for aluminum changes in the following manner when the substituent R is varied:

	Substituent								
	CH_3	C_3H_7	C_5H_{11}	C_7H_{15}	CF_3	C_3F_7	$H(CF_2CF_2)_{2\text{-}4}$	$(CF_3)_3CC_2F_4$	C_8F_{17}
ΔE (V) at 0.01 M	0.08	0.16	0.33	0.41	0.06		0.13	0.28	0.43
ΔE (V) at 0.02 M	0.12	0.21	0.37	0.46	0.06	0.12	0.22		1.0

It is clear that the protective action increases with increase in hydrophobicity of the substituent both in the alkyl series and in the fluoralkylcarboxylates. However, the lower homologues in the latter case are less effective, which is explained not only by their lower hydrophobicity but also by their weak polarizability. An increase in the number of carbon atoms in the anion has a beneficial effect on both these characteristics of the inhibitor, and the effectiveness of the fluoralkylcarboxylates increases.

The presence of a double bond in an organic anion usually increases its polarizability, but the lower olefinic carboxylates, e.g., acrylate, can depassivate metals. In solutions of chlorides they are weak inhibitors (Table

Table 3.9. The Effect of Olefinic Carboxylates, R-CH = CH-COONa(0.02 M) on the Critical Pitting Potential of Aluminium in 0.01 M Buffer Solution (pH 8.1) of NaCl ($E_{pit}^{background} = -0.33$ V)

Inhibitor	Substituent, R	HLB	E_{pit}(V)	ΔE(V)
Sodium acrylate	H	25.15	−0.26	0.07
Sodium crotonate	CH_3	24.67	−0.29	0.04
Sodium maleate [cis]	COO^-	44.25	−0.28	0.05
Sodium fumarate [trans]	COO^-	44.25	−0.31	0.02
Sodium sorbate	$-CH{=}CH{-}CH_3$	23.7	−0.27	0.06
Sodium cinnamate	$C_6H_5^-$	23.7	−0.21	0.12

3.9).[167] This is explained by the hydrophobicity of the anions and by their weak surface activity as indicated by the value of the hydrophile–lipophile balance (HLB). All the lower olefinic carboxylates are inferior surfactants, and the most hydrophilic of these are the acrylate derivatives which contain a second COO^- group. Although they should be fully ionized at an appropriate pH, the known tendency of the cis-isomers to the formation of hydrogen bonds additionally impedes adsorption. With increase in the surface activity of the compound, that is, with decrease in HLB, there is a tendency for the protective properties to improve and consequently most interest is attached to the higher olefinic carboxylates.

Sodium undecylenate $CH_3{-}(CH_2)_3{-}CH = CH\,(CH_2)_4COONa$, although not a particularly good surfactant (HLB = 21.3), is one of the most effective of the olefinic carboxylates that have been examined (at C_{inh} = 0.02 M, E_{pit} = 0.08 V). Even better protective properties are provided by compounds containing 17 carbon atoms in a hydrophobic radical which makes them, judging from the values of the critical micelle concentration (CMC), true surfactants (Table 3.10).

Despite its surface activity, sodium stearate is only a weak inhibitor. Consequently, it is not a good example of the inhibitive action of carboxylates which, in the case of aluminum inhibition, improves with increase in the number of double bonds or hydroxyl groups in the molecule. The reason for this difference in the effectiveness of the higher olefinic carboxylates may be, as we have seen in the example of sodium oleate (section 3.1.2.), the structuring of the polymolecular layer of the inhibitor, which is facilitated with increase in the degree of unsaturation of the compound. It is possible that it is connected, at least, with a partial polymerization of the inhibitor in the adsorbed layer.

A similar increase in the inhibitive properties of salts of fatty acids has been observed with iron.[195] The concentration dependence of ΔE in such solutions has an S-shape (Fig. 3.13). In the first section of low slope,

Table 3.10. Surface Active and Protective Properties of Higher Olefinic Carboxylates in Relation to Aluminium (in the same background solution as in Table 3.9)

Inhibitor	CMC mol L^{-1} $\times 10^5$	ΔE(V) Concentration of inhibitor, mol·L^{-1} 0.001	0.005	0.01
Sodium stearate $CH_3(CH_2)_{16}COONa$	6	0.025	0.065	—
Sodium oleate $CH_3(CH_2)_7CH{=}CH(CH_2)_7COONa$	15	0.090	0.320	0.470
Sodium linoleate $CH_3(CH_2)_4CH{=}CH\text{-}CH_2\text{-}CH{=}CH(CH_2)_7COONa$	75	0.260	0.530	0.730
Sodium linolenate $CH_3CH_2CH{=}CHCH_2CH{=}CHCH_2CH{=}CH\text{-}(CH_2)_7COONa$	80	0.320	0.620	0.830
Sodium ricinoleate $CH_3(CH_2)_5CH(OH)CH{=}CH(CH_2)_7COONa$	40	0.090	0.360	0.660

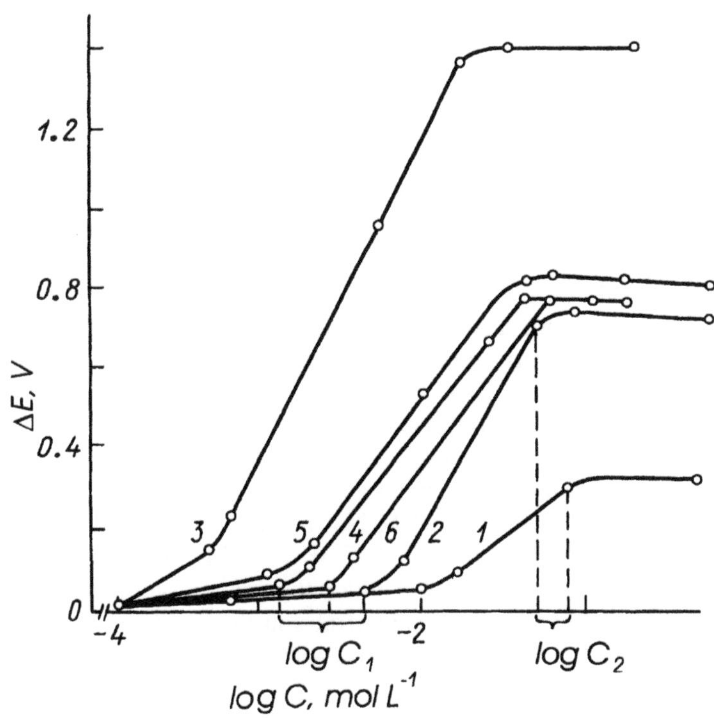

Figure 3.13. Dependence of the protective effect ΔE on Fe in a borate buffer containing 0.01 M NaCl on the concentration of sodium salts of the acids: 1, $CH_3CH_2CH_2COOH$; 2, $C_5H_{11}COOH$; 3, $C_6H_5(CH_2)_{10}COOH$; 4, $C_8H_{17}COOH$; 5, $C_{13}H_{27}COOH$; 6, $C_{10}H_{21}COOH$.

small additions of fatty acid salts (FAS) block the defect regions of the surface and the accompanying kinetic limitation of pit formation, and ΔE is found to be higher the greater the hydrophobicity of the inhibitor. However, ΔE at $\log C_{FAS} < \log C_1$ does not exceed 0.10–0.12 V and is apparently at a maximum: the higher fatty acid salts even with $C_{FAS} < 1$ mM form unstable colloid systems, e.g., sodium stearate, and so the use of the more hydrophobic fatty acid salts is considerably hindered.

On increasing the concentration of fatty acid salts, $C_{FAS} > C_1$, there was a sharp increase in ΔE, culminating in a "plateau" at C_2. Such an increase can be associated not only with chemisorption of the inhibitor but also with the beginning of the formation of a polymolecular barrier layer. In this case, there is some uncertainty in the value of ΔE because in the polymolecular or hemimicellar film that is formed with $C_{FAS} \leqslant$ CMC[196] there is a potential drop which it is not possible to estimate. This must also be associated with the absence of any clear dependence of ΔE in the $C_1 - C_2$ region on the electronic or hydrophobic constants of the fatty acid salts. However, C_2 depends on the hydrophobicity f_Σ (section 2.3. footnote) of the fatty acid salts by

$$\log C_2 = -0.52 - 0.20 f_\Sigma \tag{3.24}$$

This equation indicates the limitations in the use of organic inhibiting anions since increase in their hydrophobicity should lead to a reduction in the solubilities of the corresponding salts. The logical outcome of this situation is the search for inhibitors which, like oleate and other higher olefinic carboxylates, could form structured adsorbed layers or show an increased capability for adsorption on oxide covered metals.

However, the effectiveness of other types of carboxylates, for example, cyclic compounds, is to a large extent determined by their hydrophobicity and by the fact that the conformation of the hydrocarbon can adversely affect the inhibitor effectiveness. At the same time, attention should be drawn to the statement concerning sodium phenylundecanoate ($C_6H_5(CH_2)_{10}COONa$),[197] which in 0.01 M NaCl at pH 7.4 ennobles E_{pit} by 0.5 V even at $C = 5$ mM and which at $C = 10$ mM completely suppresses the depassivation of aluminum. Unlike the higher olefinic carboxylates, phenylundecanoate forms stable neutral solutions with $C < 0.1$ M; it is also of low toxicity and has a high passivating action towards various metals.

Good protective properties, according to Deberry,[198] are also possessed by the higher alkyl-N-acylsarcosines. Thus, the addition of 3 mM lauroylsarcosine ennobled E_{pit} of a chromium–nickel steel by 0.28 V in a quite aggressive solution, i.e., 0.2 M $NaOOCCH_3$ + 0.1M NaCl at pH 5.2. The author considered that the effectiveness of the inhibition resulted from the formation on the surface of, at least, a bimolecular adsorbed layer and emphasized the role in its formation of not only the hydrophobic hydrocarbon chain but also that of the carboxylic group. A somewhat smaller ΔE was obtained by the addition of 5-laurylthioglycollate or lauryl-glycine, but the limitations of solubility and other technical requirements are hardly less significant than in the use of the more readily available higher carboxylates.

In practice, it is rarely the case that single inhibitors will be used since mixtures often provide a synergistic action, that is, a mutual reinforcement of the effectiveness of the protective action.

3.5. COMBINED (SYNERGISTIC) ACTION OF INHIBITORS

If a metal or alloy is subject to localized corrosion in a particular environment, then the minimum concentration, C^{min}, of an inhibitor that is necessary to suppress corrosion completely is often used as a criterion of its effectiveness. An inspection of such values for the aluminum alloy D16, as shown in Fig. 3.14, indicates that these concentrations are, as a rule, higher than those of an activator present in the same system. Thus, the lines in Fig. 3.14 separate regions of C_{inh}/C_{NaCl} ratios where there is complete suppression of pitting from the regions in which pitting takes place. The slopes of the lines depend both on the nature of the inhibitor and on its chemical structure. In dilute chloride solutions, substituted phenylanthranilates, particularly isodifluorant (N–(m–difluoromethylthiophenyl anthranilate), have clear advantages, but at high chloride concentrations their use

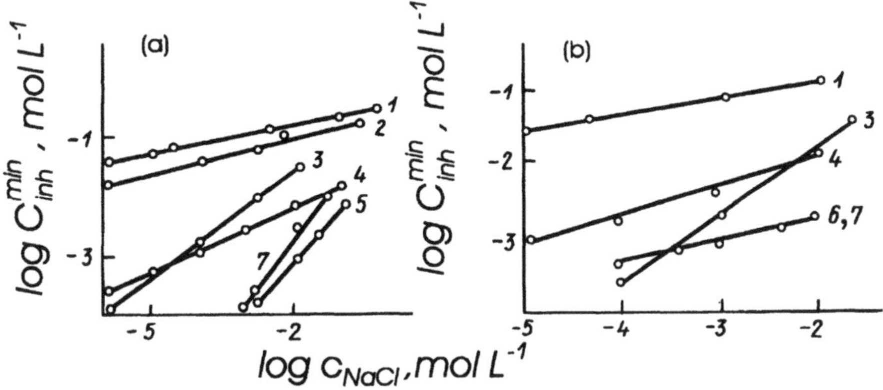

Figure 3.14. Dependence of C_{inh}^{min} for the aluminium alloy D16 on NaCl concentration in a borate buffer at pH 7.4 (a) and in water (b) containing potassium dichromate (1); sodium phenylanthranilate (2); isodifluorant (3); sodium oleate (4); sodium linoleate (5); sodium phenylundecanoate (6); IFKhAN25 (7).

is not as appropriate. Attention should also be paid to the high effectiveness of linolenate, even though the colloidal systems which it forms in concentrated chloride solutions are unstable. The low sensitivity of phenylundecanoate to C_{NaCl} makes this inhibitor promising for use in solutions with high chloride contents.

The inhibition of corrosion of low carbon steel can be achieved using lower concentrations of organic anions than would be required for an aluminum alloy. Oleate, phenylundecanoate, and isodifluorant are the most effective for the protection of steel (Fig. 3.15) although there are advantages and disadvantages with each. It should also be noted that the number of carboxylates that would replace the better inorganic inhibitors in practice is not very large. For example, benzoate, which is of low toxicity, is noticeably inferior in effectiveness to sodium nitrite, and only the introduction into its molecule of a sufficiently hydrophobic substituent which will also change the electron density at its reaction centre can improve the effectiveness.

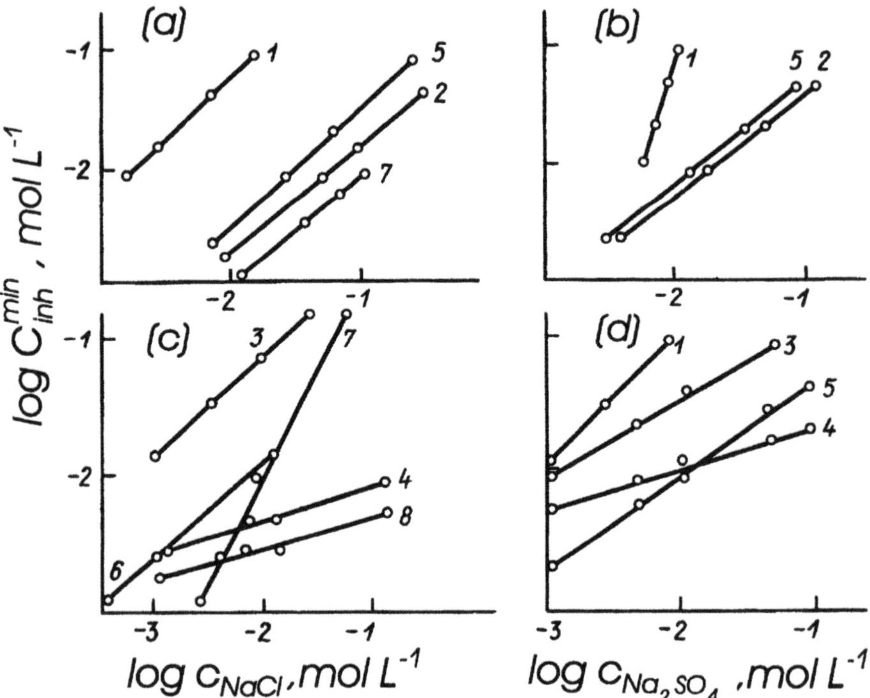

Figure 3.15. Dependence of C_{inh}^{min} for the low carbon steel St3 (C = 0.2%) on concentration of chloride (a,c) and sodium sulfate (b,d) in borate buffer at pH 7.4 (a,b) and water (c,d) containing sodium benzoate (1); sodium nitrite (2); sodium undecylate (3); sodium oleate (4); sodium phenylanthranilate (5); sodium phenylundeanoate (6); isodifluorant (7); IFKhAN25 (8).

However, new problems can then arise associated with the costs of protection or of ecological effects.

According to Antropov *et al.*,[5] the synergism of protection by inhibitors is often the result of a weakening of the repulsive forces between adsorbed particles and of an increase in their coverage of the metal surface. Such a situation can be realized by the introduction of two inhibitors of which one is anionic and the other is a surfactant of the cationic type. While accepting that cationic surfactants can inhibit the initial stages of pitting, it must be pointed out that they only slightly ennoble E_{pit} and furthermore, their use in conjunction with effective inhibiting anions is not very promising because of the formation of difficultly-soluble salts. A second route is via the introduction into an organic molecule of a basic group together with an acid group, but this is evidently only effective with the higher homologues since both functional groups in such compounds are hydrophilic, and for the provision of surface activity an increase in hydrophobicity of the hydrocarbon part of the molecule is required. Furthermore, the effective negative charge of these inhibitors is below that for monovalent anions and this also weakens their adsorbability and their nucleophilic properties. It is also difficult to expect a significant increase in the effectiveness of protection from the combined use of inhibitive organic anions and nonionic surfactants, since the latter can hydrophilize the metal surface.

A significant effect can result from certain combinations of components for example, by using anionic inhibitors, such as phenylanthranilate and oleate. Thus, protection by such a binary inhibitor, even at a low total concentration, can be more effective than that given by the better of the individual inhibitors, i.e., oleate. With a 1:1 mass ratio of these components such an inhibitor, in this case designated by IFKhAN-25,* is more effective than oleate for steel (Fig. 3.15), D16 alloy (Fig. 3.14), or other aluminum alloys. A feature of this system is that IFKhAN-25 provides good protective properties even at elevated temperatures. Thus, the protective effect of a 0.1% solution of IFKhAN-25 increases in a linear manner with temperature (up to 363K) and is stronger than that given by solutions with twice as high a concentration of phenylanthranilate. IFKhAN-25 is also effective in a flowing aggressive solution. The protective effect is retained even at temperatures above 333 K when pitting corrosion is prevented as a result of the high passivating properties of both organic anions. A further advantage of this formulation is that it provides reliable protection to low carbon steel at lower concentrations than sodium oleate and prevents atmospheric corrosion in very aggressive conditions. Thus, the rate of corrosion of the D16T

*IFKhAN from *I*nst *F*iz. *K*him. *A*kad. *N*auk (Inst. Phys. Chem. of the Academy of Sciences).

aluminum alloy in an atmosphere with 100% relative humidity in presence of deposited nuclei (chlorophos, trichloracetate) reaches 0.48 $gm^{-2}d^{-1}$, but with a prior treatment with a 4% solution of IFKhAN-25 this falls to below 0.02 $gm^{-2}d^{-1}$. This high protective action of IFKhAN-25 has led to the development of a process for protecting aircraft equipment using crop-spraying chemicals.

Synergistic protection by the higher carboxylates is a characteristic not limited to phenylanthranilate and sodium oleate. In the case of the IFKhAN-31 inhibitor, synergism, resulting from the presence of two carboxylates, is also observed with significant ennoblement of E_{pit} of aluminum or iron reached at concentrations at which the individual inhibitors show no protective action (Fig. 3.16). It is then possible to suppress pitting corrosion using IFKhAN-31 without the formation by the latter of colloidal

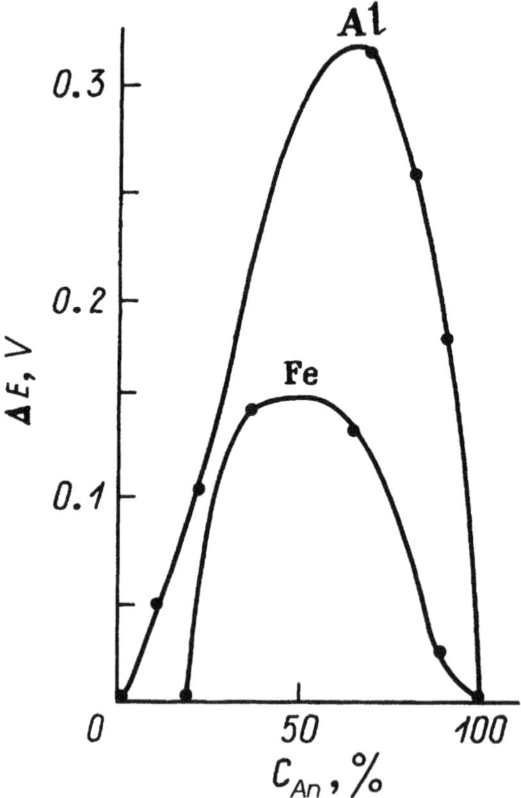

Figure 3.16. Dependence of the protection of iron (Fe) and aluminum (Al) in borate buffer at pH 7.4 containing 0.01 M NaCl on the content of carboxylate, An, in the IFKhAN31 inhibitor formulation. Total concentration of inhibitor: 0.1 mM for Fe and 0.5 mM for Al.

solutions. Furthermore, this inhibitor is less sensitive than IFKhAN-25 to hard water, which along with its low toxicity and better protective properties provide promise for its practical application.

Another route for improving the inhibition provided by anions is to change the composition of the surface complexes that they form. In this case, a marked ennoblement of E_{pit} can be achieved by introducing a relatively small quantity of hydrophobic anions. Inhibition by this approach can be understood in the light of the independence of E_{pit} on a wide range of concentrations of activator as shown in Ref. (138). This occurs even with low anion concentrations, C_{An}, if the activators are large organic anions, e.g., caproate ($C_5H_{11}COO^-$). It can be explained by the influence of the ψ' potential on the kinetics of the heterogeneous reaction leading to pitting formation. Since its value decreases with increase in $\log C_{An}$, then the ψ' effect, which as an electrostatic component, is measured by $\Delta G^{\neq}_{eff}$, can be neglected in concentrated solutions. In support of this proposition is the decrease in critical activator anion concentration, C_{cr}, (above which E_{pit} is constant, see Fig. 3.17) with increase in ionic strength of the solution and the fact that $\Delta G^{\neq}_{eff}$ at $C_{An} > C_{cr}$ ceases to depend on concentration, although the rate of pit initiation as judged from the reduction in the induction period, τ, continues to grow. Thus, there are certain limiting values of E_{pit}

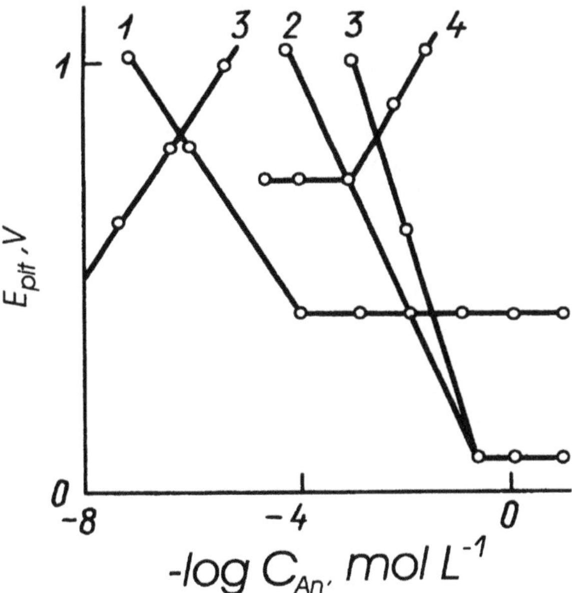

Figure 3.17. Dependence of E_{pit} of nickel in borate buffer on concentration of sodium caproate (1,4), and sodium chloride (2,3), without (1,2) and with addition of 0.05 M sodium caproate (3) or 1 mM NaCl (4).

and $\Delta G^{\neq}_{eff}$ which do not depend on C_{An} but which, undoubtedly, reflect the energy of interaction of the activator with the electrode surface.

Therefore, it is of interest to examine the depassivation of nickel in solutions with concentrations of caproate, $C_{An,1}$, above C_{cr} and with chloride $C_{An,2}$ much below C_{cr} (Fig. 3.17). It is found that in 0.05 M $C_5H_{11}COONa$ the constancy of E_{pit} breaks down even with the introduction of 10^{-8} M NaCl. Bearing in mind that the ratio of $C_{An,1}/C_{An,2} = 5 \times 10^6$, the adsorptive release of caproate by chloride is unlikely. At the same time, the chloride ion alone does not activate nickel at such $C_{An,2}$ and caproate is one of the most aggressive of the studied anions, thus signifying its high surface activity. On further increasing $C_{An,2}$, a clear inhibition of pitting developed since E_{pit} in the individual solutions of both activators was more negative than in the mixed solutions, except for $C_{An,2} > 0.1$ M where caproate can no longer be considered as an activator.

This effect of low concentrations of chloride can be explained by the fact that, although unable to displace caproate, chloride can impede the hydration of the complex by replacement of the water, which is a weaker nucleophile. Similarly, an analogous inhibition is observed if $C_{An,1}$ is varied with $C_{An,2} = 10^{-3}$ M. This is caused by the fact that components of the solvent (H_2O, OH^-) are more hydrophilic than chloride and even more so than caproate. Therefore, any of the indicated changes in the composition of the complex are capable of leading to, as already noted, a change in the controlling stage of the depassivation and even to complete blocking of active centres and so E_{pit} is ennobled. As $C_{An,2}$ is increased, the inhibiting effect grows (the ascending branch of dependence 3, Fig. 3.17), while Cl^- is not able to replace caproate (the descending part of the same dependence). In this case, pitting formation is inhibited by caproate although $C_{An,2}$ is not sufficiently high for the complete displacement of $C_5H_{11}COO^-$ and E_{pit} has not acquired the values for pure chloride solutions.

The above route to inhibition may be considered as a consequence of the formation of mixed complexes from two activators. Caproate is by no means special as an activator since other carboxylates, $RCOO^-$, will react in a similar manner; R can be H, alkyl with 1–9 carbon atoms, HC = C, $CC1H_2$, or CF_3. Thus, with $C_{An,1} = 0.05$ M and with $C_{An,2} = 10^{-5}$ M the inhibiting effect of Cl^- will depend only on the hydrophobicity (see section 2.2.) of the first anion

$$E_{pit}^{An+Cl} - E_{pit}^{An} = 0.05 + 0.17f \qquad (3.25a)$$

which is in full agreement with the mechanism of inhibition put forward for the ascending branch of the concentration dependence. For the descending branch, for example, at $C_{An,2} = 10^{-3}$ M, the electronic influence of the substituent on the reaction centre in the carboxylates, which is taken into account by its inductive constant σ^* and steric constant E_s (see section 2.3.), cannot be omitted

$$E_{pit}^{An+Cl} - E_{pit}^{An} = 0.37 + 0.15f - 0.023\,(2.6 + 0.6\sigma^* - 2.4E_s) \qquad (3.25b)$$

If carboxylates are considered as inhibitors and Cl^- as an activator, then the role of the electronic structure of an organic anion shows up even more clearly (for example, with $C_{an} = 0.05$ M and $C_{Cl} = 10^{-3}$ M):

$$E_{pit}^{An+Cl} - E_{pit}^{Cl} = 0.15 + 0.07f - 0.048\,(2.6 + 0.6\sigma^* - 2.4E_s) \qquad (3.25c)$$

It should be emphasized that inhibition according to (3.25a) is achieved by small additions of the activator which, at high concentrations, is even more aggressive than caproate. Despite the dangers of such inhibition it appears to be extremely attractive from the economical and ecological points of view.

Yet another way to improve the protective action of inhibitors of the adsorption type is by their combination with oxidizers. This has received relatively little basic study although the patent literature has many examples of such formulations. A composition based on sodium benzoate with small additions of sodium nitrite[199] can serve as the simplest example. We have shown[200] that a stronger oxidizer, sodium nitrophthalate, in very small quantities (> 0.2 mM) has a beneficial effect on the corrosion behavior of iron in neutral phthalate solutions. These phthalate ions are strong stimulators of the active dissolution of a metal, evidently as a result of the ease of transfer into solution of the surface compounds that they form. This dissolution prevents the organic anion from fulfilling its passivating function. On the other hand, its adsorption is accompanied by a fall in the surface concentration of the passivating components of the solvent which leads to an increase in the critical current density for passivation, i_p. With O_2 or other oxidizer present in the solution, the deficit of passivating components can be made up by the accumulation of OH^- resulting from the accompanying cathodic reaction. In this case, the negative influence of the lower arylcarboxylate on the passivation of iron is reduced or disappears completely. This proposition finds support in the work of Davies and Slaiman,[10] who found that i_p for iron in benzoate solutions decreased with increase in pH and C_{O_2}. They came to the conclusion that the effect of O_2 could not be explained without taking into account its inhibition of the anodic reaction, since the difference in i_p for deaerated and aerated solutions significantly exceeded the limiting diffusion current density, i_d, for oxygen.

The combined effect of nitrophthalate with phenylanthranilate on the initiation of pitting on iron merits attention. Our own work has shown[184] that with increase in temperature the generation of passivating OH^- by the reducing nitrophthalate did not result in a decrease in τ—as would be the case with adsorption inhibitors—but in an increase (Fig. 3.18). Since the adsorption of OH^- impedes activation of iron, there is the possibility that in the region of small θ_{inh} the effectiveness of protection can be improved by the presence of an oxidizer. This suggestion is supported by our observation of an increase in $\log \gamma = \log \tau_{inh}/\tau_{back}$ (at $C_{NaCl}/C_{phenylanthranilate} = 10$) with increase in pH from 6 to 8.2. In a borate buffer containing phenylanthranilate + nitrophthalate and chloride (Fig. 3.18), $1/\tau$ actually decreases with increase in temperature and the protective effect improves. Thus, the products of the associated cathodic reaction can turn out to have an influence on the rate of anodic dissolution not only in the active but also in the passive regions of potentials. It should be noted that with $C_{phenylanthranilate} < 2$ mM

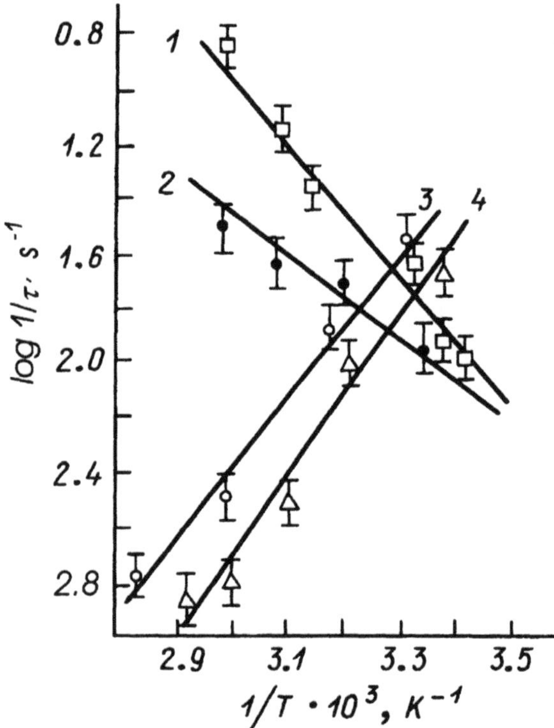

Figure 3.18. Dependence of log $1/\tau$ on temperature for iron in a borate buffer containing 3.6 mM NaCl (1) and with addition of inhibitors: 0.7 mM sodium phenylanthranilate (2); 86 mM nitrophthalate (3); 0.7 mM sodium phenylanthranilate + 0.4 mM sodium nitrophthalate (4).

the mechanism of pit initiation involves substitution of OH^- by the activator and phenylanthranilate only blocks the most defective parts of the surface. An increase in activity of OH^- with increase in temperature improves the protection to a greater extent than the deterioration by the weakening of the blocking action of the phenylanthranilate.

Corrosion tests have confirmed the conclusions obtained from electrochemical experiments. Although small additions of nitrophthalate have practically no effect on the value of the minimum concentration of phenylanthranilate for inhibition at 293 K, a noticeable synergism of protection occurs at 353 K; this minimum concentration is almost halved on addition of only 0.1 mM nitrophthalate. Analogously, the combined action of nitrophthalate and N-2,3-xylylanthranilate on the protection of steel is more effective than the latter alone. Thus, for oxidizers, the rates of reaction of which increase with temperature and cause the generation of OH^-, one can expect a beneficial effect on the effectiveness of inhibitors of the adsorption type.

3.6. ASPECTS OF THE PROTECTION OF METALS IN MIXED SOLVENTS

The protective action of inhibitors is usually found to decrease with increase in the content of the non-aqueous component in mixed aqueous–organic solutions. Although the effect of organic solvents on the corrosion and protection of metals has received much attention,[149,157,201], only a small part of this has been concerned with the initial stages of pitting. The mechanism of this process in water-rich media is not, in principle, different from that involved in the depassivation of metals in aqueous solutions. However, even a partial change in the composition of a solvent can have a significant effect on the kinetics of surface reactions. Thus, in aqueous–organic media, passive layers can be of different thicknesses and defect structures, as is particularly the case for certain aprotic solvents, e.g., dimethylsulfoxide, DMSO, and dimethylformamide (DMFA). The double-layer structure and the reactivity of anions, whether activators or inhibitors, will also depend on the solvent composition.

Despite the many difficulties that arise in the examination of such a complex system, there is no reason to depart from the concept in which pitting arises from the occurrence of a nucleophilic substitution of ligands in a surface complex. In fact, the stability constant, K_s, of complexes in aqueous–organic media generally increases with increase in concentration of the nonaqueous component, although it is rarely a linear function of the latter.[149]

It follows from the Eq. (3.14) that the nature of the solvent can influence the kinetics of the reaction, not only through the composition of the adsorbed complexes but also because of the differences in the solvation of the nucleophiles, i.e., the aggressive anion An^-, and the inhibitive anion In^-. Furthermore, substitution in the surface complex of part of the oxygen-containing passivating components of the water, notably OH^-, by molecules of an organic solvent can impede the growth of protective oxide. In fact, coulometric data that we obtained after a 15 min oxidation of iron at $E = 0.2$ V in a borate buffer containing an organic solvent (Fig. 3.19) showed a reduction in the thickness of the oxide film, with the film not detected at all in the presence of 50% DMSO.

However, the stability of the passive state is not uniquely determined by the thickness of the oxide film, as may be seen from a comparison of the E_{pit} values obtained in various solutions (Fig. 3.20). Ethylene glycol (ethanediol) prevents the depassivation of iron by shifting E_{pit} to more

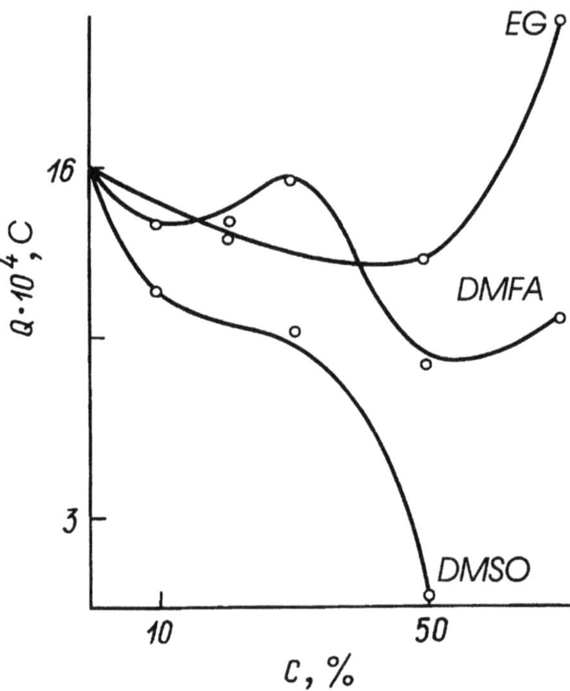

Figure 3.19. The effect of the concentration of organic solvent, C, on the oxide film thickness (as measured by the quantity of electricity used in its reduction) formed on iron in a borate buffer containing EG (ethylene glycol), DMFA (dimethylformamide), and DMSO (dimethylsulfoxide).

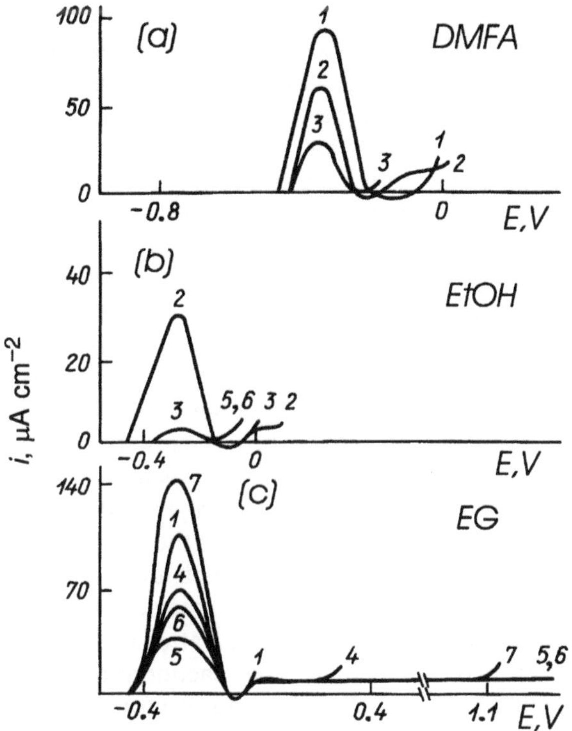

Figure 3.20. Polarization curves for iron in a borate buffer containing 0.015 M Na_2SO_4 and DMFA (dimethylformamide); EtOH (ethanol); EG (ethylene glycol) in (%): 1, 0; 2, 10; 3, 30; 4, 40; 5, 50; 6, 60; 7, 70% EG + 0.1 M Na_2SO_4.

positive values. Ethanol acts in this way only up to a certain concentration, above which it debases E_{pit}. This effect is, to a large extent, associated with a partial dehydration of the surface although the possibility of an easier desolvation of An^- which accelerates reaction (3.14) should not be excluded. A characteristic feature is that ethylene glycol with two hydroxyl groups, which therefore causes stronger solvation of anions than ethanol, produces the opposite effect and the aprotic solvent DMFA causes practically no solvation, an effect analogous to that of debasing E_{pit}.

This situation also holds true for the depassivation of aluminum, although it is difficult to tell how far the observed facts may place in doubt the widely held view on the predominance of the formation of aquo complexes over solvation by other solvents in aqueous–organic media. The cause of such an influence of the solvent can include not only the formation of mixed solvates but also the displacement of water from the electrical double layer which, in turn, will encourage the depassivation of metals. In any case, it is logical to expect an influence of the hydrophobicity of the solvent on the basic characteristics of the process leading to pitting formation. At the same time, it is clear that an assessment of the contribution of hydrophobicity would be more correctly applied to a series of related solvents that are close in their nature to water. The results obtained in aqueous–alcoholic media with a constant mole fraction (0.167) of organic solvent are of importance in this connection.[202]

Thus, E_{pit} in aqueous solution becomes more negative in the presence of monohydric alcohols but more positive in the presence of polyhydric alcohols, such as ethylene glycol or glycerol. In the series of these alcohols dissolved in water, there is a debasement of E_{pit} with increase in hydrophobicity of the alcohol according to Hansch* (Fig. 3.21). It is difficult to explain this effect solely on the basis of a better adsorbability of the more hydrophobic alcohols since it is also quite large for ethylene glycol. It should also be emphasized that pit formation occurs as a result of adsorption not of the solvent but of SO_4^{2-} or Cl^-. Furthermore, the rate of Eq. (3.14) should depend on the anion-solvating properties of the solvent. With increase in acceptor properties of the solvent—of which the acceptor number, *AN* (section 2.4), serves as a measure—the initial species in this reaction become more stable (as a result of their high solvation) as compared with the activated complex, and this decreases the reaction rate. It follows from this that an increase in *AN* of alcohols will impede pit formation. Unfortunately, values of *AN* for polyhydric alcohols are not available in the literature, but bearing in mind their tendencies to structure formation, it can be suggested that their acceptor numbers will be higher than those of monohydric alcohols because of the presence of larger associations of molecules.

Apart from the influence described above on the desolvation of An^-, the solvent can also have an effect on the kinetics of the reaction through the desorbability of OH^-, since the presence in the surface complex of more hydrophobic solvent molecules should encourage the transfer into the solution of hydrophilic hydroxyl anions. The basicity of the solvent is also important. It is a recognized fact that those alcohol-forming aqueous solutions in which depassivation is more difficult than in water; ethylene glycol and glycerol, are more acidic than water (pK_a equal to 14.18, 13.99, and 15.74 respectively.)† It can be suggested that these alcohols are to a

*Here the hydrophobicity of alcohols is expressed as the logarithm of the partition coefficient, *P*, calculated using the method of fragmentary constants, i.e., by the summation of the additive tabulated constants reflecting the contributions to the total hydrophobicity of the fragments of molecules and the interactions between them.[116]

†For comparison with the pK_a values of the alcohols, the pK_a value used for water refers to a hypothetical 1.0 M solution of water in water.

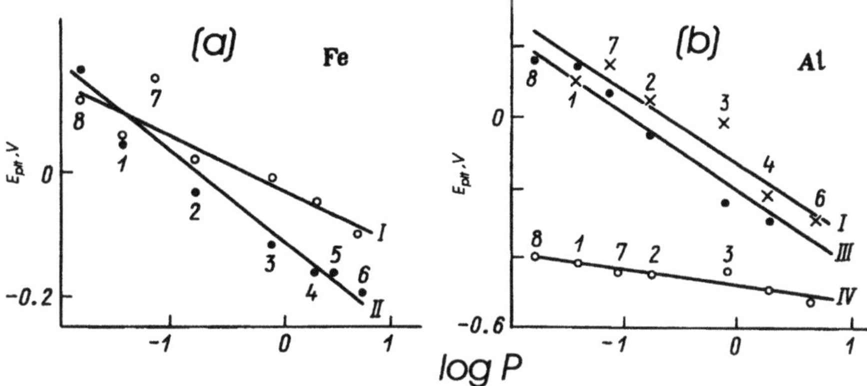

Figure 3.21. The dependence of E_{pit} of iron (a) and aluminum (b) on the hydrophobicity, P, of the organic component of the solvent in a borate buffer containing 0.015 M Na_2SO_4 (i) or NaCl (ii), 0.03 M Na_2SO_4 (III) or NaCl (IV). 1, H_2O; 2, CH_3OH; 3, C_2H_5OH; 4, C_3H_7OH; 5, isoC_3H_7OH; 6, tertiary C_4H_9OH; 7, ethylene glycol; 8, glycerol. Mole ratio H_2O: alcohol = 5:1.

large extent in the form of anions and can provide stronger competition with aggressive anions in adsorption processes.

The slopes of the lines in Fig. 3.21 essentially depend on the nature of the activator. It is probable that the extent of the change in slope (by 2–5 times, on going from Cl^- to SO_4^{2-}) is mainly caused by the values of the charge on these anions, since this will affect the energy of desolvation of the anion during its transfer from the bulk of the solution to the dense part of the electrical double layer. Thus, SO_4^{2-} is significantly more hydrophilic than Cl^- (because of the greater charge), and therefore should be more difficult to desolvate and thus, be less aggressive. On the other hand, according to this approach, E_{pit} in sulfate solutions should be more sensitive to changes in the hydrophobicity of the medium.

The dependence of E_{pit} on the nature of the activator in aqueous–organic media can also be expressed with an equation of the type (2.29) but the effect of its separate terms can differ considerably (Table 3.11). Thus, in the depassivation of iron or aluminum 50% mono- or diethylene glycol by a number of activators, the polarizability coefficient of these in some cases [Fe in diethylene glycol, (DEG) or Al in ethylene glycol (EG)] plays practically no part. The aggressivity in such cases will be determined mainly by the hydrophobicity coefficient. As this increases, pitting—as judged by the E_{pit} value—occurs more readily.

It can be suggested that in this case the squeezing out of the activator from the solution is more difficult than the formation of new, and the fracture of old, bonds on substitution of ligands. This is supported by the

Table 3.11. The Effect of the Nature of the Solvent on the Coefficients of Equation (2.29) for E_{pit} of Various Metals

Metal	Solvent	Solution composition	*a*	*b*	*c*
Iron	H_2O	borate buffer[BB]+0.05 M NaAn	0.00	0.26	0.21
Iron	H_2O+EG[1:1]	BB+0.05 M NaAn	0.093	0.05	0.02
Iron	H_2O+EG[1:1]	BB+0.1 M NaAn	0.13	0.61	0.46
Iron	H_2O+EG[3:1]	BB+0.1 M NaAn	−0.08	0.07	0.15
Iron	H_2O+DEG[1:1]	BB+0.1 M NaAn	−0.19	0.13	0.02
Aluminum	H_2O+DEG[1:1]	BB+0.1 M NaAn	0.049	0.33	0.00
Iron	H_2O+DMFA[1:1]	BB+0.1 M NaAn	−0.09	0.36	0.42

An = hal^-, N_3^-, NO_3^-, $HCOO^-$, CH_3COO^-, $C_2H_5COO^-$, $C_3H_7COO^-$, $C_6H_5SO_3^-$.

fact that the prevention of depassivation of low-carbon steel in aqueous diethylene glycol solutions of Na_2SO_4 is achieved only with the use of sufficiently hydrophobic organic anions, and also by the fact that an inhibitor such as sodium benzoate is ineffective in these media. The concentration dependencies of ΔE_{pit} in the case of sodium benzoate (Fig. 3.22) have, as a rule, an S-shaped form. However, with increasing hydrophobicity of the solvent (from water to tertiary butanol), the concentration corresponding to the inflection in these curves increases and the protective effect decreases. A constant concentration of inhibitor, 0.01 M, will correspond to different sections of the concentration curve, which makes a correct quantitative assessment of the role of the solvent difficult in this case. In the scope of our model, the change in hydrophobicity of the solvent should change the solvation of both nucleophiles—inhibiting and aggressive ions.

It follows that one must take into account that in an inhibiting solution two competing processes, the adsorption of benzoate anions and the adsorption of sulfate anions, can occur in parallel. By replacing the solvent it is possible to change the nature of the more rapid of these processes which, in the case of parallel reactions, will control the overall rate of the process. This was apparently also so in the case we studied. In sufficiently hydrophilic media (a purely aqueous borate buffer), the benzoate anion, being more hydrophobic than sulfate, is more easily adsorbed and blocks those active centres on the surface at which pit initiation is most probable. As the hydrophobicity of the medium is increased by going to mixtures of water with more hydrophobic alcohols, the solvation of the anions will decrease to different extents. The smaller change would be expected for the benzoate ion and the greater for the more hydrophilic sulfate. The adsorption of the inhibitor decreases more quickly in this case, which hinders the reaction in Eq. (3.13) and enhances the possibility of the

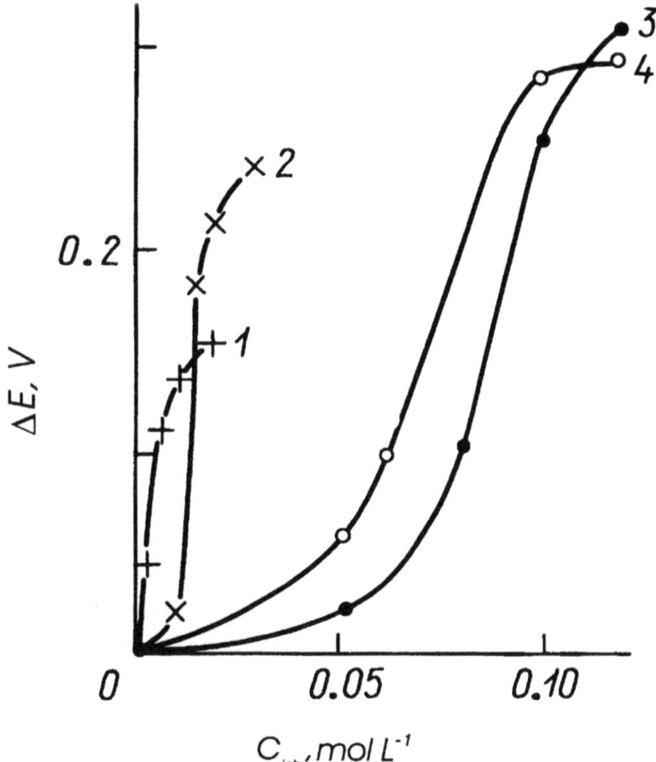

Figure 3.22. The dependence of the protective effect ΔE on iron on the concentration of sodium benzoate in a 0.015 M buffer solution of Na_2SO_4 containing 16.7 mole % CH_3OH (1); C_2H_5OH (2); isoC_3H_7OH (3); and tertiary C_4H_3OH (4).

occurrence of reaction in Eq. (3.14). In fact, as has been shown above, the increase in hydrophobicity of the solvent facilitates pitting formation in a non-inhibiting solution and lowers the protective capability of the inhibitor. Thus, in the presence of the hydrophobic propanol, E_{pit} in 0.01 M benzoate solution cannot be distinguished from that in the uninhibited solution. The concentration of inhibitor at which a noticeable protective effect is observed (hundreds of mV) in an aqueous propanol solution is approximately an order of magnitude above that of the aggressive ion, whereas in a pure water solution these concentrations are almost the same.

Similar considerations can be applied in treating the effect of the substituent in the inhibitor on its protective action in various solvents.[188]

In that work, studies were made of the inhibition of pit formation of iron in 50% solutions of ethylene glycol, ethanol, and dimethylformamide containing 0.015 M Na_2SO_4 and 0.2 M arylcarboxylate.

Compared with aqueous solutions, the concentrations of arylcarboxylate required to bring about a noticeable positive shift in E_{pit} were usually higher than the concentration of sulfate (Fig. 3.23). In the majority of cases, the dependence of E_{pit} on the inhibitor concentration had a stepwise character, but the maximum shift in E_{pit} differed for the various inhibitors, as did the concentrations at which it was achieved. In these cases, further increase in the concentration of the arylcarboxylate was practically without effect on E_{pit}. It can be suggested that with a sufficiently high concentration (0.2 M) of arylcarboxylate in the solution a high (close to $\theta = 1$) surface coverage by the inhibitor could be achieved. In this case, the inhibiting effect must be mainly determined by the strength of the bond between the inhibitor and the metal surface; consequently it will depend on the electronic characteristics of the substituent in the arlycarboxylate anion.

A comparison of the values of the protective effects (ΔE) of meta- and para- substituted arylcarboxylates indicates their dependence not only on the nature but also on the position of the substituent in the aromatic ring in relation to the carboxyl group. This shows up clearly in the case of those arylcarboxylates containing electron accepting substituents. Thus, p–Cl, p–NO_2, m–I and p–OC_2H_5 benzoates in 50% dimethylformamide are significantly more effective than benzoate, whereas their ortho isomers

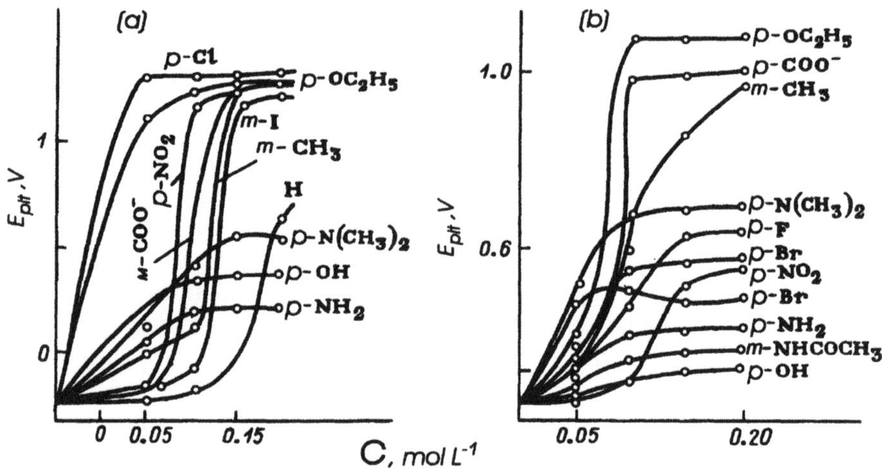

Figure 3.23. The dependence of E_{pit} of iron in borate buffered (a) aqueous-dimethylformamide (1:1) or (b) ethylene glycol (1:1) solutions of 0.015 M (a) and 0.1 M (b) Na_2SO_4 on the concentration of substituted arylcarboxylates.

effectively fail to prevent pitting. The introduction into any position of amino-, dimethylamino-, or hydroxyl groups lowers ΔE, but carboxyl and methyl groups raise ΔE, thus pointing to a controlling role of the inductive effect in the electronic influence of the substituent R in arylcarboxylates in H_2O–DMFA media. As it is known, the inductive effects of the substituent decrease as it moves further from the reaction centre, and so the ortho position is the most effective in producing changes in the electron density on the carboxyl group. Although the introduction of an electron-accepting substituent increases the polarity of an organic anion, thereby facilitating its adsorption on an electrode, it will at the same time lower the electron density at the reaction centre and, consequently, also the strength of the bond of the inhibitor with the surface. For ortho-arylcarboxylates the latter effect is so marked that they cannot resist being replaced by sulfate ions. Upon introduction of electron-accepting substituents into the meta- or para- positions, this effect is weaker and so I-, Cl-, and NO_2-benzoates have better protective properties than those of the unsubstituted compounds.

The protective effect of benzoates with meta- and para-substituents is described by a V-shaped dependence of ΔE on the inductive constants σ_I of the substituents (Fig. 3.24). Thus, for DMFA solutions, for the left-hand branch, which is composed of compounds containing either electron donors (positive inductive effect), or electron acceptors but with a small negative inductive effect, the relationship:

$$\Delta E = 0.82 - 10.45\,\sigma_I + 0.21\pi \tag{3.26a}$$

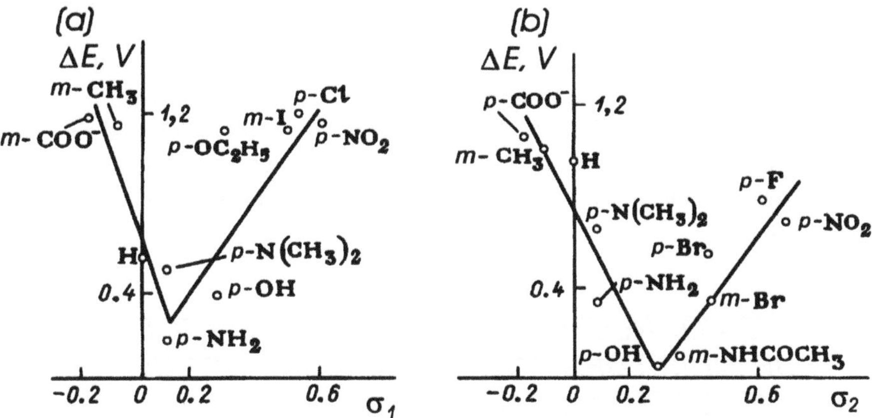

Figure 3.24. Correlation of ΔE on iron in 0.2 M arylcarboxylates with σ_I constants of the substituents in borate buffered water-dimethylformamide (1:1) (a) and water-ethylene glycol (1:1) (b) solutions of Na_2SO_4.

holds, and for the right-hand branch:

$$\Delta E = 0.49 + 1.5\,\sigma_I + 0.28\pi \qquad (3.26b)$$

It follows from the small difference between the coefficients before the π constants (related to hydrophobicity, see section 2.2.) in these equations that the effect of the hydrophobicity of the substituents is essentially the same for the entire reaction series. On the other hand, the electronic influence of the substituent on the protective properties of the benzoate is significantly higher for the case where electron-donating substituents are introduced. This is explained by the fact that these increase not only the polarity of the organic anion but also the electron density at the reaction centre and consequently, the capability to be more strongly adsorbed on the electrode, so preventing the interaction of the surface with aggressive anions.

Analogous dependencies were obtained in 50% ethylene glycol for meta- and para- benzoates (Fig. 3.24b). Due to the combination of electron-donating and hydrophobic properties of the substituent, the corresponding meta- and para- arylcarboxylates, analogously to their ortho isomers, acquire high inhibiting properties. This is generally true for the inhibition of pitting nucleation in aqueous–alcoholic media and provides a possibility for a unique test of protective agents.

Thus, moving from aqueous to aqueous–organic solutions can change the mechanism and contribution of the electron effects of a substituent and hence, the effectiveness of the inhibitors. Such changes are undoubtedly connected with the effect of the non-aqueous component in the solution in altering the conditions of solvation of the initial and final products in pit formation.

Despite the fact that the introduction of both amphoteric and aprotic solvents to a neutral medium significantly changes both the corrosion behavior of metals and also the effect of inhibitors, it can be concluded that the concept of heterogeneous nucleophilic substitution can be used to describe the depassivation of a metal and its inhibition both in aqueous and mixed media.

3.7. THE CONNECTION BETWEEN THE EFFECTIVENESS OF INHIBITORS AND THE NATURE OF THE PROTECTED METAL

Any consideration of this problem, which at the present time is still far from any quantitative resolution, must be of a partly hypothetical nature, even when it is limited only to organic compounds and pure metals. However, the importance of the electronic structure of the metal in its protection by adsorbing inhibitors cannot be overlooked. Furthermore, the frequent emphasis given to the activity of transition metals and their oxides in processes of adsorption or catalysis[42] warrants similar attention as that being given to those of corrosion inhibition.

In the corrosion of metals in acids, when the inhibitor can be adsorbed not only by electrostatic forces—as has been discussed in the works of Hackerman, Antropov, Grigor'ev, and Ekilik—an important role in its effectiveness will be played by the chemisorptive activity of the metal. Such an influence will also be expected in the protection of the oxidized surfaces, which will be present with passivated metals. In fact, as pointed out earlier, both the initiation and the inhibition of the initial stages of depassivation of metals are often governed by the chemisorption of one or another anion from the solution on to a passivated surface.

The effect of the nature of the protected metal on its sensitivity to the electronic properties of a substituent has already been examined in section 3.4.2. with substituted sodium anthranilates (Fig. 3.12). This effect is more pronounced with iron than with zinc or aluminum; inhibitors are particularly effective with transition metals. Against the background of the concepts developed above, special attention should be paid to those inhibitors of which the stability constants of their complexes with cations are known. The anthranilate anion and its N-substitutes are valuable examples of such a reaction series since the stability constants of their complexes with cations are known for important metals such as iron, aluminum, and zinc. If an assessment is made on thermodynamic principles, then iron would be capable of forming the most, and zinc the least stable complexes with anthranilic acid and its derivatives (Table 3.12). The stability of the complexes for all the ligands that have been studied varies in the order: Fe(III) > Al(III) > Zn(II), where complexes of the first two cations may be more stable as a result of a higher degree of oxidation. The actual protection by substituted sodium anthranilates is in a different order: Fe > Zn > Al. This effect can be explained by the specific formation of surface complexes and by the differences in their solubilities. In the first case, apart from ion–ion interactions, the possible occurrence of covalent bonding of the inhibitor with the surface must be considered. We have attempted to estimate the degree of covalency of the bonding in a series of complexes of these metals from the value of the shift in the bond energy of the internal electrons, E_{bond}, of the donor atom in the complex relative to that in the ligand itself.[176]

The decrease in the value of E_{bond} of ligand atoms on decreasing the electronegativity of the metal reflects the covalency of the bond of the donor atom with the metal. We have observed such a tendency in changes

Table 3.12. Stability Constants, K_s, of Complexes of Substituted Anthranilic Acids with Metal Ions and the Protective Effect ΔE of the Corresponding Anions in 0.03 M NaCl in Borate Buffer Solution at C_{inh} = 0.01 M

Substituent, *R*, Name of the acid	log $K_s/\Delta E$,V		
	Zn(II)	Fe(III)	Al(III)
H, anthranilic(AN)	3.11/0.05	5.91/0.13	5.80/0.02
C_6H_5, phenylanthranilic(PAN)	2.86/0.12	5.45/0.20	5.25/0.07
3-$CHF_2SC_6H_4$, isodifluorant	2.70/0.13	5.42/0.63	5.30/0.16
3-$CF_3C_6H_4$	2.51/0.21	5.17/0.70	4.59/0.14
$C_6H_3(CH_3)_2$, mefenaminic, (MEF)[a] (N-2,3-xylylanthranilic)	2.98/—	5.66/0.32	5.53/0.07
$C_{10}H_7$, naphthylanthranilic	2.65/0.09	5.31/0.23	5.12/0.06
$C_6H_4NHC_6H_4COOH$, o-phenylenedianthranilic	2.57/0.07	5.20/0.27	5.00/0.06
$C_6H_4NHC_6H_4COOH$, p-phenylenedianthranilic	2.51/0.07	5.13/0.26	4.92/0.06

[a]because of the low solubility of MEF, ΔE was determined only at 0.005 M

in E_{bond} of N_{1s} electrons on going from mefenaminic acid (see Table 3.12) to its complexes. The degree of covalency of the bonds in the complexes increased in the order Al–Zn–Fe, the same sequence as for ΔE. This result attaches further importance to the dependence of the protective action of aromatic acids on the electronegativity of a metal.

Furthermore, as seen from Table 3.12, the anthranilic acid anion forms more stable complexes with metal ions than its derivatives, although it is less hydrophobic than they are. In this connection, although sodium anthranilate also inhibits pit formation with many metals, the value of ΔE in solutions that are not too concentrated is usually small. On copper, for example, it is capable of initiating pitting, but the measured value of ΔE in a chloride–anthranilate solution indicates a shift of the breakdown potential, E_{br}, i.e., the potential of developing pitting, and ΔE is therefore, strictly speaking, an uncertain value. Nevertheless, we have found[201] a tendency for ΔE to increase with increase in electronegativity of the metal when this is assessed according to Pauling, χ_{π}, and to an even greater extent when using Trasatti's approach (the effective electronegativity χ_{Me}),[8] which is calculated taking the zero charge potential, $E_{q=o}$ into account (Fig. 3.25). In view of the small value of ΔE for anthranilate, the increase in the protective effect with increase in χ_{Me} for solutions of the stronger phenylanthranilate inhibitor merits attention. In this case, we did not use data obtained on copper, because of the fact that χ_{Me} was taken as equal to 2.1,

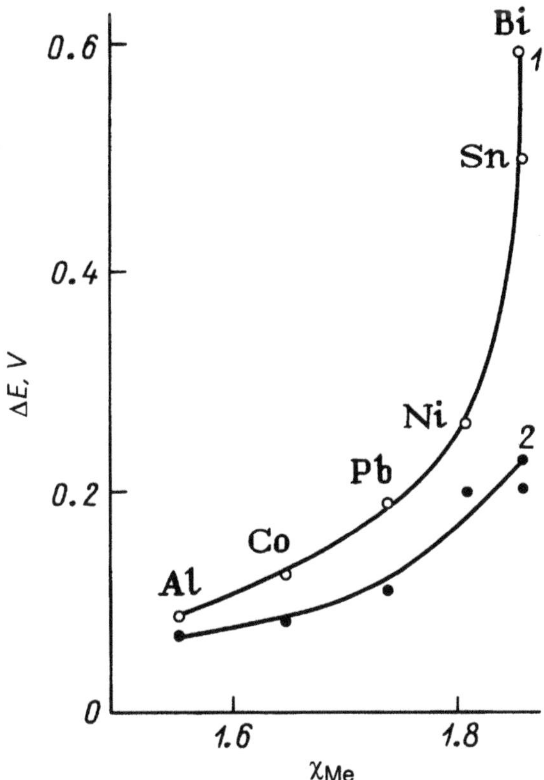

Figure 3.25. The dependence of ΔE in 0.01 M sodium phenylanthranilate in a borate buffer containing 1 mM (1) and 30 mM (2) NaCl on the effective electronegativity χ_{Me} of the metals.

and was not calculated from the values of the electronic work function and $E_{q=o}$. A complicating factor is that even a quite hydrophobic anion such as phenylanthranilate is capable of initiating, although also easily repassivating, small pits on copper.

Uncertainties and considerable scatter in $E_q = 0$ make the calculation of χ_{Me} for transition metals difficult. Trasatti considered the most reliable experimental data for nickel to give $E_{q=o} = -0.30$ V and for cobalt -0.45 V—values which have also been used for calculating χ_{Me}. However, if the value recommended for iron by Trasatti, i.e. $E_{q=o} = -0.35$ V is used, then $\chi_{Fe} = 1.44$, less than χ_{Al}, which is hardly likely, and finally, the value for ΔE in this case drops out of the correlation. Without discussing here the still unresolved question of the true value of $E_{q=o}$ for iron, it should be pointed out that in this case the value put forward by Antropov[6] of

$E_{q=0}$ = 0.0 V gives the greater value χ_{Fe} = 1.9, as expected, since organic anions provide more effective protection against pit formation for iron than for aluminum or zinc.

The observed correlations, finally, cannot be considered as strict quantitative laws, but from them it is possible to identify those metals that are relatively difficult to protect from pitting by an adsorption inhibitor. These include zinc and particularly aluminum which have high energies of interaction with water and therefore easily form oxide layers. The role of the latter in not only the initiation but also in the inhibition of pitting should not be ignored. However, the complexity of the assessment of this factor arises both from the possible deviation of the surface oxide from stoichiometry and, more particularly, from the fact that the nucleation of pitting takes place in a passive film at defects, the energetic state and potential of which differ from those for the rest of the surface. This uncertainty enters into the value of the overpotential of the pitting process, η_{pit} as determined from Eq. (2.27), in the form of a step in potential, Δ, in the passive film defects. When calculation of η_{pit} for aluminum is possible, i.e., when stability constants for complexes of its cations with activating anions, e.g., CNS^-, F^-, are known, then the η_{pit} values obtained are high (> 1.5 V). Among other metals that have been studied, high η_{pit} values are shown by nickel whereas significantly lower, but still appreciable, values are shown by tin and bismuth (Fig. 2.7). It is of interest that it is in this same sequence that the specific resistance of the oxides of these metals also decreases in ohm *m:* $Al_2O_3(10^{14})$ > $NiO(10^{11})$ > $Bi_2O_3(10^{6-8})$ ~ $SnO(10^7)$. The inhibiting effect of phenylanthranilate increases in the reverse order. It is probable that the decrease in potential at the electrode–solution boundary (due to the potential fall in the oxide) not only prevents the reaction of complex formation between the activator and the metal cation, i.e., it increases η_{pit}, but it also weakens the action of the inhibitor since the capacity to complex formation is the basis of its chemisorption. The values of stability constants of phenylanthranilate complexes (Table 3.12) are usually higher than those, for example, of the chlorides, and the hydrophobicity of the organic anion is greater than that of chloride ion. This should make surface complex formation with participation of the inhibitor easier, although there are certain difficulties in realizing this. Apparently, these difficulties also relate to the nature of the oxide of the passivating film and the potential at the electrode–solution interface. This is shown in the results of adsorption measurements of phenylanthranilate on oxidized and nonoxidized iron surfaces as reported in section 3.1.3. Although such complex phenomena require further study with broader objectives, it can be noted now that according to this mechanism, an anion which suppresses the depassivation of aluminum can also be effective for the protection of other metals of practical importance.

Even when considering pure metals, the contemporary theory of inhibition is limited by predictions that are only of a qualitative nature and cannot be made into firm generalizations.

It is still more difficult to assess the role of the composition of alloys in the inhibition of their depassivation. For example, the electronic effects of the substituent in substituted phenylanthranilate is appreciably weaker on the aluminum alloys D16 and AMg6 (Al–Mg6%) than on the pure metal. On the other hand, the alloying of iron with chromium can increase the inhibiting effect of phenylanthranilate, although the dependences of E_{pit} of iron and its Fe–10Cr alloy are analogous (Fig. 3.26).[190] At low concentrations the ΔE dependence is small and practically the same for iron and the alloy; the shift in E_{pit} to more positive values is caused only by the action of the alloying addition. However, with phenylanthranilate concentration > 3 mM, E_{pit} of iron and the alloy increase sharply and when > 8 mM, the alloy ceases to be activated by chloride right up to the potential of O_2 evolution. On iron, ΔE reaches a maximum value of 0.5 V. Thus, in

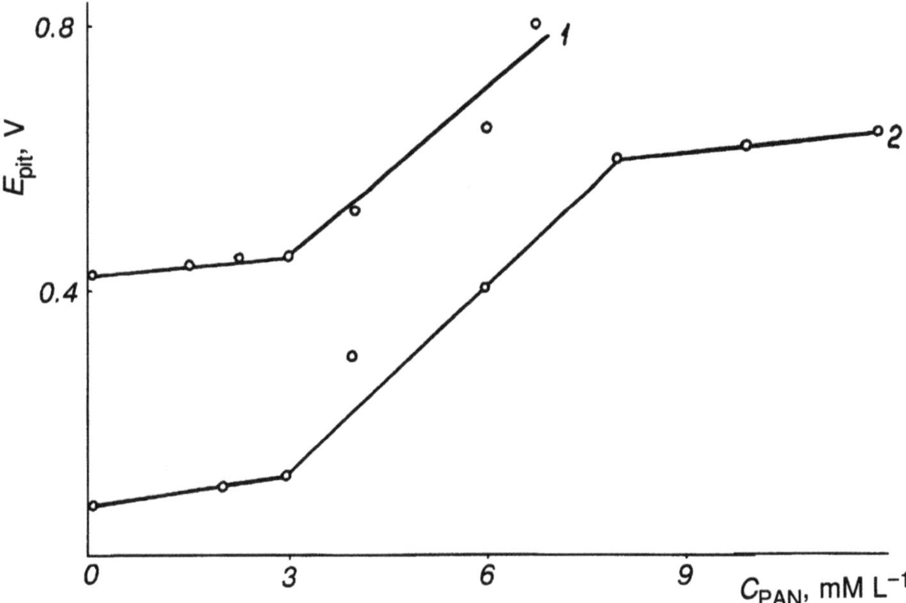

Figure 3.26. The dependence of E_{pit} of iron (2) and an Fe–10Cr alloy (1) on the concentration of sodium phenylanthranilate, C_{PAN}, in a borate buffer containing 10 mM NaCl.

the phenylanthranilate concentration region where the departing (displaced by the activator) group is the actual organic anion, the combined action of the alloying addition and the phenylanthranilate adsorption is shown quite clearly. It is possible that the adsorption of phenylanthranilate on the Fe–10Cr alloy is stronger than on iron. We have observed such a mutual effect with other iron alloys which have contained smaller additions of chromium.[201] We suggest that broader studies should be made in view of the importance of this question.

4

Corrosion Inhibitors Based on Complexing Agents

4.1. THE EFFECT OF COMPLEX-FORMING REAGENTS ON THE CORROSION OF METALS

In the preceding chapters, we have focused on the complex-forming properties of components of the medium. The consideration of these and other properties—principally hydrophobicity—allows a better understanding of the aggressive or inhibitive effectiveness of chemical compounds or ions to be reached. This is not surprising since complex formation plays a large role even in the elementary act of metal dissolution. Thus, metal cations passing into an electrolyte are always associated with solvent molecules, counter ions, or other components having free electron pairs. The aquo- and hydroxyl-containing complexes of technologically important metals are thermodynamically quite stable and are largely responsible for the importance of water in the corrosion and passivation of metals. Various compounds present in water will be adsorbed by a metal, essentially by the substitution of a solvent molecule by the adsorbate; such processes will include cases involving coordination and chemical bonding with the metal.

Common features of the adsorption are the interaction of all species at the electrode–solution boundary with the field of the electrical double layer and the dependence of this interaction on the surface charge. These aspects of adsorption were taken into account by Antropov in his scale of potentials. Furthermore, just as in the theory of catalysis, the fact that the individual properties of the active centre sometimes turn out to be more important than the collective electronic properties of the lattice of a solid body also holds for the theory of corrosion inhibition, where the need to understand the chemistry of processes taking place on a metal surface is of primary importance. In this regard, the concept of complex formation has been further developed in recent years, and has been extended to a wide range of corrosion systems.

An important conclusion arises from the ideas that we have developed on dissolution, depassivation, and inhibition of corrosion in neutral media taking place by a process of nucleophilic substitution of ligands in a surface complex. Thus, for the effective inhibition of corrosion it is not sufficient to have only a high surface activity of the additive responsible for the hydrophobicity of its aqueous solutions, or a potential capability to form complexes of low solubility. This is because in the first case a strong bond between the compound or ion with the surface may be absent. In the second case, as has been noted for the interaction of cobalt with α–nitrosonaphthol,[203] the formation of an insoluble chelate only in the bulk of the solution may even stimulate corrosion. This problem has already been examined, albeit only at a qualitative level,[148] where the contradictory nature of literature data on the effect on corrosion of a metal of compounds capable of forming soluble complexes with its ions has been pointed out. The complexity of processes taking place between a metal and a solution of a complex-forming reagent requires that these be classified into groups even if such a classification is only provisional. Thus, we have identified the three most important general cases[204]:

1. Hydrophobic ligands and the complexes formed by them are easily soluble in the corrosion medium
2. Ligands which are rather hydrophobic but providing complexes that are sparingly soluble or almost completely insoluble
3. The corrosion behavior of a metal is controlled by its interaction with complexes of this, or another, metal where such complexes are soluble in the medium and relatively stable

Although it seems that in the first case inhibition of corrosion should not be observed, this is not necessarily true. The steady-state potential of a metal E_{st} can be outside the region of equilibrium potentials, E_{eq} in which the ionization of a metal with formation of a soluble complex is thermodynamically possible. For example, if $E_{st} < E_{eq}$, then the adsorption of the corresponding ligand will not lead to complex formation but can, at least partially, block the corroding surface, change the ψ_1 potential of the electrode (see section 3.5), and so affect the corrosion reactions. As a result, there will be some inhibition of corrosion, although it can scarcely be highly effective in neutral media. Furthermore, the protection of a metal can arise as a result of the chemical conversion of a soluble complex that may be formed from interaction of a ligand with the corroding metal. The process of oxidation of the complex is a case in point. For example, tannin can accelerate the active dissolution of iron in a deaerated borate buffer (Fig. 4.1) with this activating action retained even with natural aeration of the solution. In the latter case a dark blue coloration of the near-electrode region is seen due to the oxidation in this region of Fe(II) tannate by dissolved oxygen. The potential of the iron moves into the passive region as a difficultly soluble Fe(III) tannate deposit builds up on the electrode surface. This process is complicated by acidification of the near-electrode layer associated with the complex-formation reaction. However, the stimulating effect of the tannin can transform into an inhibiting effect. It is significant that at potentials where iron dissolves in the form of Fe(III), tannin does not activate, and may even inhibit, the dissolution, as shown by an increase in τ (the induction period for pitting formation).

Another possibility for inhibiting corrosion is that associated with processes of thermal decomposition, as examined in detail by Margulova.[205] Thus, the disodium salt of ethylenediaminetetracetic acid (Na_2EDTA) (trilon B) has a high solubility in water (108 g kg^{-1} at 295 K). Despite providing an almost neutral solution (pH 5.5), trilon B at ordinary temperatures causes heavy corrosion of iron since iron complexes with EDTA anions are extremely stable and soluble. However, upon raising the temperature three processes start to occur: the thermal transformation of the complex, the formation of iron complexes of undetermined composition, and their thermal breakdown. As a result of the first process, amines can appear accompanied by alkalization of the solution by as much as three pH units

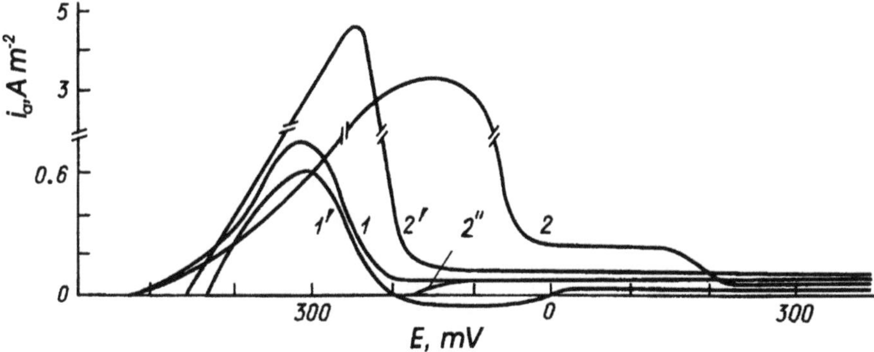

Figure 4.1. Anodic polarization curves for iron in a borate buffer without (1,1′) and with additions of 1.5 g L^{-1} tannin (2,2′,2″) with deaeration (1,2) and in conditions of natural aeration (1′,2′,2″). Curve 2″ was taken after holding the electrode in the solution for 60 h.

(presumably this will be even more at the metal surface). This by itself should accelerate the hydrolysis of the iron aquocomplexes:

$$[Fe(H_2O)_6]^{2+} = Fe(OH)_2 + 2H^+ + 4H_2O \quad (4.1a)$$

The oxidation of the hydroxide by the Schikorr reaction in the direction required for passivation is possible particularly at temperatures above 473 K where the usual controlling stage (for pure water) is accelerated[206]:

$$2Fe(OH)_2 + 2H_2O \rightarrow 2Fe(OH)_3 + H_2 \quad (4.1b)$$

Further polycondensation:

$$nFe(OH)_3 \rightarrow [FeO(OH)]_n + nH_2O, \quad (4.1c)$$

and the interaction of the polymer with Fe(II) hydroxide will lead to the formation of magnetite:

$$2[FeO(OH)]_n + nFe(OH)_2 \rightarrow nFe_3O_4 + aq. \quad (4.1d)$$

Magnetite is close in its crystal lattice structure and thermophysical properties to pearlitic steel, which it can passivate. However, the crystallographic features of the growth of such a film should be taken into account. Differences in the dimensions of crystals will lead to the appearance of "free passages" between them and to an absence of close packing. Consequently the protection provided by magnetite films in the spontaneous oxidation of a 0.2% carbon steel in water is not particularly effective.

However, the thermal decomposition of the iron–EDTA complex results in the formation of small spherical crystals of magnetite, the close packing of which prevents access of oxygen to the metal and the emergence of iron ions into the water. This process is considered to be mainly responsible for the inhibitive effect of trilon B.

Thus, the role of the complexing agent is not limited to changes in pH of the medium, otherwise the best results would be expected from solutions of tetra-substituted EDTA salts. Studies have, in fact, shown an opposite effect with the structure of the magnetite approximating that of the film formed in a pure condensate. The strong dependence of the protective properties and thickness of a magnetite film on the treatment parameters and on the steel composition, as well as the occurrence of easily-soluble secondary complexes of iron in the products of the thermal decomposition of Fe(II)–EDTA complexes point to the complexity of passivation processes. Their mechanism undoubtedly requires further investigation. We would emphasize only that complex formation taking place even at high temperatures can impede the formation of coarse crystals, thereby giving the magnetite film a closer packed and protective structure.

In the second case, many complexing agents, being easily soluble, can form complexes of various compositions and may be not only mono-, but also poly-nuclear. The latter are usually of low solubility and their formation can inhibit metal dissolution. Hydroxyethanediphosphonic acid (HEDP) is a typical example of such inhibitors, since it is capable of initiating localized corrosion of iron in those conditions when only the easily-soluble mononuclear complex HEDPFe(III) forms. The mechanism of action of such phosphonate inhibitors is quite complex and will be considered in detail later. At the same time, the existence of such inhibitors shows still further that even hydrophilic organic ligands that are capable of forming soluble complexes in certain situations should not be left out of our search for effective corrosion inhibitors. Finally, a purely adsorption mechanism of inhibition cannot be ascribed to these hydrophilic compounds, since the surface activity in aqueous solutions increases with increase in hydrophobicity of the compounds or ions. However, even for the second case, it is far from always the case that the protective layers that form are of the adsorption type.

Probably the most simple case of an effective interaction of a ligand with a metal surface is that provided by an analysis of Eq. (2.27) and by the data shown in Fig. 2.8. From this it follows that η_{pit} decreases with increase in hydrophobicity of the ligand, but this is so only up to a certain limit (which depends on the conditions and nature of the metal). Thus, in a borate buffer at pH 7.4 at $T = 295 \pm 2$ K and $C_{An} = 50$ mM, carboxylates with alkyl chain length greater than 3 do not cause pitting of aluminum, those with chain length greater than 4 do not cause pitting of iron, and those with chain length greater than 5 do not cause pitting of tin and zinc. Evidently, the corresponding compounds formed by these anions with metals are so slightly soluble that their transition into solution is either impeded or does not occur at all. It is precisely such anions that are the classical inhibitors of pitting and other local forms of corrosion. To this group belong alkylcarboxylates, benzoates and their substituted forms, the higher amino acids, etc. These noticeably increase η_{pit}, by ennobling E_{pit}, which is logically explained by changes in the energy state of the surface as a result, for example, of the blocking of the most active defects of the passivating film by a chemisorbing inhibitor. The most convenient systems for examining the mechanism of protection of metals by such ligands are those for which stability constants of the complexes with the corresponding metals are known. As already shown in Section 3.7, with the example of protection of various metals from pitting by certain substituted anthranilates, the effectiveness of inhibition depends more on the hydrophobicity of the ligand than on the stability constant of the complex. However, the limited solubilities of such inhibitors is a serious impediment to the further improvement in the protection of a metal.

As a result of this consideration, reagents that are capable of forming chelating complexes can be used. In this way, relatively hydrophilic ligands can selectively interact with some metal cations to form difficultly soluble compounds. A typical example is 8-hydroxyquinoline which is slightly soluble in water (3.6×10^{-3} mol L^{-1} at 20° C) but easily soluble in dilute alkalies in which it forms hydroxyquinolate ions. According to Sastri and Packwood,[78] the passive films formed by this compound on steel are practically free of iron oxides. The passivating function in neutral and weakly alkaline solutions is fulfilled by a film of Fe(III) hydroxyquinolate on which the ligand is adsorbed. Consequently, the mechanism of protection by such chelating reagents (to which belong many heterocyclic corrosion inhibitors that have been partly considered in section 1.4 and in detail elsewhere[119]) can be significantly distinguished from the passivating action of hydrophobic carboxylates, which as a rule, are only adsorbed on a metal or its oxide. Films of appreciable thickness—reaching several tens of nanometers—of complexes from chelating reagents often have a polymeric structure, although the mechanism of their formation remains very much unclear.

In the third case (see p. 174), the effect of ligands on the corrosion behavior can be significantly complicated when quite stable complexes with metal cations are formed. If such complexes are also of good solubility, then with the metal they will form an extremely complicated system for corrosion assessment. Thus, the stable and water-soluble complexes (of cations of the corroding metal and also dissolved cations of other metals) turn out to have a significant influence on the occurrence of corrosion processes in the system.

For example, for passive nickel placed in a neutral buffer solution of nitrilotriacetate (NTA) the Nernst equation can be used:

$$E_{eq} = E^0_{Ni/Ni2+} + 0.0295 \log a_{Ni2+} = E^0_{Ni/Ni2+} + 0.0295 \log \left[\frac{a_{NiNTA}}{K_s \, a_{NTA}} \right] \tag{4.2}$$

The complex (NiNTA) is more stable than any hydroxycomplex of nickel, its stability constant $K_s = 11.53$[207] (for comparison K_s for $Ni(OH)_2$ = 8.55). Such a marked difference in stability of the complexes provides the thermodynamic precondition for the formation of NiNTA, not only at the most defective point of the passivating film where the first pit initiates, but also across the whole surface of the electrode. The accumulation of NiNTA according to Eq. (4.2) should ennoble E_{eq} and so increase the value of E_{pit} as a result of the change in the thermodynamics of the process.

Actually, as we first observed[195] for nickel in solutions of various amino acids, complexing agents, and oxalate, there is a V-shaped dependence of E_{pit} on the logarithm of their concentrations. If the buildup of complex at the electrode surface is avoided by stirring the electrolyte or by accelerating the experimental determination of E_{pit} (by galvanostatic polarization), the dependence of E_{pit} on log C becomes linear (Fig. 4.2). On introducing into such a solution a previously synthesized complex of NiNTA the E_{pit} of nickel ennobles with a slope of 0.0295 V for the concentration dependence, in agreement with Eq. (4.2). This points to the possibility of a further effect of the inhibitor on the corrosion of the metal through a change in the thermodynamics of the system. The possibility of such a route to inhibition being realized requires further study; the protection of steel by complexes of another metal, especially zinc, has already received wide use in connection with the emphasis on ecological problems in water services. However, before examining the mechanism and ways of increasing the effectiveness of

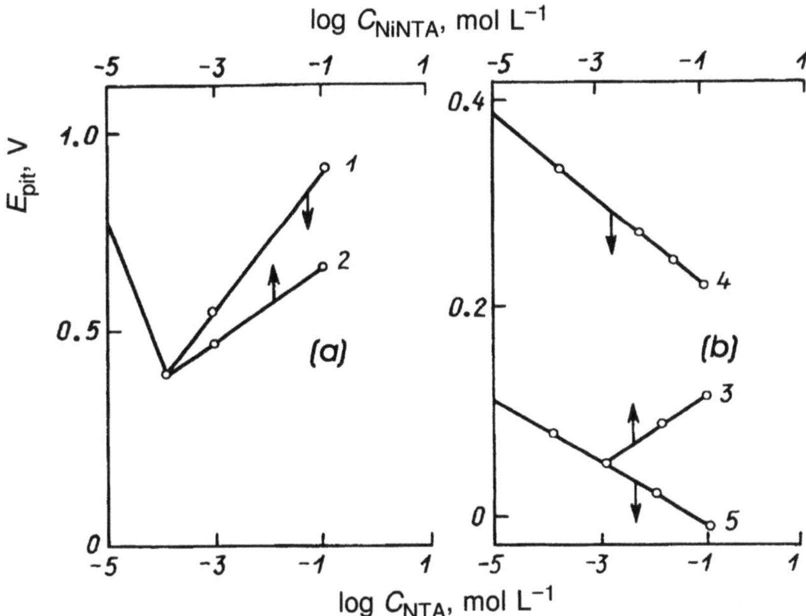

Figure 4.2. The dependence of E_{pit} of nickel in a borate buffer solution (pH 7.4) on the concentration of NTA (1,4,5) and on the concentration of the [NiNTA] complex (2,3) (in presence of 10^{-4} M NTA). Polarization curves were taken (a) potentiodynamically in a stagnant electrolyte (at 0.2 mV s^{-1}) or (b) 4, with stirring; 3,5, galvanostatically in a stagnant electrolyte.

complex-forming inhibitors, we shall analyze the corrosion–electrochemical behavior of metals in neutral solutions of phosphorus-containing complexing agents.

4.2. PHOSPHONATE INHIBITORS

Even carboxyl-containing complexing agents (NTA, EDTA, etc.), whose solubility, at least in neutral aqueous solutions, is several orders of magnitude below that of the alkyl- or aminophosphonic complexing agents,[207] are more hydrophilic than the common activators of corrosion, e.g., halide ions, alkCOO$^-$, NO_3^-, SCN$^-$. Therefore, it is not surprising that as a rule, such complexing agents do not prevent pitting nucleation, and that those cases where they act as initiators are characterized by high values of η_{pit}.[195] Furthermore, the very high values of the stability constants of their complexes make these additives thermodynamically reactive. The complexes that are formed are evidently less hydrophilic, and can therefore be more surface-active. In a number of cases they are of low solubility as, for example, with many polymeric complexes of the $[MeL]_n$ type. It is significant that, according to Dyatlova *et al.*,[207] the precipitation from solution of such complexes depends not so much on the nature of the complexing agent and of the cation making up the complex as on the metal:ligand ratio (n = [Me]:[L]). Protonated complexes, *i.e.*, those containing a complexing agent in the incompletely dissociated form (HL, H_2L, etc.), with n = 1:1 are usually less soluble than the unprotonated complexes.

Phosphonate-complexing agents containing the $-CH_2PO_3H_2$ fragment instead of, or together with, a carboxyl group have a number of features that result from their distinctive stereochemistry, their greater electronegativity, and the greater potential of PO_3^{2-} than of COO$^-$ for dentate formation. The important feature of phosphonates compared with aminocarboxyl chelants is their ability to form quite stable protonated MeHL complexes.[207] Thus, protonation of NTA breaks down the chelate structure of its complex, and its nitrilo-tris-phosphonic acid (NTP) analog forms not only an MeHL complex but also MeH_2L or even MeH_3L complex. These can exist even at relatively high pH values of 9–10. Of equal importance is the ability of NTP itself to form effectively insoluble polynuclear complexes, for instance $Ni_5(NTP)_2 10H_2O$ (pH 7.2; solubility product = 1.3×10^{-35}) or Al_3H_3 $(NTP)_2 10H_2O$ (pH 2.2; solubility product = $(4 \pm 2)\ 10^{-72}$. The high sensitivity of complex formation to changes in n and the ability to form polynuclear complexes of low solubility is also a characteristic of phospho-organic complexing agents, for example, the EDTA analog ethylene diamine tetra phosphonic acid, EDTP. No less interesting properties are shown by HEDP, one of the best known and readily available phospho-organic chelants. It differs from the classical aminocarboxyl complexing agents in, for example, its ability to form chelate rings simultaneously with several cations even with $n = 1:1$.[207] Undoubtedly, the pH of the solution and the presence of various types of complex formers in the solution has a large influence on the complex-forming properties of phosphonates and the solubility of their complexes. As a result of these features and the low toxicity of complexing agents such as HEDP and NTP, wide interest has developed in their use for inhibiting not only scale deposition but also corrosion of metals in various waters.

However, as seen from Table 4.1, in unstirred soft water phosphonate complexes, as well as carboxylate complexes, are of low effectiveness in inhibiting the corrosion of low-carbon steel, and can even stimulate the corrosion of brass.[208,209] In such solutions, the reduction in the critical passivation current density that occurs with an increase in the concentration of these complexes, changes into a stimulation of iron dissolution (Fig. 4.3), with E_{st} being displaced by 50–120mV to more negative values. This points to a difference in the mechanism of action of these complexing agents from that of inhibitors of the adsorption type, which in a borate buffer improve their protective effect with increasing concentration.

Table 4.1. The Effect of Complexing Agents[a] on the Corrosion of Steel and Brass in Unstirred Water[b,c]

Complexing agent	Corrosion rate, ($gm^{-2}d^{-1}$) Steel St.3 (C 0.14–0.25%)	Brass L-62 (Cu 62%)
—	2.30	0.14
$[HOOCCH_2]_2N[CH_2]_6N[CHCOOH]_2$[d]	2.20	0.30
$[HOOCCH]_2NCH_2N[CH_2COOH]CH_2\ N[CH_2COOH]_2$[e]	1.60	0.30
$N[CH_2COOH]_3$[f]	2.43	0.30
$N[CH_2PO_3H_2]_3$[g]	1.10	0.50
$HOOCCH_2N[CH_2PO_3H_2]_2$[h]	1.67	0.33
$[HOOCCH_2]_2NCH_2PO_3H_2$[i]	1.54	0.30
$[H_2O_3PCH_2]_2NC_2H_4N[CH_2PO_3H_2]_2$[j]	1.40	0.40
$H_2O_3PC[CH_3][OH]PO_3H_2$[k]	1.12	0.76

[a] 100 mg L^{-1}
[b] NaCl 30 mg L^{-1} + Na_2SO_4 70 mg L^{-1}
[c] Test duration, 15 d; pH 6.5
[d] hexamethylenediamine tetraacetic acid,[HMDTA]
[e] diethylenetriamine-N,N,N′,N″,N″-pentaacetic acid,[DTPA]
[f] nitrilo-tris-acetic acid,[NTA]
[g] nitrilo-tris-phosphonic acid,[NTP]
[h] glycine-N,N-di(methylene phosphonic) acid,[GP]
[i] imino-N,N-diacetic-N-methylene phosphonic acid,[IDAMP]
[j] ethylenediamine tetraphosphonic acid,[EDTP]
[k] hydroxyethane 1,1′-diphosphonic acid,[HEDP]

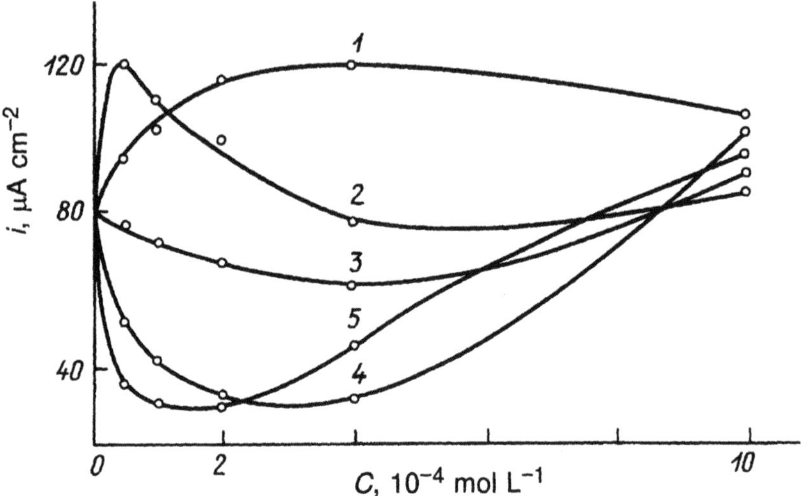

Figure 4.3. Dependence of critical passivation current density, i, for iron on the concentration, C, of complexes in a borate buffer at pH 7 containing 0.01M NaCl and 1, IDAMP; 2, NTP; 3, GP; 4, HEDP; 5, EDTP (see Table 4.1 for description of abbreviations).

In stirred, or flowing, soft water, a protective effect can be obtained at significantly lower concentrations of a complexing agent. We shall examine this with the example of HEDP.

The dependence of the corrosion rate of steel on the concentration of HEDP has a minimum at a concentration $C_{HEDP} = 2 \times 10^{-5}$ M at 20°C (Fig. 4.4a). On raising the temperature, its aggressiveness increases, but upon reaching a certain critical concentration, which depends on temperature, only localized corrosion develops (in Fig. 4.4a this region of concentrations is identified by a dashed line) with low values of mass loss. Such behavior can be explained from two features of the complex forming action of HEDP. First, because of the binding of HEDP to iron cations, the interaction of these with the OH^- formed as a result of the cathode reaction is prevented, or at least impeded. The rate of the cathode reaction increases with temperature, and in these conditions, this will lead to increased alkalinity in the near-electrode layer. Secondly, the value of the "n" ratio determines, to a large extent, the solubility of the HEDP–metal ion complexes. With $n > 2$, difficultly-soluble polynuclear complexes are formed, and with $n < 2$, soluble complexes are formed. The increase in HEDP concentration should lead to a decrease in n in the near-electrode layer and the formation of readily-soluble complexes along with the polynuclear complexes. These soluble complexes, in conditions of slow alkalization of the near-electrode layer, can stimulate the corrosion of steel as is observed at 20°C (dashed line). At higher temperatures, when the rate of reduction of O_2 increases,

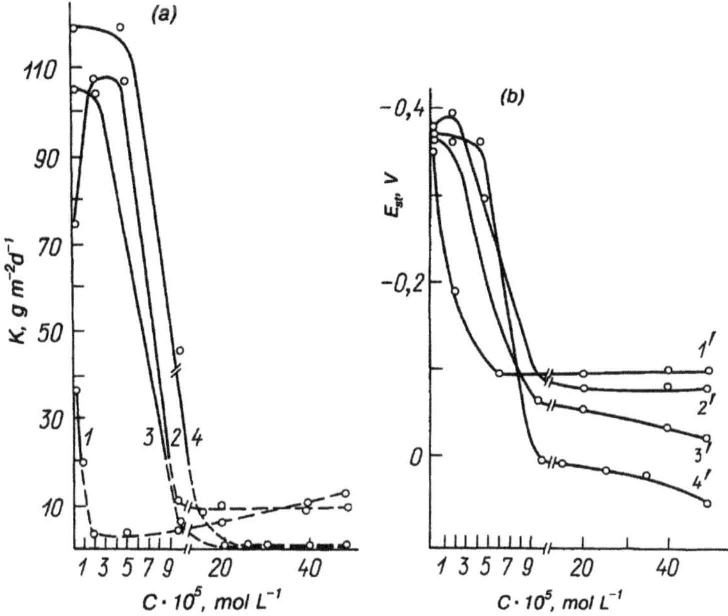

Figure 4.4. (a) The dependence of corrosion rate, K, of steel (C = 0.2%) and (b) the dependence of steady-state potential, E_{st}, on the concentration of HEDP (C) in water containing NaCl 30 + Na_2SO_4 70 mg L^{-1} at various temperatures: 1, 20; 2, 40; 3, 60; 4, 80°C. Duration of test, 8 h; specimen movement, 0.8 ms^{-1}.

the accumulation of OH^- passivates iron even within the first hour of its exposure to the solution, and the decrease in n has no effect on the corrosion of steel within a certain range of HEDP concentration. When this was greater than 20×10^{-5} (60–80°C), there was practically no loss in mass and the specimens retained their metallic reflectivity, but with numerous fine pits. The occurrence of these pits cannot be solely associated with the presence of Cl^- and SO_4^{2-} ions in the water since in a pure borate buffer at 20°C the presence of HEDP leads to formation of pits, even at −0.05 V. In fact, the potential of steel in unbuffered solutions containing HEDP moves towards this value (Fig. 4.4b). It is probable that there is a competition here between the formation of passive layers and the soluble complexes that cause pitting. Thus, at 60°C after cathodic reduction of the air-formed film, a pure borate buffer passivated a rotating iron electrode (velocity of 8 ms^{-1}), and displaced its potential to 0.0 V. When HEDP at a concentration $> 2 \times 10^{-4}$ M was added to the solution at this potential, no pitting occurred even after 8h exposure. However, even a slight amount of anodic polarization brought about pitting, thus indicating the unstable nature of the passive state of iron in the presence of HEDP. From the thermodynamic aspect, this could be explained by the possibility of the formation of a soluble complex with Fe^{3+}, even with $E < E_{eq} = -0.27$ V. Although the local depassivation of iron, as was shown above, usually takes place with relatively low values of η_{pit}, the latter increases with growth in the hydrophilicity of an activator; HEDP is sufficiently hydrophilic because of the ionized phosphonate groups ($pK_{a_1} = 1.7$; $pK_{a_2} = 2.47$; $pK_{a_3} = 7.28$; $pK_{a_4} = 10.29$; $pK_{a_5} = 11.13$).[207] In this situation, E_{st} of iron can turn out to be more negative than E_{pit}, although more positive than E_{eq}. Such a situation makes the protection unreliable and does not prevent the partial depassivation of iron in the presence of various activators.

The stimulation of the corrosion of copper and zinc that can occur in certain conditions in the presence of phosphorus-containing complexing agents can also be associated with their ability to form easily-soluble complexes with Cu^{2+} and Zn^{2+}.[209] The effect of complexing agents on the corrosion–electrochemical behavior of aluminum, which in neutral solution is usually subject to only localized corrosion, is no less complicated.

Thus, the introduction of complexing agents into a buffered (pH 7.4) 0.01 M NaCl solution debases E_{st} and somewhat ennobles E_{pit}[210] as shown in Table 4.2.

The doubly charged anions glyphosphate ($HOOCCH_2NHCH_2PO_3H_2$) ($pK_{a_1} = 2.3$; $pK_{a_2} = 5.9$; $pK_{a_3} = 10.9$) and NTA ($pK_{a_1} = 1.89$; $pK_{a_2} = 2.49$; $pK_{a_3} = 9.73$) have a weaker ennobling effect on E_{pit} than the multicharged anions, for example, NTP ($pK_{a_1} = 0.3$; $pK_{a_2} = 1.5$; $pK_{a_3} = 4.64$; $pK_{a_4} = 5.88$; $pK_{a_5} = 7.3$; $pK_{a_6} = 12.1$) or EDTP ($pK_{a_3} = 3.0$; $pK_{a_4} = 5.23$; $pK_{a_5} = 6.54$; $pK_{a_6} = 8.08$; $pK_{a_7} = 10.18$; $pK_{a_8} = 12.10$). However, no unique dependence was found of the size of the charge or the maximum dentating ability of the complexes (which varies from 4 for NTA to 10 for EDTP) on their inhibiting effectiveness. There is a tendency for E_{pit} to increase on increasing the number of phosphonic groups in the molecule with the greatest effect being brought about by EDTP, with $\Delta E = 0.12$ V. Despite this, it should be pointed out that the anodic current in the region where $E < E_{pit}$ increases in the presence of these complexing agents with the effect particularly marked in the case of the iminophosphonates (values of i = 320–340 mA

Table 4.2. The Effect of Complexing Agents (0.0015 mol L^{-1}) on the Steady-State, E_{st}, and Pitting, E_{pit}, Potentials of Aluminium in a Borate Buffer at pH 7.4 containing 0.01 M NaCl (210)[a]

	Background	NTA	IDAMP	GP	NTP	Glyphosate	EDTA	EDTP
E_{St}	−0.54	−0.67	−0.58	−0.64	−0.65	−0.58	−0.74	−0.73
E_{pit}	−0.40	−0.33	−0.33	−0.31	−0.30	−0.36	−0.30	−0.28

[a]See Table 4.1 for the full names of the complexing agents

m^{-2} being obtained in GP and EDTP and 230–260 mA m^{-2} in NTA and EDTA compared with 50–60 mA m^{-2} in the background solution). This points to a greater reactivity of the iminophosphonates, which apparently facilitate dissolution of the oxide film on aluminum. This is in accordance with the larger stability constants of the singly-charged complexes of Al^{3+} with phosphonate complexing agents compared with their carboxyl-containing analogs. It is significant that increasing the concentration of phosphonates markedly reduces the influence of their chemical structure on the passive current, $i_{p.c.}$, and also confirms the large ΔE_{st} in solutions of polyphosphonates which, for example, in 3.3 mM NTP can reach 0.2 V.

Evidently, the debasement of E_{st} cannot be associated only with the increase in $i_{p.c.}$, that is, with a formal acceleration of the anodic reaction, since these complexing agents can also inhibit the cathodic reaction. Thus, in 1.5 mM solutions of complexing agents the cathodic polarizability, P_c, of aluminum (0–50 mA m^{-2}) has the values (ohm m^2) as described in Table 4.3.

However, with an increase in the concentration of the complexing agents their inhibiting effect changes to one of stimulation of the cathodic reaction. For example, in a 0.4 mM solution of IDAMP, P_c = 10.38, but in 4.4 mM it equals 3.58 ohm·m^2. The stimulating action of NTP is still more clearly expressed, thus, for concentrations of 0.3, 3.0, and 6.6 mM the P_c values are 8.7, 4.0, and 2.8 ohm·m^2, respectively. Since difficultly-soluble compounds are formed by phosphonate-complexing agents when there is an excess of the complex-forming cations, the existence of such a transition can be explained by a change in composition of the complexes with increase in their concentration. It follows from this that to avoid the stimulation of electrode reactions by a complexing agent it should be used in a formulation with cation-complex formers.

In water containing Ca^{2+} and Mg^{2+} salts, the effect of the complexing agents on the corrosion of metals can differ significantly from that considered above for soft water. The reason for the difference lies in the possibility of the formation of complex compounds, which in turn can affect the kinetics of the electrochemical reactions responsible for the corrosion of metals. Even in unstirred water the combined presence of HEDP and hardness salts will ennoble E_{st} of steel and more markedly inhibit its corrosion than would be the case in the presence of the HEDP alone (Fig. 4.5).

Table 4.3. The Cathodic Polarisability of Aluminium (0–5 mA m^{-2}) in Solutions of Complexing Agents

Background	NTA	IDAMP	GP	NTP	Glyphosate	EDTA	EDTP
4.4	6.58	5.35	7.9	7.1	5.70	3.94	6.06

At the same time, it should be pointed out that the presence of Ca^{2+} and Mg^{2+} alone in the water will not necessarily guarantee significant protection of steel by HEDP. Again, whereas Mg^{2+}, because of the solubility of its complexes, usually does not have an adverse effect on the stability of HEDP solutions, this protection stability can be lowered by calcium hardness (H_{Ca}) as a result of precipitation of calcium complexes.

The effectiveness of the protection of steel by HEDP and Ca^{2+} is apparently at a maximum when there is a significant excess of the HEDP, i.e., when its molar concentration is 2–3 times greater than, for example, $CaCl_2$.[211] This is probably associated with the formation of a mononuclear CaHEDP complex. The relatively low solubility of this complex in water (even in strongly acid solutions) is responsible for its surface-active and inhibitive properties. With movement of the solution, there is no improvement in the protection of steel by these reagents, but with increase in calcium hardness and with the formation of polynuclear HEDP complexes, deposits will generally appear. Since the formation of the protective film on the metal is usually a rather slow process, the question of the maximum acceptable value of calcium hardness at which the stability of the inhibitor is sufficiently high in the volume of the solution becomes important. It appears that HEDP, depending on its initial concentration in the range 10–15mg L^{-1}, is effectively completely precipitated from solution within 7–12 days with $C_{Ca} > 2$ mg equiv L^{-1}. To avoid this, it is better to use not HEDP itself but its soluble complexes. This is one of the reasons why it is the normal practice to employ not the HEDP but its formulation with zinc salts to inhibit corrosion of steel in various waters. The presence in a water

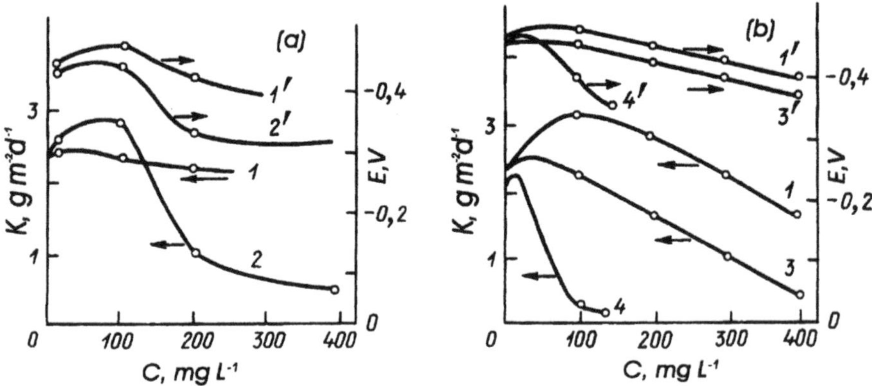

Figure 4.5. Dependence of corrosion rate (K) (1–4) and steady-state potential, E_{st} (1′–4′) of steel on concentration of HEDP (C) (1,1′) and its formulations with (a) $Ca(NO_3)_2$ in water containing NaCl 30 + Na_2SO_4 70 mg L^{-1} (2,2′); and (b) with $MgSO_4$ (3,3′) and $ZnSO_4$ (4,4′) in water containing (mg L^{-1}) Ca^{2+}, 56; Mg^{2+}, 23; Cl^-, 42; SO_4^{2-}, 100; HCO_3^-, 256; SiO_2, 6; pH 7.4.

of Mg^{2+} can play a decisive role in the stabilizing action in an HEDP solution, but only when $C_{Mg} > C_{Ca}$, an effect probably resulting from the high stability of the soluble Mg_2HEDP binuclear complex.

Thus, the presence in a corrosive medium of the natural cation-complex formers Ca^{2+} and Mg^{2+}, can affect the protective properties of phosphonate-complexing agents, which in some cases will be improved. As a rule, the presence of the hardness ions does not prevent the stimulation of corrosion of copper alloys by phosphonates, and can be the cause of the formation of deposits that are particularly undesirable in cooling systems and which can interfere with the use of inhibitors. Because of these difficulties, compositions that are most widely used in practice are those based on zinc phosphonate complexes.

4.3. THE PROTECTION OF METALS BY ZINC PHOSPHONATES

Zinc ions as inhibitors of the corrosion of steel in water have been known since the 19th century. Many examples exist of their use in combination with other inhibitors to improve the protection of metals[(199)] and the relatively low toxicity and ready availability of zinc salts have led to their wide use in practice. Phosphorus-containing compounds can also be included among such inhibitors and in fact the combination of hydroxyethane-diphosphonic acid (HEDP) and $ZnSO_4$ shows a synergistic effect in the passivation of iron (Fig. 4.6). The best inhibition of active dissolution of steel by HEDP itself is achieved at a concentration of 50 mg L^{-1}, although γ (the logarithm of ratio of corrosion rates, $K_{background}/K_{inhibitor}$) is still < 2. A zinc salt at this concentration also fails to provide passivation and in fact, only ZnHEDP can successfully provide this. The passivating properties of this complex are retained even in water of widely varying oxygen contents (1–35 mg L^{-1}). This is shown from the results obtained by chemical polarization[(212)]* which also confirm that ZnHEDP has passivating properties that are better than those of such well known inhibitors as nitrite or chromate, since the protection given by ZnHEDP at concentrations $\geqslant$ 68 mg L^{-1} ($\geqslant$ 2.5×10^{-4} M) is effectively independent of the dissolved oxygen content of the water. However, it follows from the potentiodynamic polarization curve that ZnHEDP, as also the HEDP itself, does not alter the pitting potential E_{pit}, and so there is a risk of destabilization of the passive state

*This method is based on the dependence of corrosion rate on potential where the latter is controlled by the concentration of oxidizing agent (e.g., O_2: 1–35 ppm in Fig. 4.6) in the solution and not by polarization from an external d.c. source.

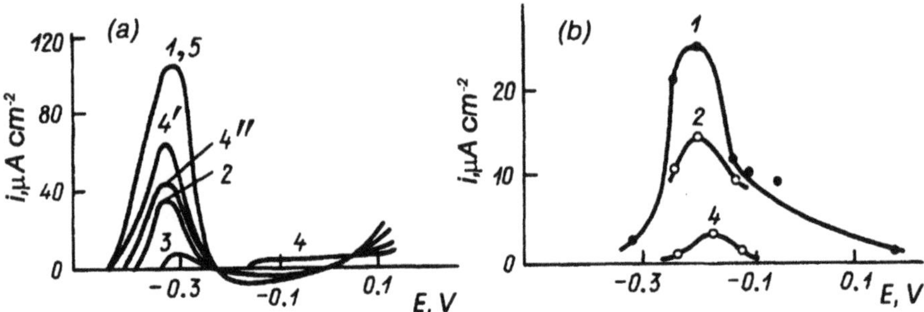

Figure 4.6. Polarization curves for steel St.3 (C=0.2%); potentiodynamic in borate buffer at pH 7.4 containing 0.03 M Na_2SO_4 (a); and obtained by 'chemical' polarization in doubly distilled water (b); in solutions of HEDP (1,4′,5), $ZnSO_4$ (1,4″) and ZnHEDP (1–4): 1, 0.0; 2, 20; 3, 40; 4,4′4″, 50; 5, 100 mg L^{-1}.

if an activator is present. Despite this, in the absence of any contact with electropositive metals (for example, copper), breakdown of the passive state of steel does not occur since the zinc phosphonate also inhibits the cathodic reaction so that the steady-state potential, E_{st}, remains below E_{pit}. Thus, depending on the ZnHEDP concentration, the polarizability of an iron cathode (at 1–20 μAcm^{-2}] can increase by almost an order of magnitude.[213]

It is reasonable to suggest that the passivation of iron by zinc phosphonates is connected with the formation of films on the metal surface that differ in nature from oxide films. In fact, x-ray photoelectron spectrographic analysis with argon ion etching of an iron surface after 16h contact with a ZnHEDP solution [200 mg L^{-1}, pH 7.6] has shown that the inhibitor is not fully removed from the film after washing with water. The film is not of any great thickness (Fig. 4.7) since the spectra taken prior to the etching clearly show a signal from Fe^0. Whereas phosphorus is found only in the outer layer, zinc is found to be present within the film—and in excess compared with its content in the inhibitor—even after the argon ion etching. Evidently, at the beginning of the formation of the film there is at least a partial breakdown of the complex arising, for example, from the replacement of Zn^{2+} by iron ions. Although the stability of the complex with Fe^{2+} with $n = [Me^{2+}] / [HEDP] = 1$ is somewhat below that of the analogous Zn^{2+} complex, conditions can be established in the near-electrode layer for the formation of a binuclear Fe^{2+} complex, or an Fe^{3+} complex which is usually the more stable.[207] The formation of these complexes releases zinc ions which react with OH^- generated by the cathodic reaction to form the slightly soluble $Zn[OH]_2$, which screens the surface and inhibits corrosion. As the corrosion rate falls, the deposition of $Zn[OH]_2$ is attenuated and the surface adsorption is then mainly of the inhibiting complex itself. Such

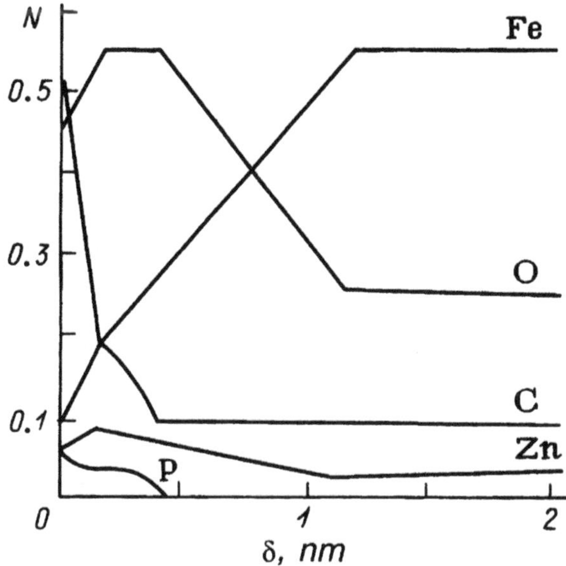

Figure 4.7. Element composition (from ESCA data) of the protective film formed by ZnHEDP (200 mg L^{-1}) on iron as a dependence on thickness.

a mechanism for the formation of protective layers by zinc phosphonates was first put forward by us.[212] In this work, we also showed by analysis of the ZnHEDP solution that the concentration of Zn^{2+} was lowered to a greater extent than that of the phosphonate after contact with an iron specimen. From this, the importance of the role of the nature of the dissolving metal cation participating in the complex can be seen since the effect of the complex-forming properties of the complexing agent on the inhibitor effectiveness will vary if the release of Zn^{2+} from its complex, and the production of the $Zn[OH]_2$ necessary for formation of the stable inhibitive complex, are impeded because of high complex stability. Actually, the passivation of iron in the presence of zinc phosphonate complexes depends on the nature of the ligand and increases in the series IDAMP–GP–NTP–EDTP* (see Table 4.1 for definitions of abbreviations).[208] It is known that the complex-forming ability of these acids with successive substitution of the acetate by the phosphonic group initially increases and then, for steric reasons, decreases, so that IDAMP > GP > NTP.[214] The stability constant, K_s, of the Zn^{2+} complex with NTP is greater than that with EDTP, but the latter more effectively inhibits the dissolution of iron.

The observation by Horner[215] that the protective action of diphosphonates $[OH]_2[O]P[CH_2]_mP[O][OH]_2$ increases only up to $m = 3$, can be explained by the reduction in stability of complexes with increase in the "separation" of phosphonic groups in the ligand since this leads to difficulties in realizing the chelating effect.[216] The inhibiting properties of the zinc complexes of diphosphonates that we have studied also depend in a complex manner on their stability. Thus, the DAPP, (dimethylamino-hydroxypropylene diphosphonic acid), complex with Zn^{2+}, which has the greatest stability [log K_s = 10.24] has the best passivating properties with higher concentrations of $ZnSO_4$*, as shown in Fig. 4.8. The minimum addition of zinc salt required for passivation is lower in the case of HEDP (log K_s = 5.7). The high effectiveness of MBAP [benzylimino di(methylene phosphonic) acid], $C_6H_5CH_2N(CH_2PO_3H_2)_2$, which occupies an intermediate position in complex-forming properties, is fully in accordance with the proposed mechanism. These results indicate that protection by zinc phosphonates depends not only on their ability to form difficultly-soluble polynuclear complexes, but also on the possibility of $Zn[OH]_2$ formation, which increases as log K_s is lowered and as the n ratio is raised.

The protective properties of a complex are also expected to depend on the electron density at the reaction centre of the ligand since an increase in electron density usually improves the stability of the zinc complex. In this case, an enhancement of chemisorption of the inhibitor can be expected in view of the stronger bonding of the oxygens of the phosphonate groups with the metal surface. As would be expected, the value of ln γ (see p. 187 for zinc complexes of aliphatic amino phosphonic acids has a non-linear dependence on the Kabachnik induction constant, σ_I^{Ph}, which characterizes the donor–acceptor properties of the substituent R (Fig. 4.9).[208] A weak complex-forming property of a compound, resulting from the low electron density on its heteroatoms does not allow significant inhibition by the zinc complex. The introduction of a stronger electron-donating substituent increases the possibility of formation of a relatively stable complex with Zn^{2+} and thereby improves the protection, but only up to a certain σ_I^{Ph}, after which the inhibitor properties of the complex again fall. In fact, the most effective complex, with the least hydrophobicity of the types studied, is that with $R = CH_3$, and consequently, the fall in ln γ is caused neither by a reduction of surface activity of the inhibitor nor by varying the chemical structure of the ligand. The most probable cause of this fall in ln γ as σ_I^{Ph} becomes more negative is the reduction in reactivity of the easily-soluble zinc phosphonate complex that is formed in this case.

We should now examine the effect of the pH on the protection of iron by HEDP and ZnHEDP since pH is the most important factor determining the possible occurrence of complexes and their reactivity, particularly in

*The curves in Fig. 4.8 correspond to the minimum concentrations of mixtures of $ZnSO_4$ and phosphonic acids which will passivate an iron electrode after removal of the air-formed oxide by cathodic reduction.

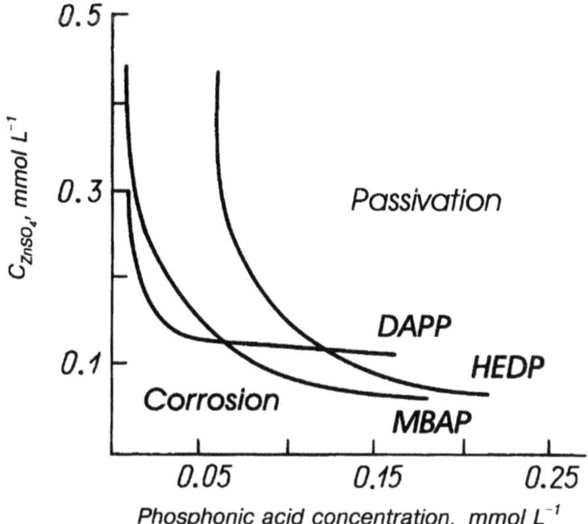

Figure 4.8. The effect of the ratio of the components of zinc phosphonate formulations on their passivating ability for iron in a borate buffer at pH 7.4.

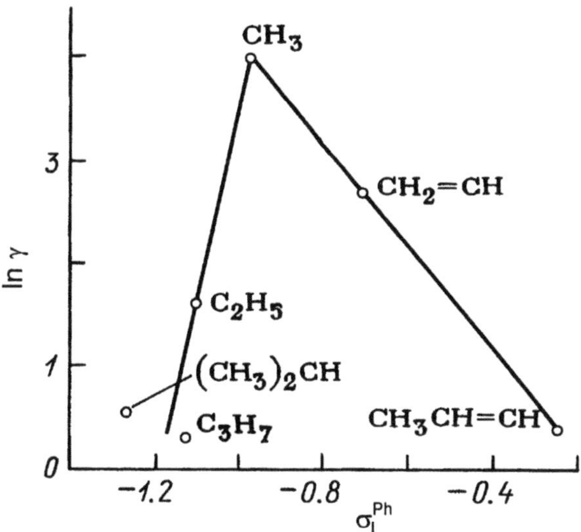

Figure 4.9. The dependence of the inhibiting properties for iron of zinc complexes of phosphonic acids of the general formula $RCH(NH_2)PO_3H_2$ on the Kabachnik induction constant of the substituent. Inhibitor concentration 1.0 mM in water containing NaCl 30 + Na_2SO_4 70 mg L^{-1}.

processes of substitution with ligands and complex formers. It follows from the pH dependence of corrosion rate at ~20°C (Fig. 4.10)[208] that the optimum conditions for protection by an inhibitor in a stagnant electrolyte are in the range pH = 5–6. Displacement of pH from this range lowers the effectiveness of the inhibitor. The increase in corrosion rate that occurs on acidification of the medium can be logically associated with an increase in aggressivity of the background solution. An increase in pH can be associated with the reduction in activity of the complex-forming cations taking part in the formation of the protective layer on the metal surface. The latter effect is explained by the formation of hydroxides, particularly those of zinc, which can occur even in the bulk of the solution. Furthermore, the ionization of phosphonic groups, the dentating ability of the ligand, and the composition of the complex all depend on pH. Thus, at pH 5, HEDP is doubly dissociated ($pK_{a_1} = 1.7$; $pK_{a_2} = 2.47$), and the formation of uncharged difficultly-soluble compounds with Fe^{2+} becomes possible so that the HEDP possesses better protective properties in weakly acidic solutions. An increase in pH will lead to further ionisation of HEDP which will gradually be transformed into a triply-charged anion. The hydrophilicity of the HEDP and the charged complexes formed is increased and the inhibiting effect reduced. On increasing the HEDP concentration, the corrosion-rate dependence on pH shows less clearly; an analogous effect is given by movement of the solution (Fig. 4.10b). In this latter case, the sharp increase in protective action of the HEDP warrants attention. This effect is connected both with the easier supply of HEDP to the metal (because of the hydrophilicity of HEDP the forces acting to remove it from the solution at the phase-separation boundary should be relatively small) and with the removal from the surface of weakly adhering Fe^{3+} compounds. These compounds, particularly in static conditions, prevent the formation of difficultly soluble compounds of Fe^{2+} with HEDP. Since alkalization of the solution accelerates the $Fe^{2+} \rightarrow Fe^{3+}$ transition, it will adversely affect the protective properties of HEDP for a stationary electrode.

Zinc sulfate cannot be studied over the whole pH range as $Zn(OH)_2$ is precipitated even in weakly alkaline solutions. The result of this precipitation is that in unstirred $ZnSO_4$ solution the protective properties are already completely lost at pH 9 (Fig. 4.10a). However, even at lower pH values $ZnSO_4$ is clearly inferior in effectiveness to ZnHEDP in corrosion inhibition. ZnHEDP significantly lowers the corrosion rate of steel over a wide pH range even with limited solution movement and at a concentration of 20mg L^{-1} (Fig. 4.10b). Although in the background (uninhibited) solution, the corrosion rate increases significantly with increase in movement of the solution, that is, with increase in the rate of rotation of the electrode, it

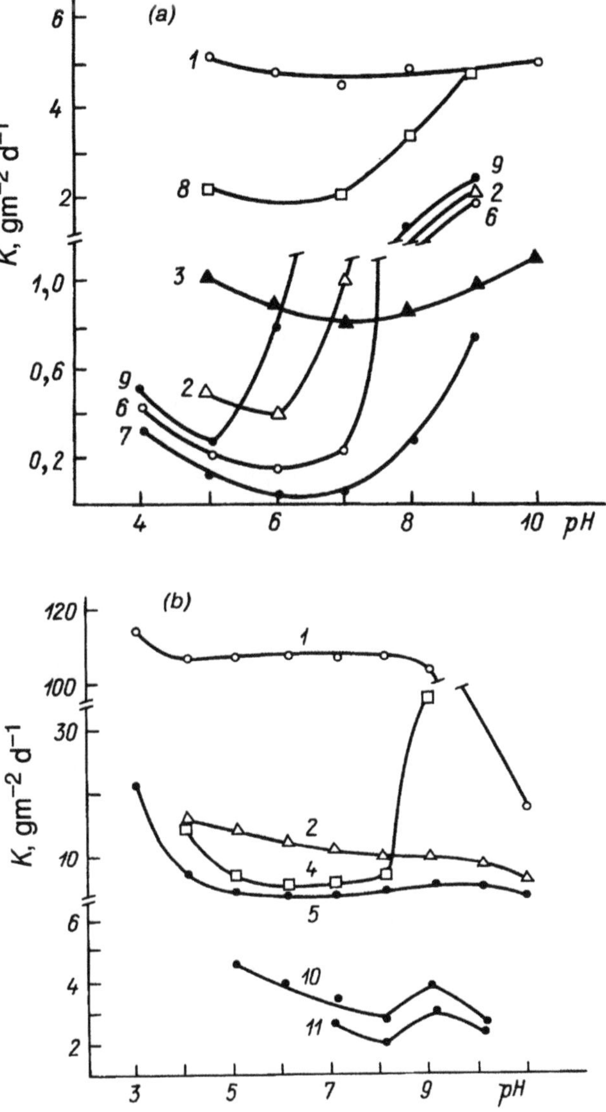

Figure 4.10. The effect of pH on the corrosion rate of steel St.3 (C=0.2%) in solutions containing NaCl 30 + Na_2SO_4 70 mg L^{-1}. 1, alone; 2,3, with HEDP; 4,8, with $ZnSO_4$; 5–7, 9–11, with ZnHEDP, (a) without rotation of the electrode, (b) with rotation of the electrode at: 1.5 ms^{-1}, curves 1,2,4,5; 1.0 ms^{-1}, curve 10; 0.3 ms^{-1}, curve 11; and with concentration of inhibitors (mg L^{-1}): 20, curves 5,10,11; 50, curves 2,8,9; 100, curve 6; 200, curves 3,4,7.

shows little change in solutions containing ZnHEDP. At a concentration of 20 mg L^{-1} at pH 8, there is practically no inhibition of corrosion in stagnant conditions; however, even a small movement of $V_{rot} = 0.3$ ms^{-1} (rate of rotation of the electrode) lowers the corrosion rate to 2 $gm^{-2}d^{-1}$. With $V_{rot} = 1.5$ ms^{-1} the value of γ is increases even more. Generally, ZnHEDP protects steel better than HEDP even in an unstirred solution, but again in this case the effect of pH on the factors mentioned above (which are particularly noticeable at low concentrations of ZnHEDP) should be taken into account.

Temperature is an important factor affecting the effectiveness of zinc phosphonates. Its influence is usually viewed from two aspects: the stability of phosphonates at various temperatures, and the sensitivity towards change in temperature of the protection of metals. Phosphonates, unlike polyphosphates, are not hydrolyzed, which gives them a certain advantage at elevated temperatures. The stability of zinc phosphonates depends very much on their composition; for example, ZnHEDP—one of the most widely used in industry—is quite stable at temperatures up to 80° C.

Analysis of corrosion data (Table 4.4) indicates a low apparent activation energy of 8.1 $kJmol^{-1}$ for the corrosion process that takes place in the background electrolyte in the absence of movement. This agrees with the recognized fact that the controlling stage is the rate of diffusion of the cathodic depolarizer, i.e., oxygen. Movement of the solution reduces this diffusion limitation but evidently, even with $V_{rot} = 1$ ms^{-1}, it is still insufficiently effective for the process to come under kinetic control (activation energy = 16.5 $kJmol^{-1}$). A more significant increase in activation energy is obtained by introducing ZnHEDP into the solution, which points to the

Table 4.4. The Dependence of Corrosion Rate of Steel St3 (C 0.15 - 0.22%) on Temperature in Water Containing NaCl 30 + Na_2SO_4 70 mg L^{-1} NaCl with and without ZnHEDP

Composition of Solution	Conditions of movement $V_{rot}(ms^{-1})$	Corrosion rate ($g\ m^{-2}\ d^{-1}$) T (°C) 20	30	40	50	60	80
Background	0	3.9	4.4	4.8	5.3	5.9	–
	1.0	37.0		55.0		85.0	117.5
+5mg L^{-1} ZnHEDP	1.0	3.25	–	25.4	–	55.4	115.0
+10mg L^{-1} ZnHEDP	1.0	0.45	–	2.4	–	30.2	115.0
+30mg L^{-1} ZnHEDP	1.0	0.0	0.0	0.0	0.0	0.0	0.0
+100mg L^{-1} ZnHEDP	0	0.4	0.8	1.2	1.6	2.1	–

important role of the chemical reactions that take place on the surface of the steel. This is especially noticeable in a moving solution when the addition of 5 or 10 mg L^{-1} of ZnHEDP increases the activation energy by 3–5 times. A further increase in concentration to above 30 mg L^{-1} guarantees full suppression of the dissolution of steel over the whole temperature range. Evidently, in this latter case a less defective layer with an adequate blocking action has formed on the steel surface.

Unlike this process on steel, the inhibition of corrosion of brass is caused by a retardation of only the cathodic process since its anodic dissolution in the presence of ZnHEDP, as with the HEDP itself, is not impeded.[209] The protective action of zinc phosphonates is retained, even in flowing water, when the corrosion rate of the alloy is increased as a result of both enhanced diffusion of the oxygen cathodic depolarizer to the metal surface and more effective removal of the corrosion products that have been formed. However, in the presence of inhibitor the protective film is formed more quickly because of the more rapid supply of inhibitor to the metal surface. This allows brass to be protected by lesser additions of the inhibiting composition. Studies in a quite aggressive water[209] with V_{rot} = 1.0 ms^{-1} have shown that zinc phosphonates based on HEDP and EDTP provided full protection of brass even at an inhibitor concentration of 30 mg L^{-1}.

On comparing the stability constant of the MeHL complex of HEDP with zinc ($\log K_s = 5.66$) with that formed with divalent copper ($\log K_s = 6.26$), it is obvious that although they are close in value the latter is somewhat larger. Consequently, in the near-electrode layer the formation of an insoluble heteronuclear complex of the $CuZnH_2L$ type can occur and this on deposition on the metal surface will retard the cathodic process. In fact, the addition of Cu^{2+} cations to a solution of zinc phosphonates leads immediately to the formation of an insoluble deposit. The release of zinc cations from the zinc phosphonates with subsequent formation of zinc hydroxide $Zn(OH)_2$, leading to the same effect, is less probable. In corrosion tests a significant alkalization of the solution (up to pH 8.6–8.8) has been observed, independent of the concentration of zinc phosphonates, which should facilitate the formation of $Zn(OH)_2$ as the concentration of the zinc phosphonates is raised. However, as seen from Fig. 4.11a the rate of corrosion of brass increases as the concentration of zinc phosphonates is increased. The formation of an insoluble heteronuclear complex of zinc phosphonates with monovalent copper is unlikely since in this case a retardation of the anodic process would be observed. Thus, a complex of the type $CuZnH_2L$ is apparently formed on the metal surface, but its deposition upon increasing the concentration of zinc phosphonates, and with the *n* ratio being used,

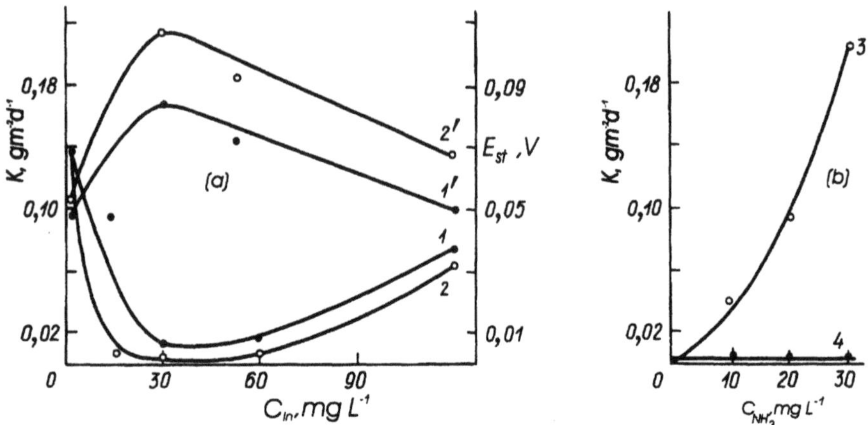

Figure 4.11. The dependence of corrosion rate, K, and of steady-state potential, E_{st}, of L62 brass (Cu=62%) in hard water on the concentration of (a) inhibitor and (b) ammonia. (a) 1,1′, HEDP + $ZnSO_4$ (1:0.5); 2,2′, EDTP + $ZnSO_4$ (b) 1, (HEDP + $ZnSO_4$) 30 mg L^{-1}; (1:0.5), 2 (EDTP + $ZnSO_4$) 30 mg L^{-1}.

is difficult because there is an excess of unbound phosphonic acid in the solution.

The formation of this complex is confirmed by the fact that it is possible to inhibit the corrosion of brass by zinc phosphonates in the presence of ammonia. When ammonia is added to a cooling system the corrosion rate of brass increases sharply as a result of the formation of copper ammoniates:

$$Cu + 4NH_3 + 2H_2O = [Cu(NH_3)_4](OH)_2 + H_2$$

As seen from Fig. 4.11b an increase in concentration of ammonia from 0 to 30 mg L^{-1} in the presence of 30 mg L^{-1} of zinc 1-hydroxyethane-1,1′-diphosphonate (ZnHEDP) leads to an increase in the corrosion rate of brass. A comparison of the logarithms of the stability constants of copper ammoniate (log K_s = 12.67) and copper 1,-hydroxyethane,1,1′-diphosphonate [log K_s for CuHL = 6.26 and for CuL = 12.5] shows that in this case there is a low probability of the formation of a heteronuclear complex. However, brass could be protected by using complexing agents for which the logarithm of their stability constant with copper is higher than, or comparable with, that for copper ammoniate. For example, such is the case with EDTP, for which log K_s for $Cu^{2+}HL$ = 11.26 and for $Cu^{2+}L$ = 23.0. In fact, 30 mg L^{-1} of zinc ethylenediamine-tetramethylphosphonate completely suppresses the corrosion of brass in the presence of ammonia (Fig.

4.11b). Evidently the formation of protective layers on brass in this case competes successfully with the formation of soluble copper ammoniates.

The behavior of aluminum or its alloys in aqueous solutions also changes significantly if a phosphonic acid is introduced together with zinc salts.[210] In the case of NTP, the zinc phosphonate complex that is formed at pH 7.36 is stable up to $n < 1.0$ with the decrease in zinc content weakening the protection of aluminum (Fig. 4.12).

A further increase in n had no effect since it led to the precipitation of a white deposit—explained by the possible formation of binuclear complexes or by hydrolysis of excess zinc. These effects have been studied with zinc nitrilotrisphosphonate (ZnNTP) (n = 1.0). In an unstirred electrolyte, an increase in concentration of ZnNTP is accompanied by increases in cathodic polarizability P_c and in ΔE_{pit} (ennoblement of pitting potential) which with C = 9 mM reaches 0.27 V. However, an increase in passive current, $i_{p.c}$ is found only in relatively concentrated solutions of the complex. Where there is intensive stirring (Table 4.5), which in the background solution leads to some ennoblement of E_{pit} as a result of the action of dissolved oxygen, the presence of even a low concentration of ZnNTP =

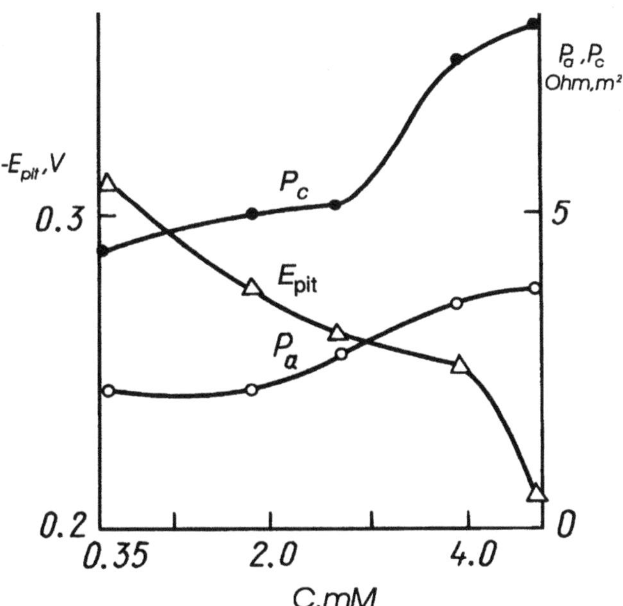

Figure 4.12. The effect of $ZnSO_4 \cdot 7H_2O$ concentration in a borate buffer (pH 7.4) containing 10 mM NaCl and 1.3 mM NTP on E_{pit}, the anodic (P_a) and the cathodic (P_c) polarizability of aluminum.

Table 4.5. The Effect of ZnNTP (1.7 mM) on the Polarizability of Aluminum (Ohm m^2)[a]

pH	Rotation of a disc electrode (rev/min)	Background solution		Background solution + inhibitor	
		P_c	P_a	P_c	P_a
7.36	0	4.4	1.7	7.0	4.0
	750	4.0	1.2	4.6	2.8
	1500	4.0	1.3	5.5	2.4
	3000	3.5	1.5	6.0	2.3
2.5	0	1.4	0.7	3.0	0.7
4.25	0	1.9	0.4	6.2	1.8
6.0	0	2.7	0.8	5.9	1.2
8.16	0	2.2	1.4	5.0	2.0
8.68	0	1.8	0.6	6.4	1.8

[a]In borate buffer, pH 7.4, containing 10 mM NaCl

1.74 mM results mainly in an inhibition of the cathodic reaction. This can be explained in the following way.

With an intensive supply of inhibitor to the surface of an electrode there are two possible interactions with ions or aquo-complexes of the dissolving aluminum: the formation of heteronuclear complexes (ligand—not coordination—saturated) or the release of zinc from phosphonate complexes less stable than those with aluminum. The absence of data on stability constants of all complexes which can form in this case and the non-equilibrium nature of the system do not allow a precise thermodynamic assessment to be made. However, since it is particularly at cathodic potentials that movement improves protection by complexes, the second possibility is the more likely. Actually, the released zinc, by interacting with OH^- generated by the cathodic reaction, will form difficultly-soluble hydroxide directly at the electrode surface. With anodic polarization there is practically no OH^- generation and the inhibitive effectiveness of the complex is low.

It follows that previously maintaining aluminum in an electrolyte before carrying out anodic polarization can increase the protective action of ZnNTP. In fact, an increase in holding time from 1 to 16 h in an unstirred 3.3 mM solution of ZnNTP leads to an increase in E_{pit} from -0.32 to -0.20 V and in P_c from 6.1 to 12.8 Ohm.m^2. A beneficial effect is provided over the wide range of pH where ZnNTP effectively retards the cathodic reaction (Table 4.5). Inhibition of the anodic dissolution of aluminum in a strongly-acid solution effectively does not occur, but as the pH is increased not only an increase in E_{pit}, but also a decrease in $i_{p.c}$, become significant (compared with the values in the background solutions). The improved protective properties of ZnNTP have been confirmed by corrosion tests with aluminium alloy D16 in such solutions.

Thus, in the background solution the corrosion rates, K, at pH values of 2.5, 4.2, 6.0, 7.4, 8.08, 8.68, 9.0 and 9.5, were respectively 1.06, 0.91, 0.84, 0.73, 0.57, 0.41, 0.33, 0.65 $gm^{-2}d^{-1}$. In a ZnNTP inhibited solution at these pH values, K had values of 0.244, 0.09, 0.073, 0.044, 0.05, 0.06, 0.08, 0.3 $gm^{-2}d^{-1}$. The increase in protective properties is still more clearly expressed with ZnHEDP, small additions of which can effectively suppress the corrosion of the alloy. Significantly, in this case there is a change in the corrosion mechanism. At pH > 8.6 the solubility of aluminum oxides increases and the tendency to localized corrosion decreases (Table 4.6). However, in the background solutions or with insufficiently high additions of inhibitor the difference between the intensity of corrosion K_{in}^* and its rate K remain quite large.

It is only by increasing the concentration of ZnHEDP that the corrosion of the alloy in alkaline solutions can be prevented over a wide pH range. In such solutions the protection is achieved by retarding both cathodic and anodic reactions, and E_{st} remains below E_{pit}: the concentration of chloride has no strong influence on the effectiveness of inhibition. Consequently, practically full protection of the aluminum alloy from localized corrosion, e.g., in antifreeze solutions based on potassium chloride, can be achieved by raising the pH to 9–10 with the simultaneous introduction of phosphonate complexes.

It is significant that in unbuffered neutral solutions of chlorides the protective action of zinc phosphonates is somewhat greater than in buffered solutions. This is connected not only with the competitive adsorption of borates but also with the change with time of the pH of the unbuffered solution resulting from the alkalization of the near-electrode layer from 7.1 to 8.4. In this case a reduction in NaCl concentration by an order of magnitude has little effect on the inhibition, thus indicating an important feature of protection by phosphonate complexes, namely, the weak dependence on the concentration of aggressive ions in the solution.

In buffered 0.01 M NaCl, an increase in temperature displaces E_{st} to more negative values as a result of easier anodic dissolution and oxidation of the electrode. The latter causes a significant increase in $i_{p.c}$ but there is no increase in the tendency of aluminum to pitting with increase in temperature, and E_{pit} can even ennoble. Despite such varying effects of temperature on the electrochemical behavior of these aluminum alloys, their corrosion, such as that of alloy D16, does accelerate with temperature.

*A value that takes into account the actual corroding area in pits etc., i.e., the tendency to localized corrosion.

Table 4.6. The Effect of pH on the Effectiveness of Protection of the Aluminium Alloy D16 in Borate Solutions Containing 0.01 M NaCl and ZnHEDP[a]

pH	Background solution		Background + 40 mg L^{-1} ZnHEDP		Background + 100 mg L^{-1} ZnHEDP	
	K	K_{in}	K	K_{in}	K	K_{in}
8.08	0.57	27.7	0.167	25	0.044	2.25
8.6	0.41	34.4	0.066	33.3	0.088	0.088
9.0	0.33	12.85	0.055	12.2	No	Corrosion
9.5	0.65	0.65	No	Corrosion	No	Corrosion
10.0	0.55	0.55	0.050	0.050	No	Corrosion
10.5	1.03	15.0	0.735	1.03	0.217	0.217

[a] K values in $gm^{-2}d^{-1}$

However, unlike the protection of low carbon steel, the effectiveness of inhibition of corrosion of aluminum alloys increases with temperature. Thus, the addition of 1.7 mM ZnNTP at 20, 35, 50 and 65°C lowers the corrosion rate by 1.1, 2.4, 3.5 and 4.8 times, respectively. This is, to a significant extent, caused by a retardation of the anodic process, increase in temperature from 20 to 80° C increases P_a by 36 times and E_{pit} by 0.2 V. The influence of temperature on the inhibition of the cathodic reaction is complex, since increasing temperature facilitates the chemical interaction of the phosphonate complexes with the surface, while at the same time reducing the generation of the OH^- ions that would assist the formation of the difficultly-soluble layers. Although the mechanism of the formation of such layers on aluminum and its alloys in neutral solutions of phosphonate complexes requires further study, it can nevertheless be suggested that these are, as in alkaline solutions, enriched in zinc coming mainly from the deposition of $Zn(OH)_2$. Thus, zinc phosphonates, depending on their composition and the hydrodynamic conditions, temperature and pH, will be capable of inhibiting both electrochemical processes responsible for the corrosion of metals.

4.4. POSSIBILITIES FOR IMPROVING THE EFFECTIVENESS OF INHIBITION

In analyzing the mechanism of action of carboxylate and phosphonate complexing agents and their zinc complexes, it has already been noticed that they do not produce the significant ennoblement of pitting potential,

E_{pit}, that would be necessary to guarantee not only passivation but also stability of the passive state. Therefore, from a mechanistic point of view, it would be logical to employ these inhibitors with anionic additives capable of moving E_{pit} in the desired direction. However, it is not easy to put this into practice since inhibiting anions, although having the advantage of hydrophobicity, are significantly inferior to complexing agents in respect of the values of the stability constants, K_s, of their complexes with metals, a property which also determines the ability of a component of a solution to be chemisorbed. The change in pH of the near-electrode layer cannot be ignored; this is fundamental to the protective action of complexing agents and zinc phosphonates, whereas in alkaline media the chemisorption capacity of inhibitive anions is, as a rule, lowered. In the conditions of a buffered environment there is, first, usually a drop in the effectiveness of the protective action of the complexing agents and their zinc complexes, and, second, a difficultly in achieving the synergism of protection by these with surface active anions.

Based on the actual mechanism of action of the carboxylate and phosphonate complexes, it should be possible to use not only the zinc cation as the complex-former, but also cations of other metals that form insoluble hydroxides. In the absence of an excessive amount of other anions capable of forming complexes or salts of low solubility with the cation, these insoluble hydroxides should be formed in the near-electrode layer as a result of the generation of free OH^- ions by the cathodic reaction as well as the electrophilic substitution by ions of the dissolving metal of the corresponding cation (for example, as in a phosphonate complex). Evidently, the inhibiting capabilities of HEDP complexes with singly charged cations cannot, because of their low stability and the high solubility of LiOH, NaOH, etc., differ significantly from those of HEDP itself. The formation in this case of insoluble heteronuclear complexes is also unlikely for neutral solutions because of the large differences between the complex-forming ability of iron ions and alkali metals.

The complexes of the majority of doubly-charged metal cations, Me^{2+}, are more stable, and the $Me(OH)_2$ solubility product is sufficiently low for these hydroxides to screen the surface of a metal. Furthermore, the K_s values that are known for these complexes are usually smaller than those for complexes of HEDP with Fe(III), and to some extent also those with Fe(II), and therefore the electrophilic substitution by iron of such ions as Ca^{2+}, Mg^{2+}, Mn^{2+}, Ni^{2+}, Co^{2+}, Zn^{2+} from their complexes is thermodynamically possible. Since the complex-formers have the same charge, the same quantity of OH^- (generated by the reduction of O_2) will be required for the formation of $Me(OH)_2$, and consequently, the pH change of the near-electrode layer will depend on the $Me(OH)_2$ solubility product. In the case of the formation of hydroxides with a relatively high solubility product (those of Ca^{2+}, Mg^{2+}, Mn^{2+}), the concentration of the complex-former should be sufficiently high for the $Me(OH)_2$ to begin to screen the surface. Actually, in moving conditions, i.e., with $V_{rot} = 0.8\ ms^{-1}$, the rate of general corrosion can be reduced by lesser additions of HEDP complexes than of HEDP itself, although full suppression of corrosion could be obtained only with a concentration at which HEDP had its maximum effect, at 0.1 mM.[217] In the region of higher inhibitor concentrations, $C_{inh} > C_1^*$, corrosion was localized and assessment of corrosion rate from mass-loss data was extremely conditional. In this situation, the minimum protective concentration, C_{min}, necessary to prevent fully corrosion of the steel is the only strict criterion of the inhibitor effectiveness. A feature of such formulations is that for an equimolar composition of HEDP and $CaCO_3$, a white deposit forms in the solution during the course of the experiment and it is not possible to obtain a value for C_{min}. Evidently, this effect is a result of the low solubility of the carbonate (solubility product $SP_{CaCO_3} = 3.8 \times 10^{-9}$) which forms in the bulk of the solution. Solutions with $Ca(NO_3)_2$ are more stable since at pH 7.0–7.5 they do not contain high concentrations of carbonate and the $SP_{(Ca(OH)_2)}$ is relatively high (5.5×10^{-6}). In such solutions, full protection of steel could be obtained, but on increasing C_{inh} a white deposit was again formed, and the rate of corrosion increased. An important requirement for inhibition is that the cation released from the complex should form difficultly-soluble compounds either in the near-electrode layer or directly on the surface, and not in the bulk of the solution. In this respect the hydroxides of metals have, because of their low solubility, an advantage over other anions since the generation at the electrode of OH^-, in addition to that in the solution, provides favorable conditions for their formation close to the surface that is being protected.

A reduction in the solubility of the $Me(OH)_2$, where Me^{2+} is the cation taking part in the composition of the HEDP inhibitor, can improve the screening of the surface and reduce C_{min} (Fig. 4.13) for the protection of steel. There is in fact a correlation between the effectiveness of such inhibitors and $\log SP_{Me(OH)_2}$

$$\log C_{min} = -3.22 + 0.063 \log SP_{Me(OH)_2} \tag{4.4a}$$

*C_1 is concentration of inhibitor at which the corrosion of iron has only a localized character, e.g., small pits after 8 h tests. On the curves of dependencies of $K(gm^{-2}\ d^{-1})$ on C_{inh}(mol L^{-1}) as for example in Fig. 4.4, where the value of C_1 corresponds to the beginning of the dashed line.

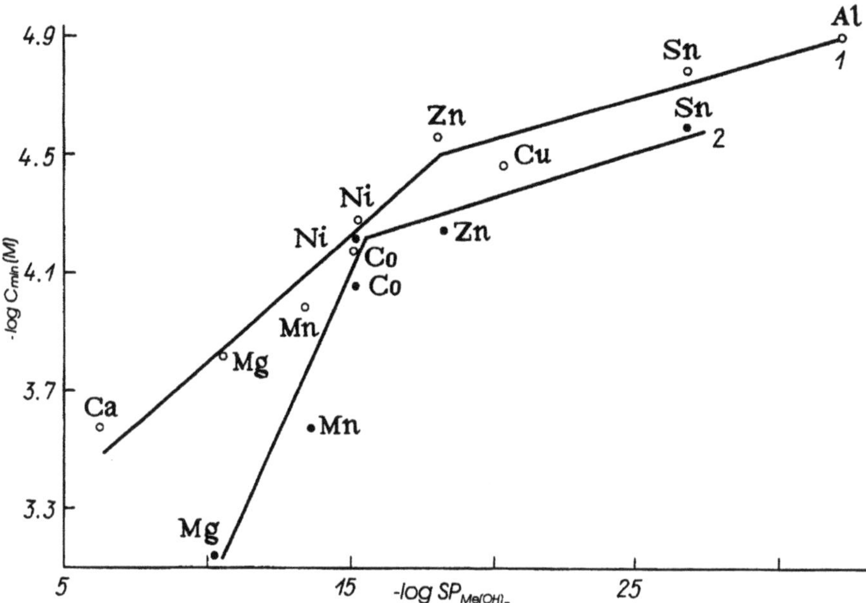

Figure 4.13. The effect of the solubility product of metal hydroxides on the minimum concentration of (1) mono and (2) binuclear complexes with HEDP for the protection of St.3 (C=0.2%) in water containing NaCl 30 + Na_2SO_4 70 mg L^{-1}.

A feature of this equation is that it is practically unchanged when AlHEDP is used

$$\log C_{min} = -3.33 + 0.055 \log SP_{Me(OH)_n} \quad (4.4b)$$

In all cases the value of the steady-state potential, E_{st}, established when C_{min} is reached is $-(0.1–0.15)$V. Taking into account the small alkalinization of the solution to pH 9 that takes place during the corrosion of steel in the presence of the phosphonate complex, this value indicates, according to the appropriate potential–pH diagram, the thermodynamic possibility that the metal-complex formers (Ca^{2+}, Mg^{2+}, Mn^{2+}, Ni^{2+}, Co^{2+}, Zn^{2+}, Sn^{2+}) will be in the form of $Me(OH)_2$. An exception is copper since $Cu(OH)_2$ in these conditions is reduced to copper. Thus, steel in the presence of a cupric phosphonate complex (at $C_{inh} > C_{min}$) was covered with a greenish-brown film having weak adhesion to the surface but leaving the surface with its metallic (steel) luster when removed.

It is found that C_{min} correlates not only with log $SP_{Me(OH)_2}$ but also with K_s of the inhibiting complexes (Fig. 4.14). If the protection of the steel

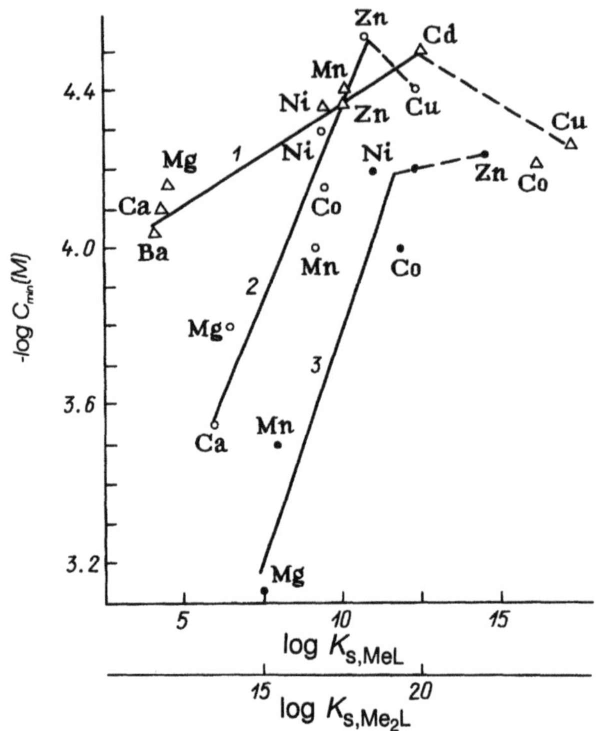

Figure 4.14. Effect of the stability constant, K_s, of mono (1,2) and binuclear (3), complexes of NTP (1) and HEDP (2,3) on the concentration for reliable protection of St.3 (C=0.2%) in water containing NaCl 30 + Na_2SO_4 70 mg L^{-1}; pH = 7.0 ≠ 0.1 (unbuffered).

were to be guaranteed by the formation of only the appropriate hydroxides, it would be expected that there would be a reduction in its effectiveness with increase in log K_s of the inhibiting complex. The effects that are observed can be explained as follows.

Apart from the reaction leading to the precipitation of the difficultly-soluble hydroxide of the metal-complex former, it is also necessary to consider the formation of polynuclear complexes, for example, by the routes:

$$MeL^{z-} + Fe^{2+} = FeL^{z-} + Me^{2+} \tag{4.5a}$$

$$m\,MeL^{z-} + (k-m)Me^{2+} = Me_kL_m^{-(mz+2(k-m)} \tag{4.5b}$$

Unfortunately, the composition and solubility of the surface complexes of HEDP are not known, but a comparison of $SP_{Me_kL_m}$ with the respective

$K_{s,MeL}$ which is known for NTP, provided a basis for proposing that the higher the stability of the MeL complex, the lower would be the solubility of the Me_kL_m complex.[218] This proposal is in full agreement with the correlations represented in Fig. 4.14; i.e., for complexes of NTP

$$\log C_{min} = -3.940 - 0.044 \log K_s \quad (4.6a)$$

and for HEDP

$$\log C_{min} = -2.770 - 0.144 \log K_s \quad (4.6b)$$

These equations show that when searching for effective inhibitors, it is necessary to take into account not only $SP_{Me(OH)_2}$, but also the K_s of the complex added to the solution.

However, it should be kept in mind that not every, or any, increase in K_s will lead to an improvement in the protection of steel by a phosphonate complex. This can be shown, for example, by comparing the protective properties of the Me_2HEDP binuclear complexes, which have greater stability, with those of the mononuclear type.[207] The K_s values of the binuclear complexes are 3–4 orders of magnitude higher than for the mononuclear, but the suppression of corrosion of steel by the former requires its presence in a greater concentration (Figs. 4.13 and 4.14). In a Cu_2HL solution, inhibition occurs in the short term test (8h) of Fig. 4.13, but corrosion can be stimulated in long-term tests by a deposit of dispersed copper which forms on the surface of the specimen within the first hour of the test.

Such increase in C_{min} of the inhibitor is observed to a lesser extent for those complex formers for which the transition from the mono- to the binuclear complexes is accompanied by a relatively small increase in K_s. Thus, for Ni^{2+} complexes this difference, $\log K_{s_{MeHL}} - \log K_{s_{Me2HL}}$ equals 2.94, and for Zn^{2+} it is 4.30, but the effectiveness of their binary complexes is practically the same.

Thus, the protective action of a phosphonate complex is determined both by the solubility product of the hydroxide of the cation-complex former and by the stability constant of the complex itself. One can suggest that there is a limiting value in the solubility product above which the effectiveness of the complexes deteriorates, despite the conditions being favorable for the deposition of $Me(OH)_n$ on the metal surface. In the protection of steel the best results have been obtained when using the mononuclear HEDP complex with Al^{3+}, this having the greatest stability constant ($\log K_{s(AlHL)} = 15.21$) and the lowest value of $SP_{Al(OH)_3} = 10^{-32}$. The binuclear Al_2 HL complex with $\log K_s = 19.33$ does not provide full protection. The best inhibitor among the HEDP complexes of metals is found to be the mononuclear zinc complex (log $K_s = 10.7$). It is of interest that the mononuclear Fe^{3+} complex (log $K_s = 16.21$) cannot provide full protection of steel even at a concentration of 2×10^{-4} M, although the $SP_{Fe(OH)_3}$ of 3.8×10^{-38} is significantly lower than that of aluminum hydroxide.

The existence of optimal values of stability constants for phosphonate inhibitors is also indicated from a comparative analysis of the effectiveness of HEDP and NTP complexes. The latter are often the more stable but in terms of their effectiveness in protecting steel can either be superior or inferior to HEDP. Typically, the complexes of metals which are characterized by relatively low values of stability constants, those on the ascending branches of the dependencies in Fig. 4.14, more effectively inhibit the corrosion of steel in the case of NTP. On the other hand, Zn^{2+}HEDP or Cu^{2+}HEDP prevent the corrosion of steel at lower concentrations than do the complexes of those metals with NTP.

To explain the position of the inflection on the dependencies, it was suggested that there is a change in the mechanism of the substitution of the cations of the inhibitor complex by the cations of the dissolving metal.[218] For example, for the nitrilo-tris-phosphonates of Ba^{2+}, Ca^{2+}, Mg^{2+}, Mn^{2+}, Zn^{2+}, Ni^{2+}, and Cd^{2+} for which $K_s \leq K_{s,Fe^{2+}NTP}$ the substitution could take place by the route in Eq. (4.5a). In the case of the similar phosphonates of Co^{2+}, Cu^{2+}, and apparently Sn^{2+}, this substitution route is thermodynamically not possible because of their high K_s values. However, with excess Fe^{2+}, or in the presence of Fe^{3+}, even if arising as a result of oxidation of Fe^{2+}, the cations of the inhibitor complex can be displaced by these cations. Such a reaction rate evidently is kinetically hindered, and the indicated complexes can prevent the corrosion of steel, but at higher concentrations. In agreement with this suggestion, an increase in temperature (up to 60° C, which accelerates the reaction), leveled out the protective action of CdNTP and CuNTP, whereas at 20° C the effectiveness of the latter was significantly lower.

Apart from the complexing agent and the cation-complex former, an important role in the inhibition of corrosion must be played by the oxidizer, the nature and concentration of which particularly affect the intensity of the generation of OH^- and consequently, the conditions for the formation of the low-solubility hydroxides. The favorable effect of higher concentration of dissolved oxygen on the protection of steel by ZnHEDP can be seen immediately in the chemical polarization in unstirred water (Fig. 4.6). The change in effectiveness of this inhibitor on going from stagnant water to water stirred at various rates is, at least partially, caused by an increase in the rate of reduction of oxygen resulting from its enhanced diffusion to the electrode. In this situation temperature has more than one effect. Thus, while it decreases the solubility of oxygen, it also increases its rate of diffusion and discharge. The latter factor in the corrosion of steel in open systems, where there is no excessive rate of flow or stirring which could fully remove the diffusion control of the cathodic reaction, usually predominates, and the rate of corrosion increases. However, it does not follow from this that the inhibiting action of complexes of low-solubility metals should increase since $SP_{Me(OH)_n}$ also changes, and the acceleration with temperature of the anodic reaction is capable of preventing both the formation of $Me(OH)_n$ as well as its deposition on a rapidly dissolving surface.

The suppression of the corrosion of steel at elevated temperatures undoubtedly requires higher concentrations of complexing agent. Therefore it should be noted that, for example, at 60°C, the temperature which is usually the most dangerous for circulating water systems from the corrosion point of view, full protection of steel can be achieved by using only the zinc and tin 1-hydroxyethane-1,1′-diphosphonates. SnHEDP provides protection at lower concentrations than ZnHEDP and protection is retained even at 80° C. Other complexes do not completely suppress corrosion at concentrations at or below 1 mM. If concentrations of inhibitors that guarantee a high level ($Z = 95 \pm 1\%$) of protection are considered (see Table 4.7) then the effect of the $SP_{Me(OH)2}$ value on the effectiveness of inhibition is retained (although, unfortunately, the collected data for $SP_{Me(OH)}$ are available only for 20° C). The higher the solubility product of the hydroxide the weaker will be the inhibitor, although in the case of Cu^{2+} phosphonates it is the metal, not the hydroxide, which is precipitated and as expected, stimulates the corrosion of steel.

It can be seen from these examples that the effect of even such important and naturally-occurring oxidizers as oxygen on the protection of steel by metal phosphonates in neutral solutions has been insufficiently studied. Still less is known about other oxidizers, although the patent litera-

Table 4.7. The Concentration of Mononuclear HEDP Complexes Necessary for 95±1% Protection of Steel in Water Containing NaCl 30 + Na_2SO_4 70 mg L^{-1}

Cation complex-former, Me^{2+}	MeHEDP concentration (in mM)	$-\log SP_{Me(OH)_2}$ (at 20°C)
Sn^{2+}	7×10^{-2}	29
Zn^{2+}	9×10^{-2}	17
Ni^{2+}	0.25	14.9
Co^{2+}	0.25	14.8
Mn^{2+}	0.38	12.4
Mg^{2+}	2.0	10

[a]at 60°C

ture has no small number of claims for the combined use of phosphonates and nitrite, chromate, or other oxidizing agents.

There are apparently two variants among the types of oxidizers for use in combination with phosphonates: either the oxidizer is reduced at higher rates than oxygen, or the reduction generates a large quantity of OH^- per unit charge transferred. An example of the first type of oxidizer is nitrobenzoic or nitrophthalic acid. An example of the second type is sodium nitrite. However, it is necessary to remember that the inhibition of the corrosion of steel by the hydrophilic complexing agent itself is possible only under the combined action of two factors, the formation of low solubility, i.e., polynuclear, complexes on the steel surface, and the alkalization of the near-electrode layer. The absence of either of these can lead to stimulation of corrosion. For example, where the formation of low solubility complexes by the complexing agent is not possible, as would be the case in solutions of trilon-B (di sodium ethylenediaminetetracetate) in which at $pH < 10$ only mononuclear complexes form,[207] no corrosion inhibition occurs and the stimulating effect even increases somewhat on raising the temperature from 20 to 80° C. No beneficial effect is obtained when this complexing agent is used with additions of oxidizers. Thus, at 80° C the corrosion rate of steel in water containing NaCl 30 + Na_2SO_4 70 mgL^{-1} is 120 $gm^{-2}d^{-1}$, and on addition of 10 mgL^{-1} of trilon-B it is 220 $gm^{-2}d^{-1}$. With further addition of 5 mg L^{-1} of nitrobenzoic acid or sodium nitrite the rates are, respectively, 192 and 188 $gm^{-2}d^{-1}$. Even increasing the concentration of such a strong oxidizing agent as nitrobenzoic acid to 80 mgL^{-1} did not inhibit the corrosion, which was then at a rate of 160 $gm^{-2}d^{-1}$. There was no reason to suppose that alkalization had not occurred during the course of the corrosion process, since even the pH of the whole solution increased by about one unit. It should be noted that with 5 mgL^{-1} of the oxidizer itself there was no such marked effect on the corrosion of the steel. Thus, sodium nitrite had practically no effect on the corrosion rate which was 36 and 130 $gm^{-2}d^{-1}$ at 20 and 80°C, respectively, and nitrobenzoic acid showed some stimulation of corrosion (42 and 150 $gm^{-2}d^{-1}$). Significant inhibition of corrosion was found with combinations of these oxidizers with HEDP (Table 4.8).

Even at 20° C, the use of the strong nitrobenzoic oxidizer can somewhat increase the degree of protection of the steel from 58 to 72%. However, the introduction of nitrite was not capable of improving the protection in this case. At 40 and 60° C, when small additions of HEDP could even stimulate corrosion of steel, the same added concentrations of both oxidizers led to 40–80% protection. However, as already suggested, the greatest effects were achieved at the highest temperatures, 80 and 90° C. Unfortunately, even in this case, the development of fine pits could not be avoided

Table 4.8. The Effect of Oxidizers on the Corrosion Rate of Steel in Water Containing NaCl 30 + Na_2SO_4 70 mg L^{-1} + HEDP

Temp (°C)	Corrosion Rate ($gm^{-2}d^{-1}$) (Concentration of inhibitor, mg L^{-1})			
	Background solution	HEDP	HEDP + nitrobenzoic acid	HEDP + $NaNO_2$
20	36	15 (2.1)	10 (2.1+5)	22 (2.1+5)
40	75	110 (5)	45.0 (5+5)	45.0 (5+5)
60	105	110.4 (5)	21.8 (5+5)	22.4 (5+5)
80	120	45 (20)	2.1 (20+5)	0.0 (20+5)
90	113	32 (5)	0.0 (5+5)	0.0 (5+5)

despite the fact that no mass loss was recorded—which would have suggested 100% protection.

In using, instead of HEDP, its complex with zinc, the alkalization of the near-electrode layer brought about by the oxidizer was at least partly reduced by the precipitation of $Zn(OH)_2$, which inhibited the rate of the cathodic reaction. As a result of the occurrence of complex competing processes, both inhibition—which may even be more effective than that given by ZnHEDP—as well as stimulation of corrosion may be achieved. The technology of water treatment using phosphonates with small additions of oxidizers has many subtleties which depend on the actual conditions of use. They present complex problems in their understanding. However, one can look forward to increased effectiveness of inhibition and further practical developments by this route.

4.5. THE COMBINED PROTECTION OF STEEL

About 40 years ago, Rozenfel'd discovered a marked increase in the protective action of some corrosion inhibitors, e.g., chromates, zinc, and calcium salts, when cathodic polarization from sacrificial anodes was applied.(219) In studying the behavior of a zinc sacrifical anode in a soft water containing NaCl 30 + Na_2SO_4 70 mg L^{-1}, he found that additions of $ZnSO_4$ and $Ca(NO_3)_2$ lowered the anodic polarizability of zinc, thus making it a more effective sacrificial anode. A mixture of these inhibitors was found to lower the steel–zinc couple current and so could reduce the consumption of a zinc sacrificial anode. The formation of a cathode deposit reduced the active surface area of the steel so that sacrificial protection was achieved with lower anode consumption. The protective properties of the deposit evidently stemmed from the fact that the cathodic polarization of the steel made the near-electrode layer more alkaline, which in turn, facilitated the formation of difficultly-soluble compounds of calcium and zinc on the metal surface. Therefore, it is clear that in water containing hardness salts the formation of the protective cathodic deposit are made easier so that the effectiveness of the cathodic protection is increased.

In fact, in unstirred artificial seawater (NaCl, 26.52; $MgSO_4$, 3.30; $MgCl_2$, 2.45; $CaCl_2$, 1.14; KCl, 0.72; $NaHCO_3$, 0.20; and NaBr, 0.08 g L^{-1}) the oxygen-limiting current density (i_d) was found to be 12 $\mu A\ cm^{-2}$, half that in soft water. On the other hand, the high concentration of aggressive ions (Cl^-, Br^-, SO_4^{2-}) lowers the effectiveness of the majority of corrosion inhibitors for steel in sea water. Therefore, there is potential attraction in the use in this environment of a combined method of cathodic protection with added inhibitors. The technique could—providing economic and ecological criteria are met—find wide use for protecting ballast tanks of ships, pipelines, or seawater-cooled equipment.

Cathodic protection in seawater differs from that in acids or soft water because in the latter cases the formation of a protective cathodic deposit has no significant role.[220] Based on this knowledge, the possibility of a combined protection of steel in seawater using small additions (< 0.1 g L^{-1}) of various inhibitors was studied.[221] When used alone, not one of the inhibitors produced any significant reduction in corrosion in artificial seawater (Table 4.9). In all cases, the potential of the steel fell upon immersion from an initial value, $E_{init} = -0.22 \pm 0.02$ V to $E_{st} = -0.475 \pm 0.015$ V, which is characteristic of values for the active dissolution of steel. This

Table 4.9. The Effect of Inhibitors on the Corrosion and Electrochemical Behavior of Steel in Artificial Seawater

Inhibitor (0.1 g L^{-1})	Corrosion rate (g $m^{-2}d^{-1}$)	E_{st}(V)	Oxygen limiting current density i_d ($\mu A\ cm^{-2}$)	Rate of potentiostatic dissolution of steel at $E=-0.475$ V (g $m^{-2}d^{-1}$)
—	1.88	−0.48	11.5	6.40
$ZnSO_4$	1.18	−0.48	6.0	0.49
Na phenyl anthranilate	1.40	−0.49	11.5	3.94
ABDM[a] chloride	1.52	−0.46	10.0	3.94
Lignosulfonate	1.68	−0.49	8.5	4.95
Tannin	1.73	−0.48	8.5	3.94

[a]Alkylbenzyldimethyl ammonium chloride

debasement of potential was due to the presence of high concentrations of Cl^- and SO_4^{2-}, which impeded the adsorption of the inhibitor and encouraged breakdown of the air-formed oxide film. The introduction of the inhibitors (except phenylanthranilate) increased the polarizability of the steel cathode and reduced i_d. The most effective was $ZnSO_4$ which provided greatly improved protection if a short cathodic polarization was applied. Thus, when a steel electrode was polarized to $E = -0.475$ V, on immediate immersion the degree of protection, Z, was increased to 92% since the alkalization of the near-electrode layer facilitated the deposition of $Zn(OH)_2$. Zinc sulphate, even at a concentration of 10 mg L^{-1}, was capable of assisting the potentiostatic protection by ennobling the potential below which corrosion is fully suppressed. Lignosulfonate and phenylanthranilate, although less effective, retained some protective role in these conditions.

For the practical application of combined protection, it is important to determine the optimum range of values of the cathodic polarization current density, i_c, for which the greatest effect can be obtained by introduction of the inhibitor. In artificial seawater, polarization of steel at $i_c = 10$ $\mu A\ cm^{-2}$ for 15 d does not fully suppress corrosion. In the presence of 1 mg L^{-1} $ZnSO_4$, $i_c = 8$ $\mu A\ cm^{-2}$ is sufficient to ensure cathodic protection. However, on decreasing i_c to 5 $\mu A\ cm^{-2}$, effective protection can not be achieved, even by increasing the $ZnSO_4$ concentration by two orders of magnitude, from 1 to 100 mg L^{-1}. It is of interest that in cathodic protection conditions, lignosulfonates, which have significant dispersing properties, lose their protective properties with an increase in their concentration and can, in such cases, even stimulate corrosion. The absence of any noticeable increase in the effectiveness of cathodic protection upon introducing ABDM (alkylbenzyldimethyl ammonium chloride) or tannin is also associated with their inability to increase the barrier properties of the cathodic deposit. Consequently, the ability of an inhibitor to inhibit the reduction of oxygen during a short cathodic polarization is insufficient for predicting its effectiveness in combined protection.

The selection of an inhibitor is complicated by the tendency of the steel to undergo localized corrosion in seawater. Thus, small additions of chromate, by localizing corrosion, increase its intensity (see section 4.3, p. 199). Galvanostatic cathodic polarization improves the protection of steel by chromate, but localized corrosion occurs even at a chromate concentration of 50 mg L^{-1} (Fig. 4.15).[222] However, the potential of the electrode even at a relatively low chromate concentration (1 mg L^{-1}) becomes more negative, which suggests that there is some cathodic protection. Unfortunately, to achieve this requires a rather long amount of time, i.e., not less than 7 d, which is thus a shortcoming in the effectiveness of such combined

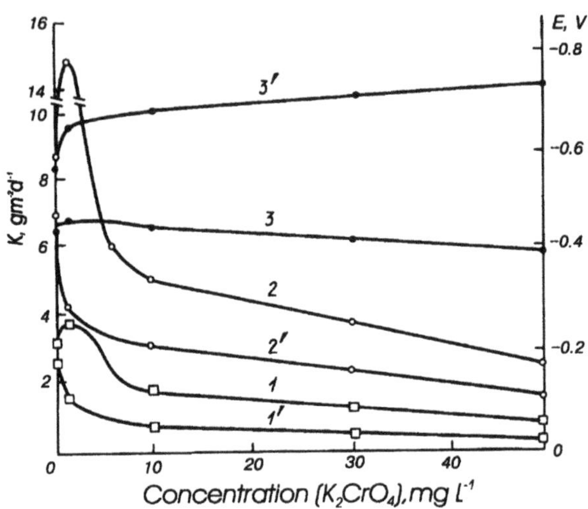

Figure 4.15. Dependencies of corrosion rates (1,1′), corrosion intensity (2,2′) and potential (3,3′) (after 15 d) on concentration of chromate in seawater without (1,2,3) and with (1′,2′3′) cathodic polarization at a current density of $i_c = 5\ \mu A\ cm^{-2}$.

protection. Furthermore, the independent use of chromate and $ZnSO_4$ for the combined protection of steel in seawater shows little promise when economic and ecological factors are taken into account. In this respect, the phosphorus-containing inhibitors are attractive since they have the ability to form difficultly-soluble protective films, as well as to impede both the anodic and cathodic (O_2 reduction) reactions. They also provide some inhibition of corrosion in seawater, even at relatively low concentrations. It is important to note that these compounds are less toxic than chromates or zinc cations.

Nevertheless, some quite complex problems of an ecological nature exist even with the use of inorganic phosphates. Because of this, it is valuable to look at the use of some phosphorus-containing complexing agents and complexes and compositions based on these.

1-hydroxyethane-1,1′-diphosphonic acid, HEDP, at a concentration of 100 mg L^{-1} weakly inhibits the corrosion of steel in artificial seawater. However, at $i_c = 2–5\ \mu A\ cm^{-2}$ a combined protection with HEDP at a concentration of 50 mg L^{-1} is more effective than with $ZnSO_4$. An undoubted advantage of HEDP over inorganic inhibitors is its ability to guarantee a high level of protection even with small polarization currents of, for example, $i_c > 2\ \mu A\ cm^{-2}$.

Some features of the action of HEDP were revealed in a study of the dependence of the corrosion rate (from the mass loss) of steel on potential in the cathodic reaction. The addition of 10 mg L^{-1} NaH_2PO_4 or HEDP to seawater sharply lowered the corrosion rate in conditions of potentiostatic cathodic polarization (Fig. 4.16). Despite the fact that a cathodic protection potential was applied 2 h after immersion of the electrode in the solution, i.e., after the steady-state potential had been established, the oxide film in these inhibited solutions was still not completely destroyed and the steel potential remained more positive by 0.15 V. At the same potential the negative charge which passed through the electrode, Q, in the presence of the inhibitor was, as a rule, higher and the corrosion rate lower, than in the seawater. However, it has to be taken into account that the protective salt layer on the steel is deposited from a solution with a relatively high concentration of Ca^{2+} and Mg^{2+} and that the reduction of oxygen is diffusion-controlled and independent of potential. Consequently, to a first approximation, the rate of formation of the salt layer causing the protective effect is determined mainly by the quantity of hydroxyl generated by the cathodic reaction, which is proportional to Q. If the inhibitor only accelerates the formation of a carbonate–hydroxide deposit, then for equal values of Q the corrosion behavior of steel in its presence cannot be distinguished from that in pure seawater. This situation occurs on adding 10 mg L^{-1} NaH_2PO_4 and at such low concentration it has to be suggested that phosphate would not inhibit the anodic reaction.

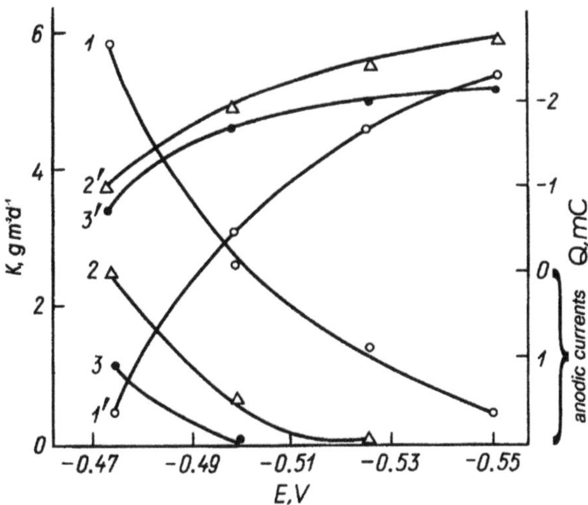

Figure 4.16. Dependence on potential of the dissolution rate of iron (1,2,3) and quantity of electricity passing through the electrode in 6 h (1′,2′,3′) in artificial seawater without (1,1′) and with (2,2′) 10 mg L^{-1} Na_2HPO_4 or (3,3′) HEDP.

In the presence of HEDP, a lower value of Q is recorded than in the case of the addition of phosphate to seawater, but it is the better inhibitor of corrosion in conditions of cathodic polarization. This is shown as a more positive potential of cathodic protection and by lower corrosion rates at constant Q (for example, 0.8 versus 2.0 $gm^{-2}d^{-1}$ in sea water at $Q = -1$ $mCcm^{-2}$). Thus, the beneficial effect of HEDP on the cathodic potentiostatic protection of steel in seawater cannot be explained only by the easier formation of the hydroxide-salt deposit on the electrode surface. Moreover, in the absence of polarization the dissolution of steel in seawater is almost unaffected by the addition of 10 mg L^{-1} HEDP the degree of protection being $< 30\%$. Thus, the increase in the protective action of the inhibitor that occurs on imposing a cathodic potential is difficult to explain by an improvement in its adsorbability on changing the surface charge since HEDP contains strongly-acid phosphonic groups, which in neutral solution are found in the form of multicharged anions. It is more probable that the protection is connected with adsorption not of the inhibitor itself but of its complex with magnesium.

In artificial seawater a zinc–phosphonate inhibitor containing equal concentrations of HEDP and $ZnSO_4$ is also more effective than HEDP. However, the synergistic protective effect of HEDP and $ZnSO_4$ that is observed in seawater with relatively high concentrations of the inhibitors can be achieved at a concentration of, for example, 10 mgL^{-1} only by subjecting the steel to galvanostatic cathodic polarization (Fig. 4.17). In this case, and with that of HEDP itself, small values of i_c may not yield a beneficial effect, and may even somewhat weaken the protective action of the inhibitor. An increase in i_c to 5 $\mu A\ cm^{-2}$ abruptly increases the protective effect, whereas cathodic polarization alone produces almost no reduction in corrosion. Combined protection is also not effective when the separate components of the inhibitor are used in such low concentrations, less than 61% protection then being achieved. This value could be raised to 90–98% by increasing the inhibitor concentration to 20–40 mg L^{-1}, but since a high degree of protection is usually not required for ballast tanks in ships, the increase in inhibitor concentration can be used more profitably for lowering i_c or for reducing the number of sacrificial anodes per unit area. In this type of application ZnHEDP is the most promising of the above inhibitors and can lower the corrosion rate of steel by 7–9 times, even at a concentration of 40 mg L^{-1} and with $i_c = 1–2$ μAcm^{-2}. It is significant that the establishment of a protective potential, E_c, depends not only on i_c but also on the duration of polarization. At an inhibitor concentration of 40 mg L^{-1} the period of time required to establish cathodic protection of steel with $i_c = 1$ $\mu A\ cm^{-2}$ reaches 10 d. In the conditions of service experienced by floating docks, when there is a prolonged period of ballasting, this regime

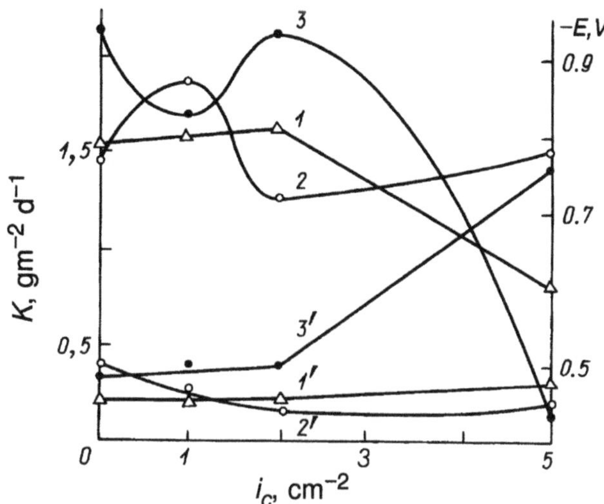

Figure 4.17. The influence of cathodic current density on the rate of dissolution of steel (1,2,3) and its potential (1′,2′,3′) in artificial seawater containing 10 mg L^{-1}: (1,1′) $ZnSO_4$; (2,2′) HEDP; (3,3′) HEDP + $ZnSO_4$, 1:1.

of combined protection is effective, since it allows a significant reduction in the number of sacrificial anodes.

Evidently, the effect of the time factor on the protection of steel is connected with features of the formation of the surface layers which change the kinetics of the cathodic reduction of oxygen. This is a characteristic of not only ZnHEDP but also of seawater. Thus, if the cathodic polarization curves are taken at the same scan rate immediately, or at 2, 17, and 60 h after immersion of the steel in seawater, i_d values of 25, 17, 11.5, and 11.0 μA cm^{-2} are obtained. Since cathodic protection of steel in seawater is obtained at i_d values greater than 12 μA cm^{-2} it can be reckoned that only the i_d value determined after holding the steel for not less than 17 h in the solution is the steady-state value. An analogous picture is observed in the presence of 50 mg L^{-1} $ZnSO_4$, for which the i_d values are respectively 25, 16, 10, and 5 μA cm^{-2}, which explains the slow establishment of E_c even with $i_c = 8$ μA cm^{-2} (about 3d). In the presence of ZnHEDP, the retardation of the cathodic reaction sets in more quickly so that after 2 h, i_d falls to 7, and after 17 and 60 h, to 5 and 4 μA cm^{-2}, respectively. In this case, the cathodic deposit could also be thinner than that obtained in seawater alone or in seawater containing $ZnSO_4$—as confirmed by X-ray microanalysis of the surface of electrodes held for 4 h at −0.8 V in uninhibited and inhibited seawater (Fig. 4.18). The cathodic deposit that is formed in seawater consists mainly of magnesium compounds, with a calcium content not above 1–2%.

This deposit is porous and of non-uniform thickness (0.1–1 μm) as a result of which the spectra also include a signal from Fe.

The alkalization of the near-electrode layer resulting from cathodic polarization of the steel should lead mainly to the precipitation of $Mg(OH)_2$ since the concentration of Mg^{2+} (0.05 g ion L^{-1}) is higher than that of Ca^{2+} (0.01 g ion L^{-1}), and the solubility product (*SP*) of $Mg(OH)_2$ is lower than that of $Ca(OH)_2$ (respectively, 7.1×10^{-12} and 5.5×10^{-6}). The precipitation of other magnesium compounds is less probable because of their greater solubility: $SP_{MgCO3} = 2.1 \times 10^{-5}$ and $SP_{MgO} = 1.85 \times 10^{-6}$. Of the Ca^{2+} compounds, the least soluble is $CaCO_3$ ($SP = 3.8 \times 10^{-9}$), and it must be assumed that this will be present in the cathodic deposit. Interestingly, the introduction of zinc sulfate only increases the thickness of the cathodic deposit without leading to the occurrence of Zn^{2+}, although, possibly it is present in the layer immediately adjacent to the metal. Mg^{2+} compounds form the basis of the film. The film formed in the presence of 20 mg L^{-1} ZnHEDP is also of non-uniform thickness and on steel there may even be areas that retain a metallic brightness. The spectra from such areas are characterized by an intense Fe peak as a result of the large contribution from the substrate in the 1 μm^2 beam area. A content of more than 1% zinc in the film is detected only in thin regions whereas phosphorus is distributed relatively uniformly (3–3.4%). Evidently in this case, zinc is

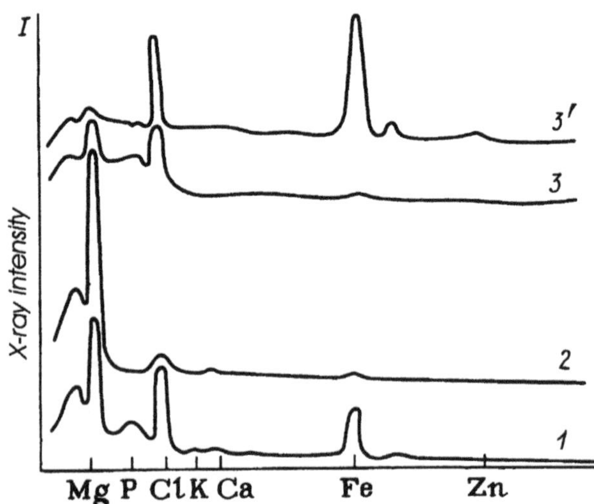

Figure 4.18. Characteristic X-ray spectra of films on the steel St3 [C=0.2%], formed in 4 h at $E=-0.8$ V in artificial seawater (1) without and (2) with addition of 10 mg L^{-1} $ZnSO_4$ plus (3,3′) ZnHEDP. The spectra of 3′ were obtained from a thinner section of the film.

deposited on the steel surface mainly in the initial period of formation of the protective film which can then be "completed" by compounds of magnesium and calcium, including those containing HEDP.

The combined protection with ZnHEDP provides a significant after-effect which is retained both during the "switching off" of cathodic polarization, as well as when the inhibited seawater is replaced by clean seawater. When a steel electrode was held for a certain time at $i_c = 5\ \mu A\ cm^{-2}$, and then the polarization switched off for 12 h, the potential began to ennoble in both HEDP and ZnHEDP solutions. Switching on the current again displaced the potential in the negative direction, although it did not reach its initial value in the presence of HEDP alone, even at 50 mg L^{-1}. Such off–on cycles gradually lower the cathodic polarization step as well as the degree of protection of the steel. For improved stability of protection it is necessary for the HEDP-containing film to be formed over a long polarization period of ~300 h.

Even at a concentration of 20 mg L^{-1}, ZnHEDP forms a protective film that guarantees rapid and complete recovery of the cathodic potential over 24 off–on cycles. After switching off the polarization, the potential would remain sufficiently negative, and only after 26 d would the film begin to lose its protective properties, although, even in this case the corrosion prevention was more effective than that provided by the addition of 50 mg L^{-1} HEDP.

The post-treatment effect depends on the duration of formation of the protective film. This has been shown from data obtained when seawater inhibited with ZnHEDP was replaced by uninhibited seawater. Films were formed over a period of 132 h at $i_c = 5\ \mu A\ cm^{-2}$ and then the electrode was transferred to uninhibited seawater. The potential was found to remain ennobled for three d after the transfer. When the time of film formation was increased to 15 d at this current density, there was no indication of corrosion occurring in uninhibited seawater after >10 d (at $i_c = 5\ \mu A\ cm^{-2}$). In practice this effect can be used for periodic ballasting of tanks or storage vessels of ships with seawater. (Significantly, the thick salt deposit that could be formed even at $i_c = 15\ \mu A\ cm^{-2}$ in uninhibited seawater over similar treatment times does not guarantee adequate protection of the steel when i_c is reduced to 5 $\mu A\ cm^{-2}$ in this environment.)

The marked reduction in concentration of inhibitor that is possible when it is used in combination with cathodic polarization reduces the ecological problem and makes it possible to use formulations containing small additions of chromate. Thus, as little as 1 mg L^{-1} of potassium chromate can increase the effectiveness of inhibition in seawater when cathodic polarization is applied. For example, zinc gluconate at a concentration of 12 mg L^{-1} was not particularly effective when used with $i_c = 5\ \mu A\ cm^{-2}$, reducing the corrosion rate by 4–5 times, with no improvement when the concentration was increased to 50 mg L^{-1}. However, in combination with 1 mg L^{-1} chromate the combined protective effect was raised to a value of 98–99%. As was the case with the other inhibitors mentioned above, chromate-containing formulations make a large contribution to the inhibition of the cathodic reaction, although with $E < E_{st}$ the anodic dissolution is also inhibited.

A greater effect than that given by zinc gluconate can be produced by using similar formulations with complexes. A binary mixture of K_2CrO_4 1 + HEDP 10 mg L^{-1} provides a synergism in the protective action during cathodic polarization, but HZnCr (HEDP + $ZnSO_4$ + K_2CrO_4 used in mass proportion 5:4:1) is still more effective. In unstirred seawater and with i_c = 5 μA cm^{-2}, a concentration of 5 mg L^{-1} of HZnCr can guarantee 90% protection of steel and concentrations above 10 mg L^{-1} provide essentially complete protection (see Fig. 4.19). Lower concentrations of this inhibitor can significantly increase the degree of cathodic protection, e.g., with i_c = 8 μA cm^{-2}, 2mg L^{-1} of the inhibitor guarantees 94% protection, versus 59% in its absence. When the low toxicity of ZnHEDP and the low Cr^{6+} content in the ballast water (0.03–0.06 mg L^{-1}) are taken into account, it can be suggested that such an inhibitor is not an environmental threat.

The replacement of HEDP by nitrilo-tris-phosphonic acid, NTP, to give the ternary formulation (NZnCr) allows a high degree of protection of steel (95.3%) to be reached at i_c = 5 μA cm^{-2}, even with an inhibitor concentration of 1 mgL^{-1}.

Combined protection can also be achieved in a flowing liquid. Since in this case there is a decrease in the diffusion limitation of the cathodic reaction, larger values of i_c are needed to allow the cathodic protection potential to be reached. However, there will be an enhancement of inhibitor supply to the metal surface which will aid in the formation of protective films. With a flow rate of 0.35 ms^{-1} and i_c = 5 μA cm^{-2} the corrosion rate of steel increases by almost an order of magnitude, compared with that in stagnant seawater (Fig. 4.19). Despite this, the suppression of corrosion, as in stagnant conditions, can be achieved by the introduction of 20 mg L^{-1} HZnCr, as a result of the considerable reduction of the cathodic reaction by the inhibitor (Fig. 4.20).

The introduction of 10 mg L^{-1} HZnCr into an unstirred solution lowers the oxygen diffusion limiting current, i_d by only a half (γ_c = 2). Stirring of the seawater significantly intensifies the inhibiting effect: with rotation frequency 94 and 295 rad s^{-1}, γ_c increases, to 7.5 and 9.2, respectively. With the inhibitor concentration raised to 20 mg L^{-1} the oxygen reduction is inhibited even more (γ_c = 2.7, 10, and 13.7, respectively). It is significant

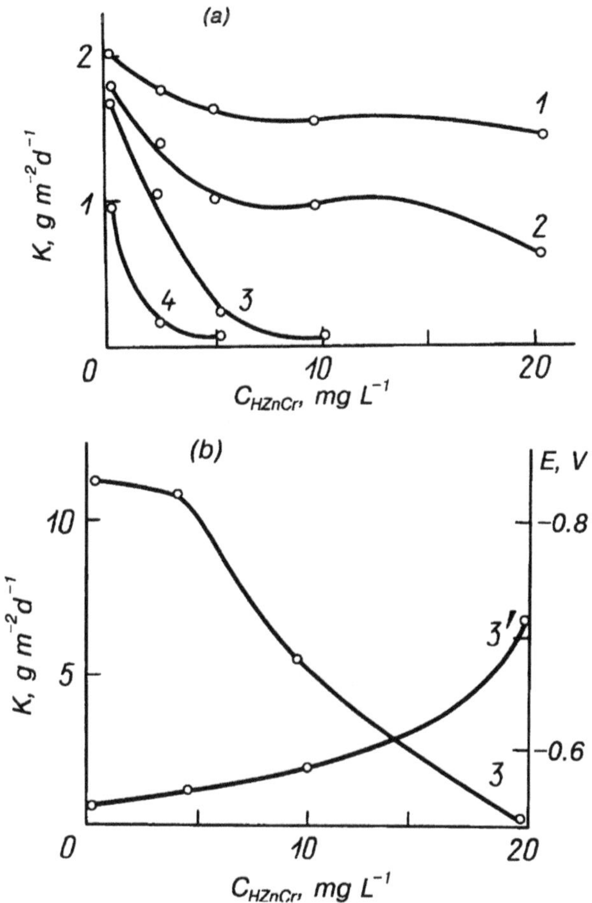

Figure 4.19. The dependence of the corrison rate of cathodically polarized steel (1–4) and its potential (3′) on the concentration of the inhibitor HZnCr in artificial seawater (a) without and (b) with stirring at 0.35 ms^{-1}, with i_c = 0.0 (1); 2 (2); 5 (3,3′) and 8 $\mu A\ cm^{-2}$ (4).

that a less negative potential is required to prevent the corrosion of steel in flowing seawater since the inhibitor more effectively inhibits the anodic reaction (Fig. 4.20) particularly near E_{st}. As these effects appear without any long period of keeping the electrode in these solutions, it can be suggested that the combined method is capable of encouraging the formation of layers that can protect the steel surface, even in conditions of subsequent use without cathodic polarization.

These ternary formulations have a strong post-treatment effect, which is greater than in the cases that have already been examined for combined

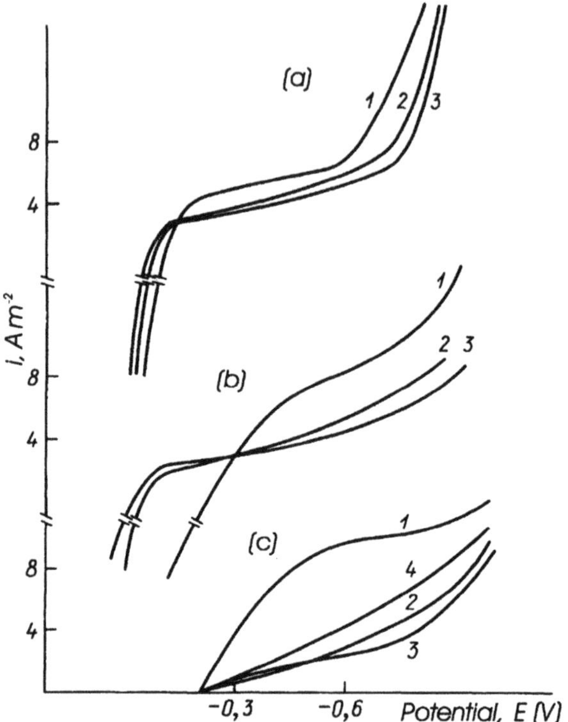

Figure 4.20. Polarization curves for a rotating steel disc in artificial seawater containing the HZnCr inhibitor (5:4:1 by mass) at concentrations of 0.0 (1); 10 (2); 20 (3); and 5 mg L^{-1} (4) and rotation frequency (a) 0.0; (b) 94; and (c) 295 rad s^{-1}.

protection using ZnHEDP. To demonstrate this, a comparison was made of cathodic deposits made with and without the presence of inhibitors. The cathodic deposits in the former case were obtained at an HZnCr concentration of 10 mg L^{-1} and i_c = 5 μA cm^{-2} since these conditions had been shown to provide practically complete protection. The cathodic deposits in the uninhibited seawater were obtained with significantly higher currents of i_c = 15 μA cm^{-2}, so that corrosion was prevented from the outset. In these conditions the potential of the steel rapidly debased (Fig. 4.21), and protective potentials in the inhibited and uninhibited seawater were reached simultaneously. On replacing both the inhibited and uninhibited seawater by clean (uninhibited) seawater and applying i_c = 5 μA cm^{-2}, the behavior of the specimens depended on the composition of the cathodic deposits. The potential of the specimen that had been protected only by cathodic polarization experienced the greater ennoblement of potential, which then debased to a value of −0.68 V only after 5 d. This suggests that

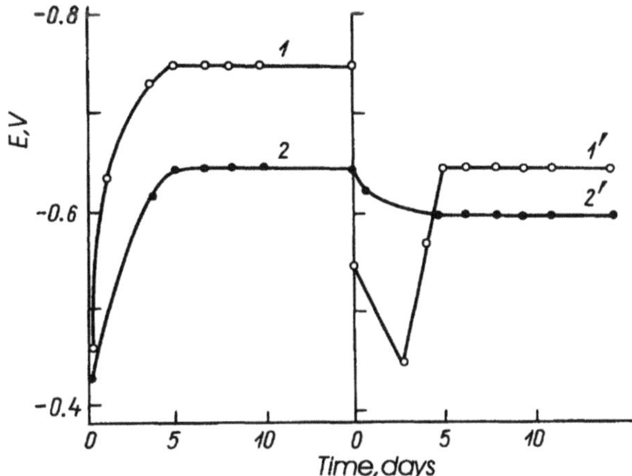

Figure 4.21. Changes of potential with time during (left) and after (right) cathodic and combined protection of steel in seawater. Concentration of HZnCr inhibitor (1,1′,2′), 0; (2), 10 mg L^{-1}; cathodic current density = 15 (1) and 5 (2,1′,2′) $\mu A\ cm^{-2}$. Curve 1 is cathodic protection only.

the earlier-formed cathodic deposit did not fully prevent corrosion when i_c was reduced. The corrosion rate when calculated over the whole test period, and even without allowing for its localized nature, was quite high—0.68 $gm^{-2}d^{-1}$. Evidently, the reduction in i_c decreased the alkalization of the near-electrode layer and led to the occurrence of corrosion in the pores of the cathodic deposit. However, the corrosion products also have protective properties and the quite compact salt deposit on the remaining part of the specimen increased the actual i_c in the pores, which with time led to a restoration of the effectiveness of cathodic protection.

The combined protection provided a more protective film. Thus, when the inhibited seawater was replaced by clean seawater the potential ennobled by 0.06 V. After formation of the inhibited cathodic deposits no corrosion occurred on subsequent exposure to clean seawater. Evidently, the inhibitor is involved directly in the composition of the film, changing its structure and improving its protective action.

X-ray microanalysis was used to monitor the formation of the cathodic deposit, the kinetics being determined from the mass increase-time behavior. In uninhibited seawater there were two characteristic regions (Fig. 4.22).[223] In the first, which covered the deposition period from 3–6 h, the rate of deposition was quite high, although falling continuously from 50 to 20 g $cm^{-2}h^{-1}$ (0.14 to 0.06 μmh^{-1}). In the second region, with times above 6–12 h, the deposition rate changed insignificantly at 5–10 μg

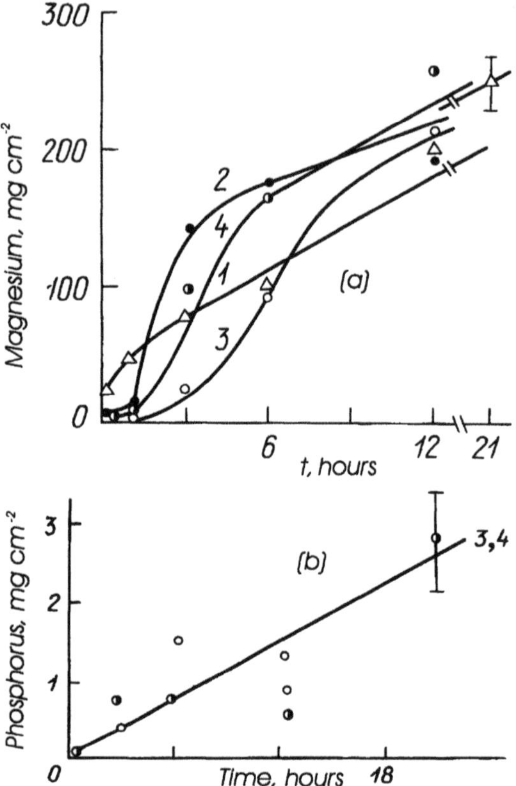

Figure 4.22. Contents of magnesium (calculated as MgO) (a), and phosphorus (calculated as P) (b), in cathodic deposits on steel as a function of holding time, *t*, at a potential of E=−0.8 V in artificial seawater without (1) and with (2) 1 mg L^{-1} K_2CrO_4; (3) 10 mg L^{-1} ZnHEDP or ZnNTP and (4) 10 mg L^{-1} HZnCr or NZnCr.

$cm^{-2}h^{-1}$. In the first region a film of 0.5–0.7 μm thickness was formed. The increase in thickness of the film correlated well with the drop in the Fe line intensity.

The deposit formation kinetics altered noticeably when potassium chromate, even at a concentration of 1.0 mg L^{-1}, was added to the seawater. At $E = -0.8$ V, chromate was reduced to Cr^{3+} which was deposited in the form of $Cr(OH)_3$ ($SP = 6.3 \times 10^{-31}$) on the steel surface and led to a reduction in the rate of the cathodic reaction. With short exposure times of less than 1 h there was almost no deposition of $Mg(OH)_2$ as the OH^- ions generated by the cathodic reaction were consumed in the formation of the significantly less soluble $Cr(OH)_3$. It was in this period that the Cr content of the film increased, with values of 0.6 and 2 μg cm^{-2} at 0.25 and 1 hour, respectively. Later, however, the chromate (apparently as a result of diffusion limitations) ceased to be reduced and a rapid deposition of $Mg(OH)_2$ then occurred (Fig. 4.22). The thickness of the cathodic deposit became rather more significant at 0.4 μm and by 3 h even the Fe signals in the spectra had weakened considerably, while those from Cr were completely absent. Thus, it would appear that the inhibitor had altered the composition of the deposit only in the layers adjacent to the metal surface.

The introduction of formulations of HEDP or NTP with $ZnSO_4$ at 10 mg L^{-1} also led to important changes in the kinetics of deposition of the protective layers (Fig. 4.22). The main, and significant, effect was a decrease by more than an order of magnitude in the rate of deposition of $Mg(OH)_2$ at times shorter than 1 h. When this rate was close to zero, one could even refer to an induction period during which the primary deposition was of zinc compounds. As in the case for Mg^{2+}, the most probable deposition would be of hydroxide rather than of carbonate ($SP = 1.2 \times 10^{-17}$ and 1.45×10^{-11}, respectively). The content of the zinc in the surface layer in the initial stage of its deposition reached 50–70 mass %. When 10 mg L^{-1} ZnHEDP was added to the water, the film thickness, calculated as $Zn(OH)_2$, was 2.5–3 μg cm^{-2} (0.01 μm). After 3 h when $Mg(OH)_2$ deposition became noticeable, the additional quantity of zinc on the surface was 0.6 μg cm^{-2}, since $Zn(OH)_2$ is barely involved in the outer layers of the thickening deposit. With increase in the time of exposure the proportion of zinc in the growing film became poorly reproducible and often it would fall to zero. Even with higher concentrations of 20 mg L^{-1} of ZnHEDP (as already pointed out) zinc at 4 h could be found only on sections of the surface that had retained a metallic luster. Consideration of all these data lead to the conclusion that the surface of the steel was only initially enriched with zinc.

Analogous results, although also indicating a somewhat lower content of zinc in the layer adjacent to the metal (0.8–1.0 μg cm^{-2}) have been obtained from analysis of the surface of an electrode held in seawater containing ZnNTP. For both ZnNTP and ZnHEDP, the Zn/P mass ratio on the surface in the initial stages (0.25 h) of formation of the protective layers was an order of magnitude above that in normal zinc phosphonate complexes. Consequently, the main quantity of the zinc was deposited in the form of $Zn(OH)_2$ and not in the complex. Unlike zinc, the content of phosphorus increased with increase in the time of formation of the film and by 6 h it had reached 5–10 mass %, when calculated as HEDP. In this case, the magnesium deposition rate was little different from that observed in clean seawater. However, it can be suggested that magnesium can be included in the film not only as $Mg(OH)_2$, but also as a magnesium complex. This is possible in view of the high content of Mg^{2+} in seawater and the good stability of its binary complex with HEDP compared with that of the normal zinc complex. Adsorption of the magnesium complex with HEDP apparently accompanies the deposition of the $Mg(OH)_2$. This, to a large extent, explains the uniform distribution of phosphorus in the cathode deposit.

The addition of 10 mg L^{-1} of the three-component formulation also changed the kinetics of the deposit growth in the first 1–3 h. Thus, the presence of chromate decreased the concentration of zinc on the surface to 0.5 μg cm^{-2}. This agrees with the views that have been expressed concerning the formation of the deposit since $Cr(OH)_3$ is significantly less soluble than $Zn(OH)_2$. The total mass thickness of these difficultly-soluble compounds was 1–2 μg cm^{-2} the cathodic deposit containing magnesium and phosphonates formed over these. The substitution of HEDP by NTP in the inhibitor led to some increase in the content of phosphorus in the outer layers of the deposit, although it remained below that in the case of the binary inhibitors. Furthermore, an increase in the concentration of the HEDP–Zn–Cr (HZnCr) formulation to 50 mg L^{-1} led to a particularly sharp increase in, the phosphorus in the deposit, which at 16 h was up to 4.5 μg cm^{-2}. Apparently, the main protective function is carried out by the barrier layer directly adjacent to the metal, although the inclusion of complexes has an additional beneficial effect on the protective properties of the cathodic deposit. Thus, a combined protection of steel in quiescent seawater can be achieved using very small additions (< 10 mg L^{-1}) of a formulation of phosphorus-containing complexing agents (HEDP, NTP) with zinc salts and potassium chromate. The small content of Cr^{6+}, amounting to 1/30 of the total concentration of the 3-component formulation, avoids ecological problems. The practical possibility of an effective combined protection of steel in flowing seawater and the strong post-treatment action indicates its possible use not only in the ballast tanks of ships but also in other cases, such as in various types of cooling systems.

5

Organic Corrosion Inhibitors for Cooling Systems

5.1. GENERAL CONSIDERATIONS

Despite experience over many years with the use of corrosion inhibitors in industrial cooling systems, the problem of the rational treatment of water is still very real in the context of the anti-corrosion treatment of pipelines and equipment. There are various reasons for this, but the most important are associated with ecological requirements and water scarcity. The intensification of technological regimes is also important, for example, the increase in temperature of heat exchange surfaces can stimulate not only corrosion but also scale formation.

Bearing in mind the electrochemical nature of the corrosion of metals in water, a primary assessment of the corrosion resistance of a cooling system could be made from the values of the steady-state potentials, E_{st}, of the materials used in its construction. For this purpose it is useful to use potential–pH diagrams. Thus, Pourbaix diagrams indicate the region in which water itself is stable, as limited by the E–pH lines for evolution of H_2 and O_2. If E_{st} of a metal lies inside this stability band in a region of passivity, then in the absence of foreign impurities, there should be no contamination of the water. For this reason, cooling systems utilizing pure water operate successfully when constructed with copper, aluminum, or alloys based on these metals.

However, cooling systems rarely consist of one constructional metal, and the stability of the different metals that may be present can be altered even without direct contact between them. Thus, aluminum and its alloys can be subject to pitting corrosion even in very pure water if copper ions accumulate in the water. The source of copper ions can be a heat exchanger made from copper alloys. On the other hand, the corrosion rate of brass in many cases can be lowered as a result of the presence of iron ions in the water, which can form protective hydroxide films on the brass surface. Thus, in a real cooling system using pure water, only one "unstable" construction material need be present to make the prediction of the corrosion behavior very difficult from the theoretical point of view. If it is also taken into account that conditions for corrosion can be established in cracks and gaps, at sites of welding or soldering, and at contacts with other materials, it becomes clear why the reliable use of cooling water requires its treatment with special reagents.

The selection of such reagents depends on the type of system and on the nature of the cooling water. Three types of system can be distinguished: once-through, open recirculating, and closed (fully sealed).

In once-through systems natural water is, as a rule, subjected to purification in order to improve its quality, then it is used for a purpose in which it may be heated by several degrees, then returned to its source. This use requires the presence of a plentiful source of water which exceeds the consumption requirement by not less than a factor of 10. It is also necessary that the temperature of the source water should not rise by more than 1° C after the water has been used. At the present time, because of the scarcity of water, increasing ecological requirements, and the complexity of the treatments needed, this scheme of water cooling is rarely used. From the economical point of view, it is limited to a distance of not more than 2 km between source and site of use and a height of not more than 10 m above the source level.

In such systems, although corrosion-resistant materials may be employed, "stabilizing" water treatments may be used. Thus, water is described as stable when, at a given temperature, it does not leave behind insoluble deposits—notably calcium carbonate—and does not act aggressively on materials used in construction. However, this is not a very precise description, particularly with respect to the condition of equilibrium between the free carbon dioxide and the calcium ions in the water, which govern these effects. More properly, the stability of water is assessed from a saturation index, usually that of Langelier:

$$J = \mathrm{pH} - \mathrm{pH}_{\mathrm{sat}} \tag{5.1}$$

where pH and $\mathrm{pH}_{\mathrm{sat}}$ are indices relating, respectively, to the pH value of the actual water and to that of water with the same composition but containing the equilibrium concentration of carbonic acid compounds. If $\mathrm{pH} < \mathrm{pH}_{\mathrm{sat}}$, that is, if the concentration of free CO_2 in the water is above the equilibrium value, then a carbonate film will not be deposited on the walls of pipes or equipment. If $pH > pH_{sat}$, calcium carbonate will be precipitated from solution and can reduce, or completely eliminate, corrosion. The object of a water stabilizing treatment is, therefore, the establishment of a layer of $CaCO_3$ on pipe walls by the introduction into the water of sufficient alkali to maintain $J = 0.5–0.7$ (any stronger alkalization could lead to turbidity of the water). The dosage of NaOH, in g kg^{-1} required at $pH < 7.7$ can be calculated from the formula:

$$A = \frac{4\,\mathrm{Alk_O}\,[CO_2](50-J)}{2200\,\mathrm{Alk_O} + [CO_2]} \tag{5.2}$$

in which $\mathrm{Alk_O}$ and $[CO_2]$ are the alkalinity of the original water in mg $equiv^{-1}$ and the content of free CO_2, respectively. The duration of the stabilization treatment should not be too long, otherwise a thick layer of $CaCO_3$ will be formed. Therefore, after the protective film has formed it is recommended that J be reduced almost to zero.

However, the Langelier index assesses purely chemical aspects whereas the nature of corrosion processes is electrochemical. The properties of the carbonate deposit, e.g., its density, strength, adhesion to the surface, and defect nature depend on the pH and composition of the near-electrode layer, the concentration of the cathodic depolarizer (oxygen), and on the nature and concentration of those anions capable of taking part in the anodic reaction. It is known[224] that even with $J > 0$, porous and friable deposits with weak protective properties form in the presence of 100 mg L^{-1} sulfate ions. The buffer capacity and flow rate of the water and the presence of cations of other metals (including magnesium), also have a marked effect on the quality of the carbonate film. The nature of the metal to be protected is also extremely important. Thus, electron microscopy has established that the rate of formation of $CaCO_3$ on a heat exchanger surface at (54−56° C) falls in the order: low carbon steel → aluminum–copper → graphite.[225] These differences are associated with the corrosion stability of metals and with the nature of the surface oxides.

Considering this, it is clearly necessary to take into account the concentration of other components of the solution, as well as that of calcium. It has been proposed that the content in waters of sulfate, SiO_2, and dissolved oxygen need to be considered, but such an approach is not without its own problems, chiefly because of the lack of information about the factors controlling corrosion. Furthermore, a low corrosivity of water may be accompanied by enhanced deposit formation, which is undesirable in the case of heat-exchange equipment. The concentration of SiO_2—a natural inhibitor of corrosion—is important; a flowing water system in which 15–20 mg L^{-1} SiO_2 are present may not require any further protection, particularly if the

SiO_2/Na_2O ratio is high. The protection of steel can then be maintained over long periods of use of the equipment, even if the concentration of silicates falls to 4–8 mgL^{-1}. Disadvantages of silicates are their tendency to form difficultly-soluble deposits in the presence of calcium and iron ions, and their high sensitivity to the depassivating action of aggressive ions, especially sulfates.

Polyphosphates also do not fully meet the requirements of inhibitors for once-through systems and attempts to find an organic inhibitor that would be more acceptable for such cases have not so far met with success. The problem lies not so much in meeting satisfactorily the ecological requirements (there are many organic compounds of low toxicity that are quite suitable for use in cooling systems) but rather in the need for an economical treatment. For example, the low toxicity of gluconates and other similar compounds obtained from plant sources would appear to offer promise, even if not used in the pure form, but as a synergistic component of an inhibitor formulation. However, this would require the development of cheaper methods of synthesis. The possibility of progress in this direction has been demonstrated by a report on the effective suppression of corrosion in water at 90° C by small additions of an inhibitor obtained by the interaction of zinc oxide with biochemically-oxidized glucose.[226]

Other organic inhibitors meriting attention include the lignosulfonates, which are sulpho-derivatives of a natural polymer, lignin, and the product from the sulfiting of cellulose. These are of low toxicity (maximum allowable concentration in drinking water in Russia = 5 mgL^{-1}), have good dispersant properties, are relatively cheap, and readily available. According to Midzumoto *et al.*,[227] the zinc salt of a lignosulfonate is comparable in its protective action with the more toxic chromate–polyphosphate formulation, which is probably the most effective inorganic inhibitor. Prospects for the use of lignosulfonates are, of course, associated with their use as the basis of synergistic formulations. The patent literature has much information on the effective increase in protection of steel by the combined use of lignosulfonates, not only with Zn^{2+} salts but also with mercaptobenzothiazole, dibutylthiourea, nitrite, polyphosphates, hydroxycarboxylates, tannins, etc. However, it should be admitted that the problem of finding a suitable inhibitor for once-through cooling water systems is still far from being solved.

In open recirculating systems cooled by evaporation, CO_2 can be lost and the water will become alkaline. Furthermore, evaporation (in the cooling tower or in the pond) will increase the concentration of low-volatile impurities. All this will facilitate the formation of $CaCO_3$, but its deposition can be retarded by the establishment of supersaturated solutions. Such supersaturation can be stabilized by the presence in the water of certain organic substances and polyphosphates which are adsorbed on the faces of crystal nuclei, so inhibiting their further growth. Depending on the nature and concentration of such substances, the degree of supersaturation of the water can vary over a wide range but with continued evaporation it will, sooner or later, become unstable and the carbonate deposits will then adversely affect the heat exchange, as well as increase the hydraulic resistance in the system.

To avoid the formation of deposits on cooling surfaces, part of the water is rejected (blowdown) and the loss compensated by "make-up" water containing lower concentrations of Ca^{2+} and CO_3^{2-}. Clearly, this is a compromise and cannot solve all the economical and ecological problems involved. With these in mind, developments have been made to aid in curtailing the blowdown by a water treatment that would allow an increase in the concentration factor of the water. This could be achieved by breaking down the bicarbonate by acidification with sulfuric acid to a pH of 6.0–6.5:

$$2HCO_3^- + H_2SO_4 = 2CO_2 + 2H_2O + SO_4^{2-} \tag{5.3}$$

The drawback of this method is that HCO_3^- is replaced by the corrosion active SO_4^{2-} which, at concentrations above 500 mg L^{-1}, can increase the attack of concrete, as in cooling towers, and lead to deposition of calcium sulfate. The use in this case of polyphosphates (especially in combination with silicates) is preferred, since this approach is often capable of preventing intensive corrosion of steel. However, such treatment is effective only with $[P_2O_5]:[Ca^{2+}] \approx 2:1$, and with low concentrations of oxygen and low flow rates. Prolonged use and higher temperatures encourage the hydrolysis of polyphosphates, which results in a reduction in their effectiveness and the formation of slimes. The conversion of polyphosphates into orthophosphates occurs even more readily in waters of higher pH values.

In the systems under consideration the deposits can be formed not only as a result of scale formation, corrosion, and phosphate slimes, but also by dusts and other substances falling into the water. These latter establish conditions that are favorable for the development of bacteria which can often lead to the formation of a gelatinous mass (slime), particularly in poorly refreshed areas. The access of oxygen is impeded by this slime layer beneath which anaerobic bacteria can proliferate and develop H_2S, which in turn can lead to significant pitting corrosion. Biological growths are formed by colonies of various micro-organisms and algae. For example, growths in the condensers of power station turbines begin with zoogenic bacteria, then filamentous and iron bacteria appear, and finally microscopic fungi and diatom algae.[228] In water conduits a large part of the biogrowths consist of mussels. Chlorination of water remains, at this time, the main method of combating the bio-contamination of cooling systems. Chlorine in water forms the hypochlorite anion, which causes a breakdown of the internal structure in living organisms leading to their destruction. A concentration of ~0.2 mgL^{-1} Cl_2 is sufficient for the destruction of the majority of bacteria and algae in clean water. However, it must be remembered that chlorine is a strong oxidizing agent, and can be rapidly consumed by secondary reactions with impurities in the water. Chlorination is therefore usually conducted on a periodic basis (1 or 2 times a day for not more than 1–2 h) but at a higher dose rate of 0.5–1.0 mg L^{-1}.

However, it should be recognized that living organisms often adapt to chlorine, that the oxidizing properties can lead to certain technical difficulties, and that despite the relative simplicity of chlorinating equipment, its use requires strict safety regulations.

In combating algae and particularly mussels, dosing of the water with copper sulfate is more effective. However, the maximum allowable concentration of copper ions is very low and their assimilation by micro-organisms occurs slowly, which unlike in the case of correctly dosed chlorination, presents an additional ecological problem. Furthermore, copper ions can cause pitting of steel and especially, aluminum alloys. For the successful operation of open cooling systems it often becomes necessary to use biocides as well as the chemicals used for controlling pH and for inhibiting corrosion and scaling. Treatment with biocides is more effective when dispersants are used. These are usually surface-active compounds, therefore it is necessary to guard against foaming, which if significant, must be countered by anti-foaming chemicals. Consequently, in selecting corrosion inhibitors it is necessary to take into account their interaction with other chemicals. Therefore, traditional inhibitors of the oxidizing type, such as chromates and nitrites, should be used with caution, not only because of ecological considerations, but also because of their high chemical reactivity. There are, however, a large number of organic compounds that can be described as multifunctional, that is, having anti-scaling or biocidal as well as inhibiting properties. When it is considered that many organic compounds do not form slimes and do not stimulate growth of algae (unlike phosphates), that they are non-toxic (especially when compared with chromates), and equally effective for different metals and against different forms of attack (unlike silicates and nitrites) then the interest in them, which has grown particularly in recent years, becomes understandable.

Finally, we turn our attention to inhibitors possessing polyfunctional properties. Thus, amine-type inhibitors, which are often used for corrosion protection of heat-exchanger equipment, are less suitable for circulating cooling systems, where inhibition of scaling is also required. Polyalkylene-polyamines or aminoalcohols, particularly triethanolamine, due to their complex-forming properties, can also be used for inhibiting deposit formation,[148] but they are of little effectiveness when used alone. Quaternary ammonium salts are more often used as dispersants, biocides, and flocculants, than as corrosion inhibitors in water. It is also known[229] that some of them [e.g., p-alkylbenzylpyridinium chloride] possess, when present with potassium thiosulfate, quite good protective properties in moving (2-3 ms^{-1}) chloride-containing water. The synergism of protection of steel in a circulating water with calcium hardness of ~10 mg equiv L^{-1} with a nitrogen-containing surfactant and 8.8 mg L^{-1} $ZnSO_4 \cdot 7H_2O$ has been reported.[230]

Those nitrogen-containing corrosion inhibitors that contain an acid group are more promising since such groups undoubtedly improve both the complex-forming and adsorption capabilities of the inhibitor. In the 1960s, good protective properties were found for N-alkylsarcosine and other amino acids, particularly in presence of sodium nitrite. Weisstuch and Lange[196] explained this as due to a combination of high surface activity and the ability for chemisorption, arising from the formation by the functional groups of 5 and 6 member chelating rings. Some compounds having such properties, such as iminodipropionic acid, the principal acid from tall oil, exhibit protective properties in cooling systems at concentrations even below the CMC (critical micelle concentration), < 10 mgL^{-1}. In studying similar compounds, which he called surface-active chelating reagents, Zecher[231] found that in industrial cooling waters N-lauroylsarcosine and tetradecan-β-aminopropionic acid provided a high ($> 90\%$) degree of protection even without the addition of zinc ions. It is important to note that amino acids are usually of low toxicity and that their solubility can be increased by oxyalkylation which, according to a Japanese patent,[232] does not weaken their good protective properties.

In recent years, polymers, polyacrylamides, and others, have been used with the same objective. For example, in a water containing Ca^{2+} 88, Mg^{2+} 24, HCO_3^- 40, Cl^- 70, SO_4^{2-} 328 mg L^{-1} (pH = 7.0), the rate of corrosion of steel was lowered from 200 to 2 mg dm^{-2} d^{-1} by addition of a zinc salt and 50 mg L^{-1} of a polymer of the general formula

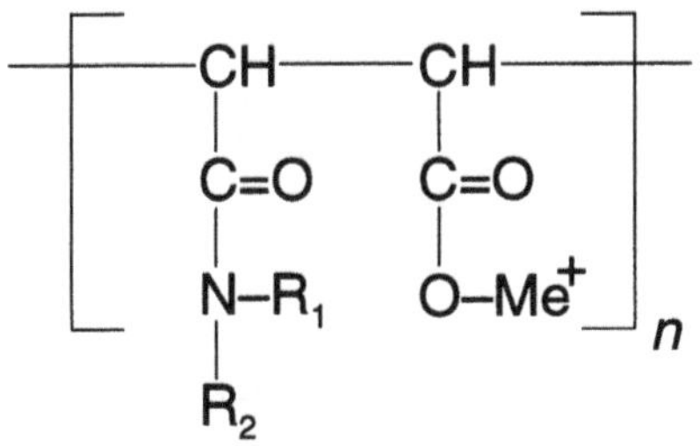

$$\left[-CH(C{=}O{-}NR_1R_2)-CH(C{=}O{-}O^{-}Me^{+})- \right]_n$$

where Me is a cation-hydrogen, alkali metal, or quaternary alkylammonium and $n = 2–100$.[233]

It may be suggested that not only amino acids, but also the products of their condensation with natural carbohydrates, for example, glucose–melanoidin, will also be low-toxicity compounds. According to Kanioternyki *et al.*,[229] melanoidin cystine is effective not only as an anti-scaling agent but also as a corrosion inhibitor giving 98.5% protection at a concentration of 100 mg L^{-1} in a flowing cooling water at 40° C.

Some amino acids and their derivatives have biocidal, as well as corrosion inhibiting properties. For example, N-acetylglycinelaurylamide at 50 mg L^{-1} completely prevented the corrosion of steel for two months, and also prevented the development of micro-organisms in a cooling system in which, in the absence of the inhibitor, corrosion occurred within one week and micro-organisms developed within three weeks.[234]

These advantages of amino acids or amido acids and their simple derivatives suggest that from these it will be possible to find a universal means of combating corrosion, scaling, and biological growths in a cooling system. However, the success in such a search will depend very much on the availability and the cost of synthesis of these inhibitors—factors which, so far, have fallen short of industrial requirements. The polycarboxylic amino acids, i.e., the classical complexing agents such as EDTA, and NTA are more readily available. These have often been used in various formulations (with Zn^{2+}, silicates, etc.) because of the good anti-scaling properties of these complexing agents. These formulations have found wide use not only in cooling systems but also in thermal power plant and heat circuits.[205]

On the other hand, the phosphorus-containing analogs, notably HEDP, NTP, and DTPP [diethylenetriamine-N,N,N′,N″,N″-penta(methylene phosphonic) acid] are well known in practice not only as scaling inhibitors, but also as corrosion inhibitors.

The studies reported in the previous chapter have shown that the introduction of these complexing agents, in the majority of cases, leads to protection of metals only in hard waters, when the corrosion inhibiting properties are essentially the result of the calcium and magnesium complexes that are formed. In practice it has been convincingly established that the protection of cooling systems can be more effectively achieved by treatment with zinc phosphonates, although in the patent literature there are many recommendations for the use of phosphonates with Mn^{2+}, Ni^{2+}, Cd^{2+}, Cr^{3+}, Pb^{4+} cations, and with the lower alkylamines.[235,236] Important factors in the use of the zinc-HEDP complex, are that zinc-HEDP is several times less toxic than the HEDP itself (maximum allowable concentration = 5 mg L^{-1}) and that the zinc salt is cheaper than the HEDP. Usually the weight ratio of $ZnSO_4$ and HEDP is close to 1, and the total concentration of the inhibitor is 5–25 mgL^{-1}. According to Burlov *et al.*,[237] in the circulating waters of petroleum refining plant with a pH of 6.5–8.0, the optimum protection of 90–93% for steel was obtained with the use of 5 mg L^{-1} HEDP and 7.5 mg L^{-1} Zn^{2+}, at which concentrations the inhibitor was non-toxic to the microflora of an active sludge in equipment for the biocleaning of water. This procedure allowed part of the circulating water to be dumped and replaced by clean water, i.e., to operate blowdown of the cooling system.

In cooling waters, the synergism of protection by phosphonates and Zn^{2+} ions is shown most clearly in relation to steel, although the protection of aluminum or copper alloys is also improved by these formulations. For the more effective protection of copper and copper alloys, additions of inhibitors of the azole class are usually recommended, benzotriazole(BTA), captax (MBT, mercaptobenzthiazole), benzimidazole, etc. Increased protection of copper alloys by BTA in the presence of zinc phosphonates has been reported.[238] The patent literature often has claims for the use of ZnHEDP or ZnNTP with gluconates, fluorosilicates, N-laurylsarcosine, benzoic and other carboxylic acids, or their salts. Attempts have also been made to lower the concentration of chromate by using it in combination with HEDP or ZnHEDP. However, more attention is now being given to compositions based on the less toxic—although more expensive—molybdates. According to Robitaile,[239] molybdate–phosphonate inhibitors also show synergistic action; they are of low toxicity and calcium molybdate is not precipitated in hard water (pH 8.5), even in the presence of 100 mg L^{-1} Na_2MoO_4. At the same time, molybdate has practically no oxidizing properties in neutral solutions and can therefore be used with many organic compounds. This can lower the cost of inhibiting formulations. An example of effective protection (98.4%) of steel in a hard water at pH 8.5 at 48° C has been given[240]; the formulation contains Na_2MoO_4 5, HEDP 3, MBT and $ZnSO_4$ 0.7 mg L^{-1}.

Phosphate–phosphonate mixtures would be expected to occupy a special place in inhibitive formulations since the mechanisms of action of the components have much in common. Thomas and Breske[241] studied about twenty low-toxicity inhibitors, i.e., without Zn^{2+} or Cr^{6+}, and selected a formulation containing polyphosphate with HEDP and BTA. In three months of operation of a pilot plant, steel corroded at an acceptable rate (0.05 mm/year) in the presence of this inhibitor, but the filters became clogged with deposits consisting mainly of iron phosphate. Algae growths were also found on the caps of separating columns which led to the need to double the NaOCl dose. The authors of this report found that such difficulties could be fully overcome and this inhibitor system was recommended for the protection of the cooling system of the plant. This same report also describes the successful use of a multicomponent inhibitor containing HEDP, BTA, hexametaphosphate, and aminomethylenephosphonic acid (AMP). In tests in waters containing Cl^- at 60–350 mg L^{-1} and hardness salts at 75–250 mg L^{-1}, the system was treated for 24 h with 200–400 mg L^{-1} of the reagents before lowering the concentration by 4–8 times. Having obtained satisfactory results, particularly in hard water, the authors then treated an industrial circulating system with this formulation with the further addition of 30 mg L^{-1} of a polyacrylate dispersant to enhance the anti-scaling properties. Unfortunately, the report of that work does not give the ratio of the components, but this can be approximately assessed from the patent literature. Thus, in one USA patent[235] results are given from tests with 9 similar formulations, of which the most effective in a hard and aggressive water was an inhibitor consisting of hexametaphosphate:NTP:polyacrylic acid in the ratio 20:8:1.7 parts in water. At a total concentration of 25 mg L^{-1} (calculated as the solid material) the degree of protection of steel in a heat exchanger at 52 ± 0.5° C reached 99%, which in presence of 2 mg L^{-1} H_2S or 75 mg L^{-1} hydrocarbon pollutant remained effectively unchanged (97–98%). In another patent[242] there is a recommendation to add 1 part of hydrolyzed polymaleic anhydride to a mixture of hexametaphosphate and NTP or HEDP (15:4); this additive allows the formulation to be used in alkaline waters (pH 8.0–9.0). In the absence of polyphosphates the ratio of the components is altered, for example, according to the patent,[242] NTP:polymaleic anhydride = 3:2 with an overall concentration of inhibitor of 25–50 mg L^{-1}. It should be noted that such "phosphate-free" and "zinc-free" inhibitors are relatively expensive and therefore are sometimes supplemented by other low toxicity inhibitors, such as lignosulfonates.[243]

There remains a real problem in reducing the cost of water treatments that use phosphonates without lowering their effectiveness and ecological acceptability. The theory and practice of cooling systems show that the introduction of Zn^{2+} within reasonable limits is beneficial not only for the corrosion protection and stability of the inhibitor, but also for ecological safety, since in forming the protective layer the Zn^{2+} is consumed to a greater extent than is the phosphonate. Furthermore, the maximum allowable concentration of zinc phosphonates in water is noticeably higher than that of the actual phosphonic acids or phosphates.

On this basis, a series of corrosion and deposit formation inhibitors based on ZnHEDP has been developed in our institute for use in circulating systems. Of these, the IFKhAN-R formulation has found the widest practical use at the present time. This inhibitor, as seen in the results of rig tests at the Redkinsky test plant (Table 5.1) is as effective as ZnHEDP in protecting steel, even when used at lower concentrations, and is also a more stable and economical inhibitor. This has allowed a complex program

Table 5.1. Results of Rig Tests of Inhibitor in the Cooling Water of the Redkinsky Test Plant

		Corrosion rate of a 0.2% C Steel ($g\ m^{-2}d^{-1}$)							
	Concentration	293 K flow rate (ms^{-1})				323 K flow rate (ms^{-1})			
Inhibitor	($mg\ L^{-1}$)	0.5	1.0	1.5	2.0	0.5	1.0	1.5	2.0
ZnHEDP	0	5.17	7.69	10.13	10.84	7.10	9.63	14.0	14.6
	50	0.019	0.016	0.028	0.031	0.43	0.15	0.40	0.38
IFKhAN-R	30	0.026	0.012	0.030	0.033	0.26	0.09	0.43	0.11
	50	0.011	0.008	0.011	0.017	0.10	0.13	0.07	0.17

to be developed for water treatment of an ammonia production cooling system. This was first introduced in 1981 at the Azot plant in the town of Ionava, and then later in several other undertakings.

In the Azot plant, the total volume of the water in the circulation cycle is 1000 m^3, with a usage of 3400 m^3h^{-1}. The feed water and circulating waters contain, respectively, in mg L^{-1}: Cl^-, 10.5, 17.8; SO_4^{2-}, 48.4, 93.2; NH_3, 0, 14.8, NO_3^-, 10.1, 76.0; Cu^{2+}, trace, 0.18; dry residue, 142.0, 515.6, and hardness 1.8, 4.5 mg equiv L^{-1}. Before the introduction of the inhibitor formulation, the blowdown was at a maximum and despite the feed water's high pH of 10.4 and supply of feed water (150 m^3h^{-1}), there were significant variations in pH around the value of 7, and in the NH_3 and NO_3^- concentrations, which led to highly corrosive conditions. These, in turn, resulted in the appearance of corrosion products on the steel pipework, and of oxide and hydroxide deposits (up to 86% Fe_2O_3, 4% Cu, and 25% organic material) which lowered the efficiency of heat-exchange equipment.

The water treatment was introduced in three stages. Biocidal treatment was conducted for two days by periodic dosing with Katamin AV (a quaternary ammonium salt similar to an alkyl benzyl dialkylammonium chloride) at 10 mg L^{-1} and then, to remove and break-up the surface layers of deposits, 10 mg L^{-1} IFKhAN-R was introduced, but without the inclusion of the zinc component. Finally, to passivate the system zinc sulfate was added and the concentration of the IFKhAN-R raised to 20 mg L^{-1}. This accelerated the development of the protective film on the steel surface. After the zinc concentration in the water reached 3 mg L^{-1} the dose was reduced.

When the system had reached the operating regime, with 12–15 mg L^{-1} IFKhAN-R, 1–3 mg L^{-1} mercaptobenzthiazole was introduced to improve the protection of brass, and 3–5mg L^{-1} Katamin AV added once a month. As a result of this treatment the corrosion rate of the steel—depending on the flow rate of the water—was reduced from 1.7 and 1.66 to 0.024 and 0.069 mm/year, and that of brass from 0.3 to 0.005 mm/year. The prevention of deposit formation stabilized the temperature drop in the heat exchange equipment, which led to increased productivity and significant economic effects.

IFKhAN-R has also been used in water with high hardness (> 10 mg equiv L^{-1}) and salt content (> 1.5 g L^{-1}). In this application it was superior to ZnHEDP in ecological and economical terms, although it did not always provide highly-effective corrosion protection when used at low concentrations.

The IFKhAN-36 formulation that we have developed also has low sensitivity to water hardness and to increases in concentrations of aggressive salts (Cl^-, SO_4^{2-}).[244] In a soft mineralized water (Fig. 5.1), complete suppression of corrosion of steel, rotating at 0.8 ms^{-1}, was achieved at the same concentration as that of ZnHEDP, but at lower inhibitor concentrations IFKhAN-36 was undoubtedly superior. In water containing 18 mg equiv L^{-1} $CaCl_2$, in which ZnHEDP becomes unstable (e.g., with a loss of 60% of the HEDP in one day), the IFKhAN-36 was noticeably more effective and for protection of steel required only 2 mg L^{-1}, an order of magnitude less than ZnHEDP even in brief test periods. The special additive which is contained in IFKhAN-36 provides a stabilizing action which allows it to be used in waters with higher levels of Ca^{2+}, and this inhibitor is thus more effective than ZnHEDP in hard water. Apparently, the calcium compounds participate directly in the formation on the steel of the protective film that is produced in the presence of IFKhAN-36. This was confirmed by comparative tests in a hard water containing a high concentration of HCO_3^- along with the calcium ions. In this case, without inhibitors, the amount of corrosion was small and of a localized character, with a calcium carbonate scale deposited on the steel surface. Only the introduction of IFKhAN-36 prevented simultaneous corrosion of the steel and precipitation of deposits. However, in this case it is not possible to establish the form in which calcium is present on the inhibited metal surface as it is difficult to distinguish between CaHEDP and $CaCO_3$ in the analysis of thin surface layers.

The superiority of IFKhAN-36 over ZnHEDP is even more noticeable at the elevated temperatures that occur when water passes through heat exchangers. Thus at 60° C, full protection of steel in a hard water [18 mg equiv L^{-1}] was not obtained, even at concentration of 100 mg L^{-1} of ZnHEDP, whereas only 20 mg of L^{-1} IFKhAN-36 was required to provide full protection. Important properties of IFKhAN-36 include the more rapid formation of the protective layer on steel—after which its concentration can be appreciably lowered—and the ability for effective protection of already corroding steel.

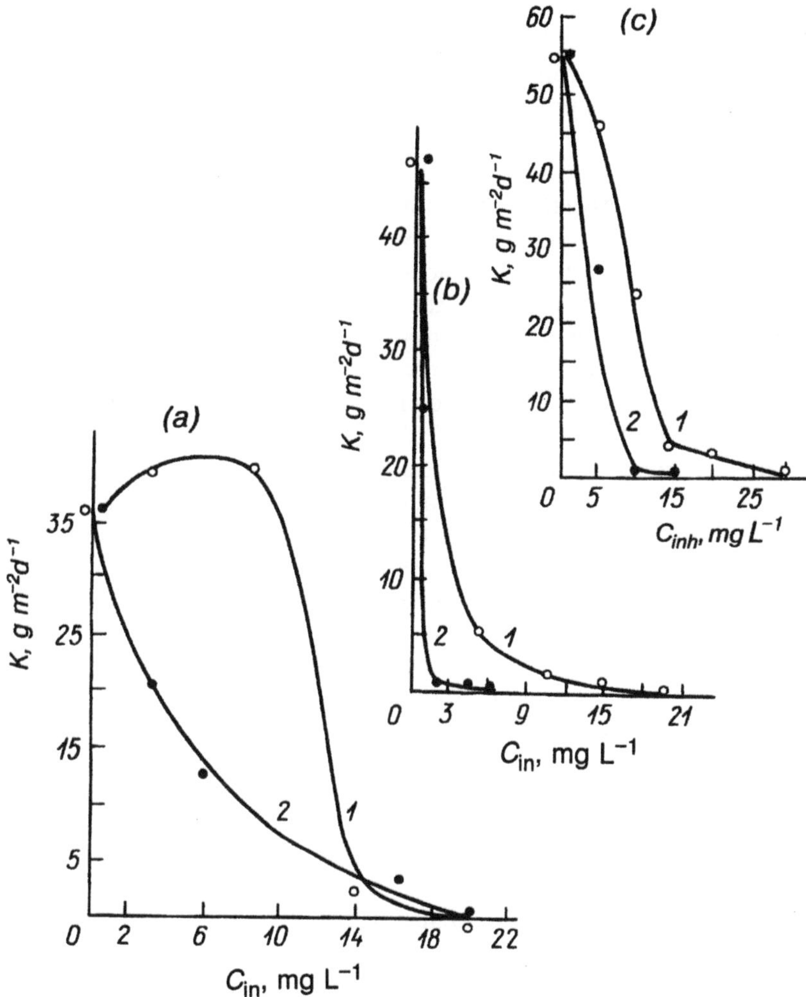

Figure 5.1. Dependence of the corrosion rate of steel on the concentration of ZnHEDP (1) and IFKhAN 36 (2) at specimen rotation rates of 0.8 ms^{-1} in water containing: (a) NaCl 150 + $NaSO_4$ 350 mg L^{-1}, (b) 18 mg equiv L^{-1} $CaCl_2$, (c) $CaCl_2$ 389 + NaCl 167 + $NaHCO_3$ 588 + Na_2SO_4 740 mg L^{-1}.

The examples of IFKhAN-R and IFKhAN-36 demonstrate the improvement in the stability of an inhibitor that can be achieved by the selection of a suitable synergist. Work in this direction is in progress in many laboratories and companies. The ecological problem has prompted a tendency to use complexing agents containing lower quantities of phosphorus—the carboxyphosphonates, of which 2-phosphono-butane-1,2,4-tri-carboxylic acid (PBTC) has been the most studied.[235,245] As with HEDP or NTP, this acid prevents deposit formation and can intensify its anti-corrosive action when present in a formulation containing zinc salts, ligno-sulfonates, alkanolamines, phosphates, nitrites, various azoles, and certain polymers. Marshall and Greaves[245, 246] report that zinc carboxyphospho-nates are more effective than the traditional zinc phosphonates, particularly in conditions of elevated pH (up to 9.5). Their anti-scaling properties are also associated with a threshold action, i.e., the disruption of the growth and adhesion of scale crystals to the metal surface, thus making possible a rapid flushing out of deposits from the system. It is of interest that a mixture of phosphonate and carboxyphosphonate can be still more effec-tive. Thus, if a carboxyphosphonate prevents carbonate deposition in recir-culating water up to a hardness of 10 mg equiv L^{-1}, then its mixture with phosphonate can guarantee this situation up to 20 mg equiv L^{-1}.[247] According to a U.S. patent,[248] a formulation consisting of (in mg L^{-1}) NTP 3, PBTC 4, and tolyltriazole 4 in a cooling water (pH 9.0) reduced the corrosion rates of steel and admiralty brass after 72 h from 2.8 and 0.015 to 0.06 and 0.005 mm/year, respectively. The combined use of PBTC, HEDP, and Zn^{2+} has been proposed in Japan[249], where more complex formulations containing PBTC or other polycarboxymono-phosphonates had been previously patented. However, PBTC is not included in the list of compounds for which a maximum permitted level in industrial water supplies of the (former) USSR has been established, and the extent to which other carboxyphosphonates will meet the ecological requirements is still not clear. This applies also to a series of other phosphonates, in particu-lar those put forward by Keller and co-workers[250] and reported to be more effective than HEDP or NTP.

The use of various phosphonates with polymeric dispersants, for ex-ample, of the carboxymethylcellulose type,[251] polycarboxylates, or other water-soluble non-toxic polymers[235] is undoubtedly one of the most promis-ing and realistic directions for improving the ecology and economy of water treatment in recirculating cooling systems.

Attention should also be paid to attempts to avoid completely the use of phosphorus-containing compounds—including phosphonates. In view of this, an absence of nitrogen in the composition of inhibitors would avoid the breakdown of the ecological balance in water storage systems and is therefore of special interest. The usual carboxylates (salts of substituted benzoic or acetic acids) can form relatively insoluble salts (soaps) when present in hard waters at concentrations that are high enough for effective inhibition. For this reason, they will be effective only in soft recirculating waters.

It should not be forgotten that the maximum permitted concentration of these compounds is usually below the concentrations that are necessary for obtaining high levels of inhibition. The relatively high cost of these inhibitors is no less a problem. Thus, at present, the possible uses of some of these compounds remain unclear and must await the development of synergistic formulations, examples of which have been given in Chapter 3.5.

Non-toxic and relatively easily available polyhydroxycarboxylates (e.g., gluconates, citrates) usually show their best inhibiting properties when used in formulations with other inhibitors.[148] The presence of a large number of functional groups in these prompts the suggestion, as in the case of tannins, that the protection in many cases is associated with the complex-forming properties of the organic compounds. Thus, synergism of protection by gluconates is often observed with the same compounds that give synergism with complex-forming agents, with Zn^{2+} and Ca^{2+} ions, silicates, phosphates, etc.

The mechanism of action of other polyfunctional compounds, the water soluble polymers and oligomers, is less clear. They are also usually recommended for use in formulations. For example, the polymer based on maleic anhydride with a molecular weight of 300–5000, as already mentioned, is not only used with phosphorus-containing inhibitors. According to a Japanese patent[252] the addition of such a polymer (referred to as Berklin 200) at a concentration of 5 mg L^{-1} with $ZnSO_4$ and lignosulfonate to water (with carbonate hardness 0.5 mg equiv L^{-1}, pH 7.1 and 20 mg L^{-1} Cl^- flowing at 0.5 ms^{-1} at 50° C) lowered the corrosion rate of steel from 8.7 to 2.8 mg d m^{-2} d^{-1}. Reports have also referred to the effective protection of steel by a formulation of a copolymer of a monoallyl ester of polyethyleneglycol with a saturated carboxylic acid and a zinc salt.[253] However, references in the patent literature to the satisfactory protective properties of these polymers when used alone, for example the copolymer of maleic anhydride with isobutylene, also merit attention.[254] Some polymers, e.g. carboxymethylcellulose, have often been used in recent years for reducing the resistance to flow of liquids in a pipeline. A secondary effect is that the reduction in the rate of mass transfer between the pipe wall and the turbulent liquid flow[255] allows these materials to be used as corrosion inhibitors.

In comparing three types of polymers for their inhibiting and adsorptive properties Sedhamed *et al.*[256] found that these properties increased in the series polyacrylamide → carboxymethylcellulose → polyhydroxyethylene. However, when studies were made in our laboratories of these and other water-soluble oligomers and polymers of various chemical compositions it was found that their protective action towards steel was, as a rule, shown only in a narrow concentration range, above which stimulation of corrosion could occur. It is obviously necessary to conduct further studies to reveal the mechanism of action of these high molecular weight water-soluble compounds.

Finally, the need to exploit various routes for the water treatment of cooling systems has to be recognized, since their variety in terms of water composition and ecological conditions emphasizes the requirement for a wide range of inhibitors of corrosion and scaling with dispersants and biocides.

5.2. ASPECTS OF THE USE OF INHIBITORS IN BRINES

The problem of the inhibition of corrosion of metals in waters with a high salt content is extremely extensive and concerns not only those systems that use recirculating water. For example, the problem requires solution in the protection of equipment used in the extraction, storage, and transport of oil,[152,257] in geothermal systems, in seawater ballasting of ships, and in various types of aqueous antifreeze. In this last example, the problem is aggravated by a number of factors. First, one of the principal construction materials used in cooling equipment, and which has adequate corrosion resistance, i.e. copper, is becoming relatively scarce. Secondly, the use of the traditional inhibitor for protection of steel or aluminum alloys in salt solutions—chromate—is undesirable from the ecological point of view. Thirdly, the development of the northern geographical regions and the combating of icing of airports, motorways, etc., places new emphasis on the use of concentrated salt solutions which will require the use of non-toxic and effective corrosion inhibitors. Finally, the ecological requirements for the composition of antifreezes or de-icing fluids do nothing to reduce the economic aspects of the problem, which remain substantial despite the considerable amount of research directed towards its solution.

The vast majority of the solutions that are used as freezing point depressants are based on sodium chloride and calcium chloride, which are most aggressive to steel at a chloride concentration of ~2.5%.[21] Concentrated $CaCl_2$ brines (27–33%) present less of a corrosion hazard than the corresponding NaCl solutions, and are quite widely used in practice. Furthermore, the eutectic point, which specifies the concentration and temperature of the brine at which the composition of the liquid phase does not change on solidification, differs significantly for $CaCl_2$ and NaCl.[258] In the former case, the eutectic temperature is −55° C at a concentration of 29.9%, whereas for NaCl these values are −21.2° C and 23.1%. Not only the economics, but also the low-toxic properties and the high thermophysical properties of chloride solutions, as well as their incombustibility, lack of explosive hazards, and non-volatility make them preferred to other heat-carrying fluids.

In regard to corrosion, it is important to consider whether the system in which the brine is circulating is closed, i.e. with all components hermetically sealed, or open. In the first case, the maintenance of a pH of 9.5–10.0 and the use of an oxygen scavenger can be sufficient for the corrosion protection of low carbon steel, although the corrosion will not be completely suppressed.

Unfortunately, open brine systems are often encountered, and the saturation of these with atmospheric oxygen leads to the occurrence not only of corrosion, including pitting, of low carbon steel, but also to the corrosion cracking of stainless steels, for example 12Kh18N10T (18:10 Cr–Ni, Ti-stabilized). With chromates not being acceptable, the search has intensified for an effective corrosion inhibitor for steel among the phosphorus-containing compounds. Inorganic phosphates, and notably hexametaphosphate, have been in use for a long time. In the 1940s it was reported that 16 mg L^{-1} hexametaphosphate reduced the corrosion of steel in a $CaCl_2$ brine by 15–20 times.(23) Later, phosphonic acids were studied for this purpose, although Duprat *et al.*,(259) having investigated salts of oleylaminopropylenamine and NTP as inhibitors of steel corrosion in 3% NaCl, pointed to the dangers of localized corrosion despite the ability of the inhibitor to retard both electrode reactions. It has also been shown(260) that even at a concentration of 100 mg L^{-1}, in a brine containing NaCl 42, $CaCl_2$ 14, Na_2SO_4 1 and $MgCl_2$ 17 g L^{-1}, some phosphonates could provide a relatively high (88–94%) protection of steel, even with H_2S saturation of the brine and a temperature of 90–95° C.

In our own laboratory, we have established that phosphonates can be used successfully in brines intended for refrigeration plants that have components in low-carbon steel or the aluminum alloys AMg–5 (Mg 5%) or D16 (Cu 3.8/4.9, Mn 0.3/0.9, Mg 1.2/1.8%).(210) In this work we took into consideration that inhibitors of the adsorption type would not be effective in brines because of the high concentration of activating anions [Cl^-] and the possible formation of insoluble compounds between sufficiently hydrophobic anions and Ca^{2+}. Therefore, comparisons were only made of the protective properties of compounds with high-complexing capabilities: NTP, HEDP, glyphosate, and lignosulfonates.

Because of the low content of oxygen in unstirred brine (Fig. 5.2), the corrosion rate of steel and aluminum alloys was relatively low. The introduction of inhibitors, of which the most effective was NTP, provided good protection of the steel, although the corrosion rate of the aluminum alloy remained significant. As has already been pointed out, the phospho-

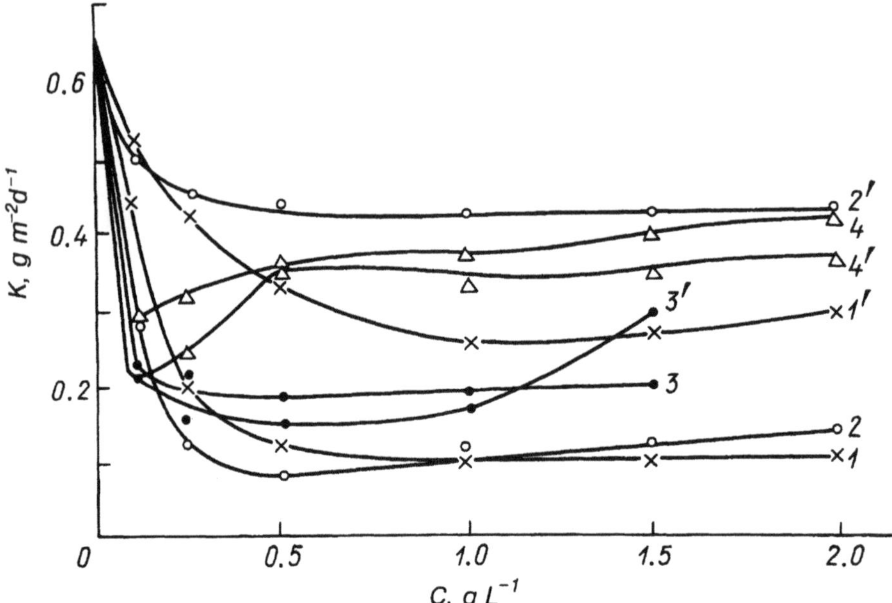

Figure 5.2. Dependence of corrosion rate of steel St3 [C 0.2%] (1–4) and of the aluminum alloy D16 (1′–4′) in 27% $CaCl_2$ on the concentration of NTP (1,1′), HEDP (2,2′), lignosulfonates (3,3′) and glyphosate (4,4′).

nates were less effective than their complexes, or compositions, with Zn^{2+} or Sn^{2+}. Thus, even in stirred brine with excess air (Table 5.2), in which the rate of corrosion of the aluminum alloy reached 28 $gm^{-2}d^{-1}$, these complexes could reduce the corrosion rate by 20–25 times. Of particular effectiveness was the inhibitor IFKhAN-36 (see section 5.1), which effectively fully suppressed the corrosion of the alloy by the formation of a light matte film on its surface. This inhibitor could be used even at low concentrations. Thus, at 25 mg L^{-1} it provided 94% protection in 33% $CaCl_2$.

Another possibility for the effective inhibition of corrosion in brines is the use of sucrose derivatives. According to Sukhotin and Borshchevskii,[261] the introduction of 0.6–1.0% of calcium saccharate into a brine could reduce the corrosion of steel by two orders of magnitude. Proposals have been made for inhibitors for brines consisting of a complex ester of sucrose and myristic acid[262] or a composition of di- or oligosaccharides (not less than 1 g L^{-1}) with sulfur-containing compounds (for example, thiocyanates, thioamides, thiourea).[263] The latter are often recommended as corrosion inhibitors for brines, but are most suitable for the oil extraction industry or for de-icing fluids. In the case of deicing fluids the ecological

Table 5.2. The Protection of the Aluminium Alloy AMg-5 in 33% $CaCl_2$ at a Flow Rate of 1 ms^{-1} in 30 h Tests with Inhibiting Complexes

Composition and inhibitor content (g L^{-1})	Corrosion rate ($gm^{-2}d^{-1}$)	Appearance of specimen surface
No inhibitor	27.96	Crumbling, black deposit
0.5 NTP+0.5 $ZnSO_4$	2.4	Matt film, minimal signs of corrosion
0.5 HEDP+0.5 $ZnSO_4$	1.0	Matt film, minimal signs of corrosion
0.5 NTP+0.35 $SnSO_4$	1.6	Black film, minimal signs of corrosion
0.5 HEDP+0.35 $SnSO_4$	1.5	Black film, minimal signs of corrosion
0.09 Glyphosate+0.08 $SnSO_4$	1.1	Matte film, minimal signs of corrosion
0.15 IFKhAN-36	—	No signs of corrosion

aspect is particularly important and, bearing in mind their low toxic properties, the possible use of lignosulfonates deserves attention. For example, according to a US patent[264] the chlorides of alkali or alkaline-earth metals can be mixed with lignosulfonate in the dry state or used as solutions at pH 5–8.

In considering de-icing compositions it should be noted that urea can not only lower the freezing point of water when present alone, but also show this property in formulations with other compounds (formamide, calcium chloride, and others). With the optimum ratio of these freezing point depressants and with phosphorus- or sulfur-containing corrosion inhibitors, it is possible to devise formulations which satisfactorily meet present day requirements.

5.3. AQUEOUS–ORGANIC ANTIFREEZE

Aqueous–organic solutions provide another example of antifreeze systems and are also often used as hydraulic fluids. The low corrosivity of ethylene glycol (ethanediol) towards most of the metals and alloys that are used in these applications, together with its satisfactory specific heat and thermal conductivity properties, which increase in aqueous solutions, make these solutions highly appropriate for use as antifreeze liquids, for example, in internal combustion engines. However, it was noted some time ago that oxidation of ethylene glycol will lead to the formation of organic acids (formic, acetic, etc.) and that buffering of the antifreeze (usually by addition of a borate) will be necessary to lower the resulting corrosive effects.[21] However, as shown in section 3.6, there is, even in buffered neutral solutions, a danger of breakdown of the passive state by anions of various carboxylic acids—including the lower members of the series. Furthermore, some anions cause pitting at the relatively positive potentials that may arise when steel and aluminum alloys are polarized to such values by contact with more electropositive metals or alloys (for example copper, brass, stainless steel) which often occur in cooling systems. The introduction into ethylene glycol–water solutions of sodium benzoate, particularly when combined with nitrite, can ensure the protection of cast iron, solder, copper, lead, and brass,[265] although this inhibitor provides only weak protection of aluminum and is usually used at relatively high concentrations. The development of more effective corrosion inhibitors for such systems will need to take into consideration those economic, and ecological problems that are associated with the use of internal combustion engines, compressors, solar energy conversion, waste heat, or hydraulic systems. Aqueous–organic solutions requiring the presence of corrosion inhibitors are also used in chemical production and absorption refrigerator systems.

The search for effective inhibitors for aqueous ethylene glycol antifreeze has been conducted both by investigating the role of the chemical structure of prospective organic compounds, for instance, arylcarboxylates,[188] and also by developing synergistic formulations from readily-available and low-toxicity inhibitors. In particular, and as already pointed out in 3.6, sodium benzoate in ethylene glycol–water [1:1] solutions, unlike aqueous solutions, is as effective as many of the substituted benzoates. It has also been established that among the ortho-arylcarboxylates, because of features of the mechanism of the transfer of the electronic influence of the substituent to the reaction center, inhibiting anions with multiple bonding groups are more effective in suppressing the corrosion of low-carbon steel in unbuffered aqueous ethylene glycol solutions of sodium sulfate [0.015M]. Such anions include those of acids with electron-donating substituents: toluate (CH_3), anthranilate (NH_2), phthalate, or terephthalate (COO^-). The minimum concentration of these necessary for complete suppression of the corrosion of steel is only 1.5–2 times less than in the case of benzoate. The use of substituted anthranilates is more effective if these also contain hydrophobic substituents. Thus, if the minimum concentration of benzoate required for protection is 25 mM, then for phenylanthranilate this is reduced to 5, for mefenaminate to 3, and difluorant to 2 mM.

Still more effective formulations based on these inhibitors are capable of preventing corrosion in aqueous–ethylene glycol antifreezes at concentrations that are an order of magnitude below that of sodium benzoate alone. An example is the use of sodium benzoate with benzotriazole, in the ratio 2:1–10:1 in a 50% ethylene glycol–water solution. This not only prevents the corrosion of copper but also, synergistically, that of cast iron, for which the corrosion rate is lowered 65–90 times at 70°C, as well as that of an aluminum alloy (Si 5.15, Cu 3.0, Fe 0.68, Zn 0.46, Mn 0.30%).[265]

Studies have also been made of cinnamic and of aliphatic dicarboxylic acids in such systems. Butler and Mercer[266] provide recommendations for the addition of 4.5% sodium sebacate and 0.15% benzotriazole to ethylene glycol. The authors point out that this inhibitor system also prevents scale formation on heat-exchange surfaces and stabilizes the pH of the antifreeze. Benzotriazole or its derivatives, and particularly tolyltriazole, are also used in other inhibiting formulations. Another effective inhibitor of copper corrosion, mercaptobenzothiazole (MBT), or its sodium salt, is also often recommended.[23,258,267,268]

Commercial antifreezes often contain other functional additives (foam dispersants, dyes, neutralizing agents etc.). In the former USSR, wide use is made of antifreezes based on aqueous ethylene glycol and designated TOSOL A-65 and A-40 (commercial products based on benzoates and azoles—with freezing points of −65 and −40° C). These antifreezes reduce the corrosion of copper, the L68 (Cu 68%) brass, grey cast iron, solder and aluminum alloy, and they are used in light- and heavy-duty engines. However, there are indications of inadequate protection of metals and alloys in the TOSOL A-40 antifreeze,[269] which is particularly associated with its deterioration during prolonged use of an engine. Toxicity of TOSOL antifreezes has also been noted.

These deficiencies have prompted attempts to develop antifreezes superior in properties to TOSOL. For example, Zhuravlev *et al.*,[270] provide a report of an ethylene glycol based antifreeze suitable for all types of nationally produced automobiles having liquid cooling systems—designated EOZh antifreeze. This contains a buffering additive, a dye, froth suppressant, and a corrosion-inhibiting formulation that is more effective than that in TOSOL, particularly with regard to the protection of aluminum alloys.

Despite the wide range of inhibited aqueous ethylene glycol antifreezes, there is a continuing need in many countries to find more effective corrosion inhibitors. For example, in analyzing the corrosion properties of two inhibited commercial aqueous ethylene glycol antifreezes, Davies and Prigmore[271] concluded that one, containing an organophosphate inhibitor, did not guarantee prevention of corrosion of solders of low tin content (Pb + Sn 2.5 + Ag 0.5%) and could also produce a phosphate-containing layer of low thermal conductivity on heat exchange surfaces. This could lead to local overheating, necessitating withdrawal of the equipment from service. The second, containing silicates, although better than the first in protecting solder, was rapidly denuded of inhibitor, which consequently weakened the protection of aluminum alloys. These and other deficiencies of commercial antifreezes have led to the introduction of more effective inhibitors—often of undisclosed compositions—that are claimed to be of universal action,[272] and to the broadening of the range of non-inflammable aqueous–organic liquids. Thus, increasing attention has been given to esters of ethylene glycol, to other glycols and glycerol since these show good compatibility with many metals and sealing materials and can also act as the basis of hydraulic fluids. However, it is still not clear whether success in this direction will mean a complete move away from the use of ethylene glycol in antifreezes to chemicals that are more effective at lower concentrations. The prospects for the latter approach have been demonstrated in our laboratory with the development of the IFKhAN-T antifreeze, which differs from TOSOL in having a lower viscosity even at low temperatures—3 cSt at 20°C and 100 cSt at −60°C—better specific heat, better thermal conductivity, and improved thermal stability. The use of highly effective organic corrosion inhibitors in this antifreeze allows, despite their low concentration (0.2%), protection of ferrous and non-ferrous metals and alloys over a wide temperature range. This antifreeze does not solidify, even at −75°C; it is non-flammable, non-explosive, and can be used for cooling not only internal combustion engines but other equipment as well.

6

The Protection of Metallic Components During Assembly, Transport, and Storage

6.1. ANTICORROSION COMPOSITIONS FOR DEGREASING AND CLEANING METAL SURFACES

In the manufacture or repair of various types of machines, mechanisms, or components and devices, there is usually a need for careful cleaning of the metal surface since contaminants may not only adversely affect the appearance of the item, but may also interfere with subsequent technical treatments. For example, it is not possible to deposit high-quality metallic or paint coatings on a surface that has not been degreased. Furthermore, the operations of soldering, cold-working, or finishing often require the removal of not only preservative oils and greases, but also oxide films. It is also often necessary to remove metal working compounds, polishing liquids or pastes, metallic dusts, abrasives, etc. from a metal surface.

All contamination can be conditionally divided into three levels.[273] A light contamination is usually found after storage between treatment processes. In this case, the surface can be covered by contaminants (dusts, grease) in quantities up to 1 gm^{-2}. Mechanical treatment of metallic components can leave a thin layer of grease or liquid emulsion coolant, as well as metal shavings. This degree of contamination, up to 5 gm^{-2}, is classified as intermediate. Greater contamination, described as heavy, most often arises after heat treatment of a metal, long-term storage, or with the use of preservative greases. Heavy pollution includes thick layers of corrosion products, the removal of which requires the use of pickling acids or cleaning pastes (as also used for carbon deposits). However, this section of Chapter

6 is limited to those compounds which remove contaminants that are lightly bound to the surface by deposition, and not as a result of oxidation or corrosion.

The compounds described here are used for removing greases of vegetable or animal origin, mineral oils, polishing pastes, abrasives, etc. For a long time, organic solvents had been foremost among these chemicals, but in recent years they have been generally replaced by the less toxic and less flammable aqueous- or emulsion-type compositions. Combustible solvents, that is, solvents that will ignite from a spark and which can burn after removal from the surface (e.g., benzene, white spirit, kerosene) have, apart from speed and simplicity of the degreasing action, a number of advantages. Thus, they do not corrode metals and alloys and they rapidly evaporate from cleaned surfaces. The majority of these solvents are easily flammable with flash points not exceeding 66° C in the open and 61° C in a closed cup. The use of these chemicals for degreasing is acceptable only below room temperature and with careful ventilation. These requirements represent serious problems not because of safety considerations, but also because the degreasing action cannot be intensified by raising the temperature.

In solvents which do not ignite easily—those which burn only in the presence of a flame—degreasing can sometimes be accomplished more rapidly and effectively. For example, the ability of trichlorethylene vapors to remove grease, tar, and oil is 40 times greater than that of benzene. Since the vapors of chlorinated hydrocarbons are significantly heavier than air, a layer of vapor will form above the liquid and can be of a sufficient thickness for the degreasing to occur not in the liquid, but in the vapor phase. As a result, the component remains dry. However, this is possible only with a light contamination of the surface, in other cases a prior treatment with liquid solvent will be necessary. Such treatment will be required for the high-quality cleaning of components. In vapor-phase cleaning there will be a build-up of contamination in the solvent, but the component is cleaned by effectively pure condensate which, as it forms, will remove contamination from the surface. In this situation, components at a low initial temperature and of a high mass are cleaned more effectively since the condensation on the metal surface takes place over a longer period.

The relatively short life of solvents of low flammability makes their periodic regeneration necessary. This does not present any great difficulty in the case of trichlorethylene, a fact which contributes to its wide use. Components made from any metal—except aluminum and magnesium and their alloys—can be degreased in trichlorethylene, although in the case of ferrous materials stabilization of the solvent with amines is usually required. In the presence of light, or in contact with hot objects, trichlorethylene can break down to form corrosive compounds. Another drawback of chlorinated solvents as degreasing media is their significant toxicity, which is often connected with a narcotic action of the organic vapors on the human body. It should be pointed out that in low flammable solvents in some conditions (powerful flame source, hot vapor-air mixtures), and especially in the presence of hot contaminants, an explosive situation can develop—this is the case particularly with chlorethylene and tetrachlorethylene.

These various deficiencies of organic solvents point to the need for aqueous degreasing compositions. The action of these depends on the fact that the saponification by alkalies of oils and fats lowers their adhesion, and hence of entrained mechanical contaminants, to the surfaces of metals. The intensity of the saponification of fats increases with increase in the alkalinity of the solution; however, alkalinity should not be too high since the soaps that are formed in concentrated alkali solutions are of low solubility. Depending on the form of the contamination and the cleaning technology, the alkalinity of the degreasing solutions vary within a wide range (pH 10–14). In most cases this can be achieved by a change in the concentration of the hydrolyzing alkali salt, i.e., calcined soda or potash.

To remove mineral oils from the surface where they are chemically resistant to the action of alkaline solutions, it is necessary to provide a high emulsifying capability to the degreasing medium. This property can be supplied by additions of anionic or non-ionic surfactants. For environmental protection, preference is given to those agents that can be biologically degraded in reservoirs or easily removed in the purification of ground waters.

Because of the high electrical conductivity of aqueous solutions, cleaning and degreasing can be carried out electrochemically; this is often more effective than other methods. The method is based on the fact that polarization of the metal reduces the adhesion of a grease film, and hydrogen bubbles evolved during cathodic polarization will further assist in the removal of contamination from the surface, thereby improving the degreasing action.

Electrochemical degreasing is widely used, particularly in the preparation of surfaces for galvanic coating, with the most effective cleaning being provided by direct current polarization. Alternating current of densities up to 7 Adm^{-2} is also used, especially in those cases when hydrogen charging of the metal cannot be accepted. However, it is necessary to guard against foaming, since excessive froth can prevent the escape of gases from the solution, which could then lead to the danger of explosion. Thus, the requirements for surfactants to be used in electrochemical degreasing include a low tendency to foaming and stability in the presence of the electrical current. For these reasons the introduction of soaps or salts of multivalent metals is not recommended.

In recent years there has been a noticeable increase in the use of ultrasonics for intensification of the degreasing of metallic surfaces. In an ultrasonic field there is a marked increase in the rate and quality of degreasing and, unlike the electrochemical method, it is possible to degrease components of complex profiles without particular difficulty. The action of ultrasonics is based on the fact that in the propagation of the sound wave, zones of alternating reduced and increased pressure form in the solution. As a result of the reduced hydrostatic pressure, the forces acting in the liquid become greater than the intermolecular interactions and the liquid structure is disrupted. This is accompanied by the formation of bubbles of gases already present in solution in the liquid. In the next period, when the hydrostatic pressure is high, the bubbles collapse, leading to the formation of an impact wave with a strong localized action. This phenomenon, known as cavitation, explains the speed and quality of degreasing of metallic surfaces in these conditions. This method also has a secondary effect in that the hydrodynamic forces will remove any emulsifying pollutants from the metal surface.

In the selection of the regime to be used in ultrasonic cleaning, it is necessary to remember that the effectiveness depends on the frequency of the vibrations. Thus, low frequencies bring about a strong cavitation with a low sound intensity; vibration of the components is achieved because long waves propagate more deeply into the materials, thus enhancing the cleaning action. On the other hand, a high frequency treatment allows energy to be focused. Usually, effective cleaning of a surface is achieved with a distance between the component and the source of sound of not more than 7–8 cm and an ultrasonic frequency of 20–25 kHz.

In the degreasing of aluminum or magnesium surfaces the ultrasonic treatment time should not exceed 30–60 s, in order to avoid erosion damage, with the optimum distance from the sound source equal to 7–10 cm.[273] Better results are obtained at temperatures of 80–85° C because the cavitation effect is increased as a result of the lower surface tension of the cleaning solution. However, increasing the temperatures up to the boiling point is not desirable since the vapor pressure in the cavitation bubbles increases and reduces their rate of collapse, which in turn lowers the effectiveness of the hydraulic impacts.

In using water as the solvent in ultrasonic treatments, it is necessary to take measures to protect metals from corrosion, particularly as many components of cleaning solutions are aggressive. Furthermore, after cleaning, the components may be stored for quite long periods in unfavorable conditions, where they may be subject to atmospheric corrosion. Measures that can be taken to cope with these problems include passivation in special solutions and wrapping in paper impregnated with volatile corrosion inhibitors. These procedures can be appreciably simplified if the cleaning solution itself has good protective properties. For this purpose, special inhibitors which can form stable protective films on the metal surface can be introduced.

The most promising inhibitors are those which form difficultly-soluble compounds with metal ions so that the metal surface is screened from the corrosive action of the cleaning solution. These inhibitors include salts of phosphoric and silicic acids, which are widely used in cleaning solutions, as well as various chelating reagents (benzotriazole, benzimidazole, mercaptobenzthiazole and others). The thickness of the films produced on the metal will vary from one to tens of nanometres and will depend on the composition of the medium and the nature of the metal. Inhibitors of this type are usually selective in their protective action and are seldom of equal effectiveness towards all the various metals that are likely to be encountered in cleaning processes. Thus, trisodium phosphate is quite effective as an inhibitor of the corrosion of ferrous metals, but can stimulate the corrosion of non-ferrous metals, particularly copper alloys. Benzotriazole or mercaptobenzthiazole are effective inhibitors of the corrosion of copper, but insufficiently reliable for protection of zinc or galvanized steel in cleaning conditions. Such selectivity of inhibitors follows from the mechanism of their protective action since the process of complex formation is sensitive both to pH and to the type of metal.

In aqueous solutions used for the degreasing and cleaning of metal surfaces, inhibitors are often present not only to prevent corrosion but also for other purposes. For example, trisodium phosphate is used as an alkalizing component for neutralizing acidic contaminants, for saponifying fats and oils, and for emulsifying pollutants (in which case additions of sodium silicate are often made). It is known that the emulsifying properties of inorganic anions increase in the series $HCO_3^- \rightarrow H_2PO_4^- \rightarrow CO_3^{2-} \rightarrow SiO_3^{2-} \rightarrow PO_4^{3-}$. The non-toxic properties and availability of phosphates lead to the wide use of detergent media based on these compounds. However, those detergents that do not contain surfactants should be considered first. These are widely used, but mostly in flowing processes or in electrochemical methods. The concentration of trisodium phosphate in these can reach 100 g L^{-1} since no special emulsifying agents are present. Thus, for the degreasing of components that are not heavily contaminated in a mechanized cleaning plant, use is made of a solution containing trisodium phosphate 80–100 and water glass 10–15 g L^{-1}. To remove light oils from aluminum alloys a cleaning solution containing trisodium phosphate 40–50, NaOH 5–10 and water glass 3–5 g L^{-1} can be used. In both cases, the temperature is held in the 50–70° C range and the treatment time is no more than five min.[273] Sometimes the concentration of the trisodium phosphate, which is aggressive towards copper and zinc, can be lowered from 100 to 15 g L^{-1} by adding Na_2CO_3 and NaOH at 15 g L^{-1}, and raising the temperature to 95° C.

Careful washing of metal surfaces after degreasing in these solutions is necessary to remove residues of the cleaning solution. This not only prevents subsequent corrosion, but also avoids the development of white spotting of galvanized surfaces after drying. However, washing with water significantly weakens the protective post-action of the inhibiting components of the solution, so in some industrial processes the components are again protected by a thin film of oil for storage between the various manufacturing treatments. In some cases, there is a recommendation to avoid washing with plain water and to use a dilute cleaning solution based on orthophosphates and silicates at a total concentration of 8 g L^{-1}.

Our own experiments have further shown that by adding inhibitors to the cleaning solution, it is possible to retain protection of metals during storage between manufacturing treatments after the final washing (even when this is done in flowing water) and drying. This can be achieved by the use of the (Russian) IFKhAN-1M formulation, which prevents the appearance of corrosion not only on steel and cast iron but also on copper, brass, and galvanized steel. This formulation consists of cheap, readily-available, and non-toxic compounds and can be used in machines with moving cleaning fluids, although best results are obtained in ultrasonic cleaning. The introduction of such a technology into a number of bearing factories has led to an increase in the level of cleanliness of ball and ring bearings following the mechanization and automation of the work, and has improved the hygiene and safety conditions of the plants. This general increase in production has been accompanied by significant improvement in economies.

In degreasing non-ferrous metals and alloys it is necessary to consider not only the corrosive action of phosphate ions but also the alkaline nature of the solution, aluminum alloys being particularly sensitive to alkalinity. Thus, Doufin *et al.*,[274] having studied the effect of alkaline solutions on the corrosion of four aluminum alloys, concluded that the introduction of corrosion inhibitors was a necessity. They proposed the use of calcium tartrate and citrate. However, this can be accompanied by the deposition (with time) of calcium carbonate.

The use of organic inhibitors to avoid this problem does not guarantee complete protection, although the corrosion rate can be significantly lowered by up to 70–75%. Best results have been obtained by addition of a mixture of silicate 4–5 and aluminum sulfate 0.5–0.6 g L^{-1} to the alkaline solution.

However, it is possible to reduce the corrosion of aluminum in alkaline solutions by small additions of organic inhibitors, in particular, zinc phosphonates. In many cases, organic inhibitors are found to be more effective than potassium dichromate, which was recently still being used in cleaning solutions as well as for the passivation of components after degreasing. Naphthenate should be mentioned among these organic inhibitors since it can suppress corrosion of aluminum and copper alloys in alkaline media. For the protection of copper alloys it is better to use inhibitors of the azole class, particularly mercaptobenzthiazole, with mixtures of their water-soluble derivatives with an inorganic thiosulfate to provide a synergistic action. The use of benzotriazole, aminobenzotriazole, and other inhibitors of copper corrosion of this type[(199)] in polyphosphate solutions has also been reported. These formulations sometimes provide good protection of other metals, for example, mercaptobenzthiazole for carbon steels and aluminum alloys.

The cathodic degreasing of surfaces of non-ferrous metals is less dangerous from the corrosion aspect because the component is electrochemically protected, its time of contact with the aggressive solution is significantly less than in chemical degreasing, and the cleaning solution can be used at lower concentrations. For example, copper and its alloys can be degreased at 50–70° C in a solution of Na_3PO_4 (25–30 g L^{-1}) and the same quantity of Na_2CO_3 in 1–2 min at a current density of 3–10 A d m^{-2}.[(273)] When degreasing zinc by this protocol, the sodium carbonate is fully replaced by trisodium phosphate. The degreasing of aluminum alloys, zinc, lead, magnesium, and cadmium can also be carried out by the electrochemical method in a solution of trisodium phosphate 30–40 + sodium carbonate 5–10 + water glass 3–5 g L^{-1} at temperatures, in the absence of a surfactant, of up to 80° C.

To avoid hydrogen charging of metals and the deposition, in some conditions, of contaminants as a result of their cataphoresis, anodic degreasing (in addition to the cathodic process) has also been considered. However, the danger of corrosion damage is increased with this approach. For example, black spots may appear on pressure cast zinc–aluminum–copper alloys as a result of the selective dissolution of the alloy. The addition of combinations of inhibitors to the alkaline solutions has therefore been proposed. In this approach, one inhibitor would increase the rate of the cathodic process (an oxidizer) and the other would broaden the range of passive potentials of the alloys.[(275)] Such a formulation can also be used for ultrasonic cleaning at a concentration of 45 g L^{-1}, instead of the 70 g L^{-1} used in the electrochemical method. After mechanical grinding, polishing, and a prior degreasing in trichlorethylene (for example), the components are treated in a solution containing an additive of composition: sodium carbonate 6–9, trisodium phosphate 6–9, water glass 6–9, monosodium phosphate 6–9, alkali 5–7, oxidizer 5–7 and an inhibitor of the amine type 55–57%, making up the 100% of the additive, excluding water as the solvent. The ultrasonic treatment is then conducted in this aqueous solution at 60–80° C for 1–4 min. Cathodic degreasing is conducted at the same temperature with a current density of 3–7 Adm^{-2} followed by an anodic current density of 1–4 Adm^{-2} for an electrochemical treatment time not exceeding 2–4 min. This treatment results in a transparent, compact, protective film which, if necessary, can be easily removed by hydrochloric acid.

Even with the flowing and the electrochemical methods of degreasing, when the formation of foam is undesirable, surfactants are often added to phosphate-based alkaline solutions. The effect of surfactants on the corrosion behavior of metals and alloys depends on their structure, as well as on the charge and the nature of the metal. An important role is played by the concentration of the surfactant, particularly when it is of the anionic type. For example, sodium alkylsulfates at low concentrations reduce the corrosion rate of steel in alkaline solutions but increase it at higher concentrations. Whereas steel in the absence of alkylsulfates is usually passivated in solutions of pH 10–11, with surfactants present it is necessary to raise the pH to 12.[276] The stimulating action of the alkylsulfates on the corrosion of steel is associated with the possibility of a direct participation in the elementary stages of the anodic reaction, according to Ekilik et al.[276] It has to be pointed out that combating corrosion caused by such aggressive surfactants is not easy because of the difficulties in displacing them from the surface by inhibitors. It is therefore not surprising that many of the inhibitors known to be of the adsorption type (amines, salts of carboxylic acids) are found to be of little effectiveness in these circumstances. The best protection is, in fact, found with the use of inhibitors that take part in a chemical reaction with the metal ions, e.g., silicates, phosphates, and salts of hydroxycarboxylic acids, since the films formed by these provide a significant diffusion barrier to the aggressive components of the medium.

Anions of hydroxycarboxylic acids that are bound to metal ions (such as iron) in soluble complex compounds accelerate corrosion, but in some cases it is possible to form (as a result of saturation of the near-electrode layer) a phase film on the surface that will inhibit the process of metal dissolution. Depending on the relative effects of these factors, hydroxycarboxylic acids can either stimulate or inhibit corrosion. On this basis, one can explain the increase in protection of steel that occurs in surfactant solutions on going from acid to normal salts of hydroxycarboxylates or when the solution is strongly aerated.

It must be noted that in relation to both iron and aluminum, the better inhibiting properties are provided by complexes of hydroxycarboxylates with magnesium, calcium, and zinc. This is obviously connected with the fact that in certain conditions ions of the corroding metal will release zinc or calcium from the complex, and so lead to the formation of insoluble hydroxides of these metals. These hydroxides are deposited on the corroding surface and markedly raise the effectiveness of the protection. For example, it has been reported that calcium gluconate in 20% alkylsulfate solutions, even at a concentration of 0.002 g mol L^{-1} lowers the rate of dissolution of steel by 36 times.[276]

From the example of the alkylsulfates, it can be seen that the use of surfactants that are aggressive to metals can complicate the corrosion protection in degreasing and cleaning processes. Thus, if all other properties are equal, it will be of advantage to use those surfactants which, even if they do not have protective properties, at least do not stimulate corrosion. In the selection and development of chemicals for these processes, it is also necessary to bear in mind that present-day conditions require strict adherence to regulations concerned with protection of the environment from pollution. Many surfactants, for example, tetrapropylene-benzosulfonate are quite resistant to oxidation and can accumulate in rivers, where they may lead to significant problems. Usually, the production of surfactants that are less than 80% biochemically oxidized ("biochemically hard") is not recommended; therefore, at the present time only "biochemically soft" surfactants are of potential use in washing formulations.

Relatively non-toxic surfactants are represented by the salts of unsaturated carboxylic acids, in particular, oleic acid, which are excellent inhibitors of the atmospheric corrosion of not only ferrous, but also non-ferrous metals, particularly aluminum alloys. Of the surfactants of various types studied in our tests, sodium oleate was found to be the best inhibitor of aluminum corrosion, although some protective effects were observed from olefinic-sulfonates, the dispersant NF, the cationic ABDM-chloride, or the alkamone OS-2. It is significant that the non-ionic surfactants Sintamide-5, OP-7 (see Table 3.5), or Brokason-228 had practically no effect on the corrosion behavior of aluminum. Their use in washing formulations will therefore probably not provide any reliable passivation of aluminum alloys. Nevertheless, Sintamide-5 has found wide use in washing formulations because of its biological softness and high degreasing properties.

A washing solution containing Sintamide-5 3–5 + trisodium phosphate 20–30 + Na_2CO_3 8–10g L^{-1} has been used for the ultrasonic degreasing of items[273] at a temperature of 50–60° C, and a treatment time of 10 min. After degreasing, the articles were washed in warm water for 3–5 min and passivated in a solution of sodium nitrite (10–20 g L^{-1}) with an addition of triethanolamine (2–10 g L^{-1}). This solution is also used at 22–28° C for the degreasing of phosphated components before painting. However,

Sintamide-5 is best known for its use in the MS-8 washing formulation. The minimal foaming experienced with these solutions allows them to be used for degreasing not only by immersion, but also by spraying methods. Drawbacks of MS-8 include its weak passivating action in relation to certain steels, and its aggressivity towards copper alloys. To overcome these problems the introduction of corrosion inhibitors has been suggested. According to our data, the addition of 50 mg L^{-1} benzotriazole will effectively prevent the staining of copper. At 100 mg L^{-1} it will also prevent staining of brass. However, the presence of heterocyclic chelating agents in MS-8 can impede the protection of steel.

To replace Sintamide-5 for washing powder, the proprietary formulation 'Labomid' used Sintanol DS-10, which is a mixture of polyethyleneglycol esters of synthetic primary fatty acids of the C_8–C_{10} fraction. Labomid-101™ or -102™ (trademarks of washing powders) can be used in any spray machines, but are usually used in those for cleaning components during repair. However, the temperature of the solution should not be lower than 70° C, otherwise there will be considerable foaming. Labomid-203 is recommended for use in ultrasonic baths and in equipment with circulating solutions, particularly for the removal of tar contamination at 80–100° C. The combination in this formulation of anionic and non-ionic surfactants is often used in other washing solutions, as for example in the ML™-type, to broaden the use of the solution towards other contaminants. Washing with water after degreasing of items with these media is not recommended, although good results can be achieved by using the inhibitor IFKhAN-21, a formulation of carboxylates and azoles. It is suitable for corrosion protection of ferrous and non-ferrous metals. Thus, treatment with a dilute (<1%) solution of this inhibitor provides reliable protection from corrosion in warehouse storage conditions for 15 days or more. The Labomids often cause the staining of copper and its alloys, so a corrosion inhibitor (benzotriazole, mercaptobenzthiazole, etc.) should be included in these formulations.

Phosphorus-containing surfactants described as Oxyphos B™ and Estephat-383™ have been used with success in washing media. The latter is included in the TMS-31 formulation which is recommended for the degreasing of ferrous and non-ferrous metals, including magnesium alloys.[273,277] At a temperature of 60–80° C, 6–8% solutions of TMS-31 remove preserving greases, mineral oils, and similar contaminants in periods of 3–15 min, the time depending on the degreasing method used. For intensifying the degreasing as well as improving the anti-corrosion action of phosphate washing formulations, recommendations have been made for the use of 3–5 gL^{-1} of triethanolamine salts of mono- and dialkyl esters of phosphoric acid and primary fatty alcohols of the C_{10}–C_{13} fraction or their ethoxylates containing 4–7 moles of ethylene oxide.[278]

As already pointed out, the strong foaming tendency of surfactants is undesirable in many applications of degreasing and cleaning, therefore frequent attempts are made to produce phosphate solutions with minimal foaming, but which will still contain surfactants for improving their effectiveness. Apparently, the best results are obtained with polymeric surfactants. For example, in a USSR patent,[279] it has been proposed that additions of carboxyethylcellulose, 2–3%, and a nitrogen-containing block copolymer of ethylene oxide with propylene oxide (7–8%) should be made to solutions containing sodium tripolyphosphate (30–40%), trisodium phosphate (20–23%), polyhydroxyethylated synthetic fatty C_{10}–C_{18} alcohols (7–8%), and Na_2CO_3 (17–20%). Such additions, as well as decreasing the foaming action, also increased the cleaning effectiveness and throughput of components.

The use of complexing agents is another way to increase cleaning effectiveness, and is particularly useful for removing contaminants containing inorganic particles such as metal oxides or salt. Such complexing agents are usually used with inhibitors in washing solutions. There is, in fact, a very wide range of application of these formulations, from cleaning large-scale equipment and cooling systems to the removal of oily emulsions and the degreasing of metal surfaces before the application of paint and metal coatings. An inhibited solution of ethylenediamine tetracetic acid (EDTA) at pH 3–11 and at temperatures up to 90° C is often used for large-scale equipment.

As a rule, washing solutions are multicomponent. Thus, Vertolin 74™ is recommended for protection between technical operations and for the conservation of metallic parts. This formulation contains, apart from a polyhydroxyethylated monoethanolamide of synthetic (C_{10}–C_{16}) fatty acids 4–6%, a polyhydroxyethylated fatty alcohol 8–10%, tributylphosphate 6–10%, the disodium salt of EDTA 2–3%, as well as known corrosion inhibitors such as oleic acid 8–10%, triethanolamine 8–10% and borax 3–4%.[280] This composition does not cause corrosion of LS-59 brass, the aluminum alloy ML-5 (anodized and sealed with VL-725), and is not aggressive to bronze or cadmium coatings. Oleates of ethanolamines are also included in other washing solutions containing complexing agents.[281]

In the preparation of metal surfaces for galvanic coating, use is made of polyaminocarboxylic acids or their salts, for example: hydroxyethylated alcohol 3-43, mono-, di- or triethanolamine salts of carboxylic acids 4-27, ethanolamine, the disodium EDTA or the sodium salts of nitrilo-tris-acetic acid or tri-polyphosphoric acid 10–70, the disodium salt of sulfosuccinic acid 1.5–20, and sodium citrate 10–70 with water up to 100 (mass %).[282]

Better cleaning of a surface before the deposition of galvanic coatings can be realized by electrolytic degreasing, as in a solution containing > 1 gL^{-1} alkali, Na_2CO_3, or trisodium phosphate, not less than 1 gL^{-1} EDTA or monohydric alcohols, gluconic or citric acid and > 0.1 gL^{-1} of derivatives of acetylenic compounds.[(283)]

Hydroxyethane 1,1′ diphosphonic acid, HEDP, is also recommended for washing solutions. Thus, in a USSR patent[284] a method has been proposed for the electrochemical cleaning of metal surfaces using a 3–6 V treatment for 5 sec, in a solution containing NaOH 0.6–20%, HEDP or its salt 0.001–7.0%, and hydroxyphenoxypolyethylene 0.06% at 85° C. The corrosion inhibitors recommended for protection of iron in this solution are gluconates, which as hydroxycarboxylates with complex-forming properties provide better cleaning and degreasing of metals. Solutions with complexing agents are also effective in chemical cleaning, particularly for the removal of inorganic contaminants.

The methods for providing a passivating wash of a steel surface after degreasing in alkaline solutions merit attention.[(285)] A typical formulation contains triethanolamine 5–15, 40% solution of sodium EDTA 15–20, acyclic saturated alcohol 10–30, non-ionic surfactant 2–20 and water 20–50%. The treatment involves immersion in a 1–3% bath at 30–60° C for 1–3 min or spraying with the solution. Surfaces passivated by this method are then dried in warm air.

Anti-corrosion washing formulations do not always contain complexing agents or phosphates. In such cases, providing an appropriate surfactant is used, the processes of degreasing and cleaning will present fewer problems. For example, there will be less danger of corrosion of non-ferrous materials. Bearing in mind that olefinic sulfonates are quite good inhibitors of the corrosion of aluminum alloys, it could be expected that washing solutions based on their magnesium salts would provide protection, as well as cleaning of these alloys. In such solutions it is usually the practice to introduce buffering agents, such as borates. Compositions of this type are used in the form of pastes which are particularly effective when the metal surface must not only be degreased, but also carefully cleaned from corrosion products or oxide films, or polished to a bright finish. An example is a paste for cleaning silver which contains oleic acid 1.5–10, 25% ammonia solution 2–14, Al_2O_3 10–50, chalk 2–30, carboxymethylcellulose 0.5–6 and water 30–70 parts by weight. For relatively long retention of the bright finish, the paste should contain a corrosion inhibitor so that thin protective-phase films are formed.

Attempts have been made to retain the good cleaning properties of organic solvents by modifying them so that the contents of toxic, flammable, and explosive substances (for example, hydrocarbons) can be significantly reduced. Success has been achieved with the use of emulsions of organic solvents in aqueous alkali in which, first, the minute drops of hydrocarbon have a large surface area and, second, the high alkalinity of the solution promotes saponification and removal of fatty pollutants. Phosphates, carbonates, and silicates are often used in these emulsion formulations since they have good dispersant properties. It is desirable that a surface-active emulsifier should be dissolved in the organic solvent used in such formulations, since this will significantly improve the wetting of the metal by the emulsion. These solvents include kerosene, benzene, and xylol.

From the anti-corrosion aspect, such formulations are best used at pH 7–9 since steel is darkened at lower pH values, and corrosion of aluminum alloys takes place in more strongly-alkaline solutions. Effective degreasing in these emulsions takes place at 50–60° C, which, when benzene is present, hinders their use in practice. However, if OP-7 (see Table 3.5) is replaced by a more effective surfactant, noticeable improvements can be obtained in the properties of the emulsion.

The use of kerosene in emulsion formulations lowers the fire hazard. In this respect, emulsion cleaners are less dangerous than organic solvents, particularly when the emulsification is with the heavier fractions from oil condensates. The presence of a hydrocarbon phase means that oil-soluble corrosion inhibitors can be used, particularly calcium sulfonate. To improve the washing properties, the formulations may contain, apart from the hydrocarbon solvent (1–1.3%) and inhibitor (0.01–0.03%), also salts of fatty acids (0.8–1.0%) (which have inhibiting properties), primary C_3–C_5 alcohols (0.01–0.03%), and other additives. These and similar formulations are stable over a period of 1 month in cold water, and even at room temperature, can wash oils, oil-based cutting fluids, various pastes, and other contaminants from surfaces (the presence of which could lead to corrosion). There is no need for drying after their use since the thin protective films that will have been formed will provide long-term protection from corrosion.

Oleic acid or its salts, with their high anti-corrosion and surface-active properties, are also used in emulsion formulations. An example is, 9 parts of water with 1 part of a concentrate containing kerosene 89%, oleic acid 7%, triethanolamine 3%, and tricresol 1%. Sometimes emulsion cleaners for steel, and cast iron components are formulated so that after treatment the metal surface develops water-repellent properties, so preventing corrosion in handling and also retaining a bright appearance during periodic contact with soapy water, fruit juices, etc. Usually these emulsion cleaners contain not one, but several, corrosion inhibitors. These emulsions can be used not only for degreasing, but also for cleaning surfaces of enamel, bituminous mastics, etc. For example, there has been a proposal to use for this purpose a 1.5% solution of the following composition: an organic solvent of the ethylene or propylene glycol type, and its ethers with methanol, ethanol, propanol, phenol, or butylcellosolve 100, hydroxyalkylated amide of a hydroxycarboxylic acid < 100, alkanolamide 10–300 parts by weight, a surfactant that is a type of soap of oleic or hydroxyoleic acid, and the sodium salt of sulfosuccinic acid. The surfactant is added in a quantity, depending on the content of amide of the hydroxycarboxylic acid, in a proportion up to 30:1.[286]

Degreasing and cleaning are the most widespread, but not the only, technological operations involved in the preparation of metallic items in which corrosion inhibitors have the dual effect of increasing the effectiveness of the actual operations and guaranteeing subsequent protection of a component during storage between operations.

6.2. INHIBITORS IN COOLING LUBRICANTS AND POLISHING COMPOSITIONS

Significant static and dynamic loads act on metal surfaces during machining, pressing, grinding, and polishing processes. The high temperatures which consequently arise in the treatment zones and the rapid renewal of surfaces mean that adsorption and other physico-chemical phenomena can differ significantly from those for "usual" surfaces and are important factors that will influence the corrosion resistance of the treated items, not only during the particular engineering process but also after its completion (on exposure to atmospheric conditions). For these reasons, corrosion inhibitors have been introduced into cooling lubricants. Cooling lubricants were previously mineral oils with good lubricant properties with additions of anti-oxidants, anti-scuffing agents, cleaning, anti-wear agents, etc. The corrosion inhibitors that were used included oil-soluble compounds which provided protection even during subsequent storage periods between processes. However, oils have low cooling properties, and also significant volatility, which adversely affects the hygiene of metal working areas. Oils are also inflammable and relatively expensive.[287] Because of these problems, increasing use has been made in recent years of water-based lubricating coolants, which not only have better cooling properties, but are also less dangerous to the health of operatives, less inflammable and, generally cheaper than the oil products. Their drawbacks are tendencies to bioactivity and foaming, which makes it necessary to add biocides and anti-foaming agents and to employ filters in the water supply. The role of corrosion inhibitors in these compositions—as will be examined in this section—is particularly important. Inhibitors should suppress corrosion of the plant and the components being treated, not only during the actual treatment times, but also during periods between operations; they should not have any adverse effects on paint coatings, rubber seals, or other non-metallic parts of the equipment nor on the microbiological resistance or long-term stability of the coolant. The hygienic and ecological aspects of inhibitors are particularly important and in fact, it is normal practice to disinfect storage vessels and pipework and to refill with fresh liquid every 3 months.

It is probable that the most widely-used inhibitors of steel corrosion in recent times for these aqueous fluids have been sodium nitrite and triethanolamine (0.3–0.5% of each), borates, and inorganic phosphates such as tri-sodium phosphate. To improve the lubricating properties of these additives, they are often used with water-soluble polymers (polyacrylates and polyethylene glycols having molecular weights up to 6000), copolymers of ethylene oxide, and propylene, etc. It is found that the addition of the usual phosphates to some fatty-acid salts not only guarantees corrosion protection but also has a beneficial effect on the life of the cutting tool. On the basis of thermoanalytical studies, Osten-Sacken *et al.*[(288)] concluded that phosphate-containing water-based cutting fluids could form protective surface layers at their decomposition temperatures—thus protecting iron from oxidation—and breaking down only at 630–820° C. Such layers function as lubricants at high rates of mechanical treatment.

It is also the case that sulfur, chlorine, and nitrogen compounds that are components of aqueous cutting fluids can facilitate the cutting process and decrease the tool wear. Although these compounds do not form thick films of reaction products with iron on the metal surface,[(289)] they are capable of increasing the concentration of carbon on the surface and in sub-surface layers of the metal. So it appears that, in many cases, organic corrosion inhibitors will act as polyfunctional additives in aqueous cutting fluids. Furthermore, the best inorganic corrosion inhibitor, sodium nitrite, fails to satisfy the increasing health, hygiene, and ecological requirements and is already being replaced in triethanolamine-containing aqueous cutting fluids by organic anions.

Testing of the effectiveness of organic inhibitors for aqueous cutting fluids is often carried out by the "cast iron chip" method. In this method, freshly-prepared steel millings are placed on a polished and degreased steel plate and wetted with 3 cm^3 of the fluid under test. The development of corrosion is then noted, either with the millings in place or after their removal, following exposure of the plate in a humidity cabinet for 24 h. This method provides only a preliminary assessment of the effectiveness of the inhibitor, which should then be confirmed by other methods more relevant to the intended use of the cutting fluid. For example, for polymer-containing aqueous cutting fluids there are optimum slip speeds, critical loadings, etc. Often the concentration of the inhibitor in a cutting fluid will have to satisfy the required health and safety requirements.

Antiseptic properties also play an important role in the selection of an inhibitor. Usually the most corrosion-active bacteria in cutting fluids are the sulfate-reducing type, as well as certain fungi. The use of strong bactericides or fungicides in cutting fluids is not without problems; therefore, the optimization of the composition of a cutting fluid is a quite complex task and will depend on the actual uses of the fluid. Thus, it is necessary to have a wide range of inhibitors available that can satisfactorily meet all the requirements of the fluid.

Sodium benzoate, because of its ready availability and low toxicity, has been used for a long time in aqueous cutting fluids, but its effectiveness as an inhibitor is not always adequate, and other arylcarboxylates known to have better inhibiting properties have been considered. Watanabe and co-workers[(290)] found that salts or adducts of triethanolamine with bromo- or iodobenzoic acids have not only anti-corrosion, but also good antiwear properties. In later communications, they drew attention to the beneficial effects on the properties of the cutting fluids of additives such as compounds of triethanolamine with other carboxylic acids (linoleic, ricinoleic, undecylic, oleic) or its esters with, for example, monoethylsebacate. Various derivatives of C_7–C_9 or C_{10}–C_{16} fatty carboxylic acids (their soaps or condensation products with triethanolamine, glycols, etc.) are often used as the basis of emulsion (usually 3–5%) cutting fluids since, in addition to their anti-corrosive action, they have good wetting properties and are relatively non-toxic and easily available. As would be expected, the higher unsaturated carboxylic acids, which are present in significant quantities in the majority of vegetable oils, provide poor protective properties when they or their esterification products are used in emulsion cutting fluids.

The wide use in such aqueous cutting fluids of triethanolamine and its derivatives is the result not only of their anti-corrosion and wetting actions but also of some bactericidal effects. However, they are often of low effectiveness and attempts have been made to improve the biocidal properties of cutting fluids by selecting supplementary additives which also have corrosion inhibiting properties. Examples of such additives include benzotriazole and its derivatives, partially esterified succinic acid and bis-sulfonamidocarboxylic acid. Low stability towards biocontamination is frequently a cause of the deterioration of the anti-corrosion properties of aqueous cutting fluids; therefore, well known bactericides such as aldehydes of carboxylic acids are often recommended to be added to aqueous or emulsion cutting fluids. For example, Ref. (291) suggests a proposal to inhibit the corrosion caused by a cutting fluid used in machining cast iron by the addition of 50 mg L^{-1} of the aldehyde of glutaric acid with the pH maintained at 9.0.

The wide range of organic corrosion inhibitors available for neutral media include compounds capable of improving the technological properties of a cutting fluid to be selected. Surfactants containing phosphorus, or both phosphorus and sulfur, have been known for some time as anti-oxidants, anti-wear and anti-seizing additives in oils. Apparently, the dialkyldithiophosphates are the most widely used, both in these, and in oil cutting fluids. However, difficulties with their use in aqueous compositions arise not only from their low solubility in water but also from their low stability towards hydrolysis and thermal oxidation—factors which can increase the corrosivity of the cutting fluid. Nevertheless, when specific requirements of cutting fluids are taken into account, it is possible to obtain a significant effect by additions of phosphorus-containing surfactants. For example, the introduction into the (Russian)ET-2 (Russian Standard TU 38 101599-75) emulsion cutting fluid of sodium salts of the alkylphosphates of C_{10}–C_{20} secondary fatty alcohols can improve the life of the cutting tool by almost an order of magnitude.[292] Derivatives of phosphonic acids, in particular the products of the interaction of aminophosphonic or carboxyphosphonic acids with polyethylenepolyamides and zinc oxide (hydroxide),[293] have also been recommended for increasing the anti-corrosion action of cutting fluids for metal cutting lathes. There are also examples of the use of chelating reagents in various cutting fluids, although the mechanism of their action in such complex systems has not been studied in any depth. Nevertheless, it has often been stated that such cutting fluids are characterized by increased stability under load as well as by improved protective properties.[294] Phosphorus-containing inhibitors can be used in formulations with other surfactants in the abrasive polishing of ferrous components. For example, in preparing surfaces for the deposition of galvanic coatings, the proposal has been made[295] for the use of a new emulsion based on oxalates and a phosphorus-containing passivator, which is mixed with the abrasive and then used in barrel polishing. Generally, these formulations that are used for grinding and polishing the surfaces of metals and alloys often contain the same components that are included in cutting fluids for machining or pressure treatments. Furthermore, these formulations, depending on their use, will have other functional additives (including abrasives, thickeners, and odorants for improving the smell). Often the components forming the basis of polishing solutions include such well known corrosion inhibitors as oleic, tartaric, or citric acids and their salts, including those with triethanolamine. It is therefore not surprising that metallic surfaces treated with these solutions acquire an increased corrosion resistance. However, the introduction of chelating reagents with corrosion-inhibiting properties into polishing pastes or solutions deserves close attention. This is particularly important in the polishing of non-ferrous metals and alloys where reflectivity and good electrical conductivity of components and assemblies are required.

To achieve this requirement, polishing compositions for copper, zinc, aluminum, or their alloys should contain 1–5% 8-hydroxyquinoline, benzotriazole (or its derivatives)[296] whereas for polishing silver or silver coatings the best results have been obtained with the use of sulfur-containing compounds such as thioamides, mercaptoesters, magnesium mercaptide, disulfides, and so on. Silver objects polished with such compositions do not tarnish, even in quite severe conditions such as in the presence of moist H_2S, which is particularly aggressive towards silver. This provides one more example of the fact that organic compounds capable of complex formation with a metal in various situations are capable of forming surface layers that may be more resistant than oxides to the corrosive action of an environment.

6.3. TEMPORARY PROTECTION FROM ATMOSPHERIC CORROSION BY INHIBITED AQUEOUS DISPERSIONS

A combination of technological operations with reliable post-treatment protection from atmospheric corrosion avoids the need for additional protective measures if the periods between further technical processes or transport do not exceed 2–5 months. However, in particularly severe conditions of transport, with unfavorable storage conditions, or with times of up to one year or more between processes, various methods of temporary protection are necessary. The possibilities and achievements in this field which have been made with the use of organic corrosion inhibitors do not come within the scope of this book. Such questions have been examined in many review papers and books, see Refs. (297–299). Moreover, special attention should be given to the ecological aspects of the use of many types of temporary protectives in the light of the new requirements for coatings prepared from aqueous dispersions. Corrosion inhibitors make it possible to use water as the solvent instead of the traditional organic solvents used previously for polymers and waxes. In the absence of inhibitors, the corrosive nature of water precludes its use in paint coatings, for example, particularly when these are applied to ferrous metals. Inhibited aqueous dispersions of polymers can be effective in the painting of surfaces that have already begun corroding. An example is the formulation based on the latex described as VKLVD-65D, and tannin, which can be used on steel having a rust layer up to 100 μm thick. According to Ref. (300), this guarantees protection for six months, and in the case of welded components the presence of this primer has no adverse effect on the quality of the weld. We have also shown that by introducing organic corrosion inhibitors into a latex–tannin rust-converter, the service life in industrial atmospheres of the protective coating, both on corroded and clean steel surfaces can be increased by 2–3 times, i.e., to more than a year.[301]

The use of colloidal solutions for passivating treatments on ferrous and non-ferrous metals has been well established. Salts of sufficiently hydrophobic carboxylic acids, such as oleic acid (as already indicated in preceding chapters) possess not only high surface activity but also the capacity to be strongly adsorbed on many metals and oxides, as well as to form structured and even polymerized layers. Thus, a film formed by a short immersion of metallic components in a 5–10% dispersion of sodium oleate containing 0.5% Na_2CO_3 can maintain ferrous metals, tin, nickel, and the aluminum alloy D16 in a passive state for five years in an industrial atmosphere. However, to achieve this it is necessary to avoid mechanical damage or wetting by direct precipitation. Storage in sheds or wrapping in paper wetted on one side with the passivating solution and then in a barrier packing should provide suitable conditions.

Such a method of protection has little technological value; therefore, in practice temporary protectives are used that are less sensitive to mechanical damage of the coating, such as wax or polymer dispersions. For example, the use of alkenylamine oleate and a water dispersing polymer, obtained by free radical co-polymerization (dimethylaminoethylmethacrylate, methacrylic acid, etc.) can provide temporary protection of ferrous and non-ferrous metals by coatings which are resistant to damp atmospheres but removable, when necessary, with water, but not with organic solvents.[302] In particular acrylic copolymers are frequently recommended for a range of formulations in view both of the physico-chemical properties of the film formers and of their decorative properties. Not only oleates, but also corrosion inhibitors such as benzoates and nitrobenzoates of amines, including hexamethyleneimine, guanidine phosphate, salts and amides of carboxylic fatty acids, and derivatives of morpholine or benzotriazole, may be used. The selection of an effective inhibitor for such temporary protectives very much depends on its compatibility with the binder: some aspects of the action of inhibitors in primer and paint coatings have been examined in Ref. (303). In light of this, the development of ecologically-clean and highly-effective temporary protectives based on aqueous dispersions of copolymers is still far from a satisfactory solution and further detailed studies of the mechanism of inhibition are required.

In recent years, inhibited wax compositions have acquired increasing importance. These materials combine many of the useful properties of inhibited thin-film coatings and preservative oils. They form thin lamellar films on metal surfaces which are resistant to the action of solar radiation and atmospheric deposits, and which possess sufficient fluidity for self-healing (as in the case of oil films) when their continuity is disrupted. Wax compositions are capable of penetrating into micro-gaps and cracks and are more effective in their anti-corrosion properties than the majority of oils, even when these are inhibited. Wax coatings are less susceptible to accidental removal by touch, they are not sticky, and have good decorative properties; they do not crack in frost conditions and have a drop point that ensures reliable protection of metallic components, even in the tropics. The presence in such compositions of specially-selected inhibitors in combination with the hydrophobicity of the wax component is capable of guaranteeing a strong anti-corrosion after-effect. These properties of inhibited wax compositions have attracted worldwide interest. Practically all the leading companies working in the field of temporary protectives have waxes in their range of products. Evidently, in the light of present-day requirements, particular attention will continue to be given to aqueous wax dispersions.

There are three basic components of these aqueous wax dispersions: (i) wax, or wax-like substances, (ii) corrosion inhibitors and (iii) surfactants, capable of emulsifying and stabilizing the dispersion. The most important factor in producing a water dispersable wax is the selection of the corrosion inhibitors. Thus, the correct choice of an inhibitor should (without adversely affecting the barrier properties) provide a capability for after-action and improvement of the technological parameters. However, although methods for the selection of oil-soluble corrosion inhibitors are frequently discussed in the literature, inhibitors for the aqueous phase of water dispersible waxes have received insufficient attention. The need for the use of water-soluble inhibitors in making aqueous preservative compositions can be illustrated with the example of the VVDK dispersion.(304) (These VVDK dispersions are aqueous wax dispersions based on higher aliphatic amines).

As follows from the data of Table 6.1, the use of oil-soluble inhibitors—as in the VVDK-1 dispersion—increases the protective properties of the coatings above those provided by the uninhibited composition (the VVDK dispersion), although it does not prevent the development of corrosion micropits during the stage of protective film formation. The introduction into the composition of 1% of the water-soluble inhibitor (IFKhAN-25), completely suppresses the corrosion of steel in these conditions.

Table 6.1. Results of Tests of the Protective Properties of Aqueous-Wax Coatings Applied to a 0.2% C Steel[a]

Dispersion used for forming the coatings	% of specimen surface corroded	Number of pits on the specimen surface
VVDK	14	0
VVDK-1	0	5
VVDK-1-IFKhAN-25 [1%]	0	0

[a]Test conditions: full immersion in 3% NaCl for 24 h.

However, the majority of the inorganic compounds traditionally used for inhibiting aqueous solutions are of little use in designing aqueous wax dispersions. Chromates, nitrites, silicates, and phosphates are often not capable of suppressing corrosion during the formation of the coating below concentrations that would destabilize the dispersion. Therefore, it is preferable to replace these by organic surfactants and inhibited formulations with synergistic action, although in this case it will be necessary to avoid overdosing with the water-soluble inhibitors. The introduction of even the most effective inhibitors, i.e., surfactants in excessive quantities, can lead to a general hydrophilization of the coating obtained with aqueous wax dispersions, and consequently to a reduction in the anti-corrosion characteristics of the coating. This situation concerns not only the inhibitors but also the water-soluble emulsifiers and stabilizers, without which it would be difficult to make stable aqueous wax coatings that are not susceptible to coagulation and aging. Furthermore, significant concentrations of water-soluble surfactants in the composition of protective coatings can promote their re-emulsification, that is, the loss of the barrier properties of the preservative. The optimum concentrations, from the anti-corrosion point of view, of water-soluble surfactants do not always guarantee the production of stable, non-coagulating preservatives. The use of ammonium salts of higher carboxylic acids as simultaneous emulsifiers and inhibitors could solve the problem. The hydrolysis of these compounds can lower the concentration of water-soluble substances in the protective film during the drying process. A drawback of this approach is the strong ammonia odor of the preservative, which not only presents an ecological hazard, but also severely limits the region of its potential use. A much more rational approach is to use an auxiliary solvent. Thus, the inclusion into the dispersion composition of 1–2% of certain organic solvents guarantees a high stability without producing any difficulties from the flammability or ecological aspects.

The introduction of inhibitors into aqueous wax dispersions retards not only the anodic, but also the cathodic, reaction. Severnyi *et al.*,[305] found this with the example of coatings formed with the IVVS dispersion, which was 3–4 times more effective than a previously used commercial wax dispersion. In this case the best results were obtained with coatings containing an inhibitor of the amine type, which are recommended not only for the protection of metals from atmospheric corrosion but also for the prevention of aging of agricultural items made from resins and polymers for periods of up to one year in open storage.

Amine-type inhibitors are referred to in many patents, for example, in Japan, aqueous wax dispersions[306–307] have been recommended for the protection of the internal surfaces of doors, wings, and other components and sub-assemblies of automobiles. In particular, coatings obtained from aqueous wax dispersions containing 50% solution of a metallic salt of a sulfo-acid 100, oxidized wax 100, amine 10 and water 315 parts by mass, have provided full protection to steel in a salt spray cabinet for 1400 h. The dispersion of this composition in water can be conducted in equipment, guaranteeing a high rate of passage of the solution (830 kmh^{-1}) through a fine orifice, which is apparently achieved as a result of the absence of strong emulsifiers. This last fact can hinder the production of stable highly-dispersed aqueous wax preparations and can also impede subsequent re-emulsification of the coating. The optimum composition of an aqueous wax dispersion and the technology of its production are usually confidential matters, so it is difficult to draw conclusions about the preference for amines over other types of inhibitors, particularly as the higher sulfonates of the alkaline earth metals or oleylsarcosines that are often used also show good protective properties.

In the former USSR, several protective aqueous dispersions have been proposed and tested. These go under the trade names of Akvamin, ZVVD-13, NG-224, etc.

Table 6.2. Technological Characteristics of the Aqueous-Wax Inhibited Dispersion IFKhANPOL'-1 (IMPOL ZR)

No	Technological characteristic of the dispersion	Description	Notes
1	External appearance	Viscous opaque liquid free from stratification and mechanical contamination	
2	Color	From white to cream	Color can be altered
3	Density	0.95–0.99 g cm^{-3}	
4	Viscosity	20–50 s [using a Russian (VZ-4) type] funnel	
5	pH of aqueous phase	8–9.5	
6	Content of dried residue	Not less than 20%	As supplied
7	Ability to be wetted by water	Unlimited	
8	Storage conditions	In a heated room (10–30° C) in a hermetically sealed container	A short term in cooling of up to 24 h down to −30° C is acceptable
9	Method of application	Brush, roller, air, or airfree sputtering, electrophoresis	

One of the more effective inhibited aqueous wax dispersions has been developed by our laboratory in cooperation with Polish researchers. It is described as IFKhANPOL′-1 and some characteristics are given in Table 6.2. Coatings of IFKhANPOL′-1 have no adverse effect on the majority of paint coatings in industrial use. The coatings are not sticky, are frost resistant, and have a drop temperature of >60° C. Fungicides in the composition guarantee the coating free from biodeterioration.

One of the valuable properties of this composition is its ability to prevent the corrosion of metals, even when applied to already corroding surfaces. A thin film of IFKhANPOL′-1 deposited directly onto rust can immediately retard, and even fully prevent, the further development of corrosion.

The combination of the anti-corrosion and technological properties of IFKhANPOL′-1 means that it can be recommended for use in the anti-corrosion maintenance of automobile and aviation equipment (both in civilian and agricultural use), in metal treatment, in machine tool and ship construction, and in metallurgical and other branches of industry for inter-operation and warehouse storage, including large-scale components and sub-assemblies. Bearing in mind the possibility for modification of the external appearance of the protective coatings (color, thickness, opacity, brightness) they will also be suitable for use in cosmetic treatment in the repair of automobiles. They provide reliable protection from corrosion of steel, cast iron, aluminum, and copper alloys. IFKhANPOL′-1, depending on the coating thickness, can be used in conditions of open air storage and transport of components for periods up to 10–12 months.

References

1. Pourbaix, M., *Atlas of Electrochemical Equilibria in Aqueous Solutions* (Pergamon, Oxford, 1966).
2. Markovic, T., *Proc. 2nd European Symposium on Corrosion Inhibitors (2SEIC), Ann. Univ. Ferrara, N. S.,* Sez. V. Suppl. No 4.(1966) p. 377, Vol. 1.
3. Antropov, L. I., *Introduction of the Scale of Phi Potentials and Its Use for Studying the Kinetics of Electrochemical Reactions* (in Russian) (Nauka, Leningrad, 1981).
4. Damaskin, V. V., *J. Electroanal. Chem.* **65,** 799 (1975).
5. Antropov, L. I., Makushin, E. M., and Panasenko, V. F., *Inhibitors of the Corrosion of Metals* (in Russian) (Tekhnika, Kiev, 1981).
6. Antropov, L. I. and Pogrebova, I. S. in: *Summaries of Science and Technology,* Vol. 2 (in Russian) (VINITI, Moscow 1973), pp. 27–104.
7. Frumkin, A. N., *Zero Charge Potentials* (in Russian) (Nauka, Moscow, 1982).
8. Trasatti, S., *J. Electroanal. Chem.* **64,** 128 (1975).
9. De, C. P., *Nature* **180,** 803 (1957).
10. Davies, D. E. and Slaiman, Q. J. M., *Corros. Sci.* **11,** 683 (1971), and **13,** 891 (1973).
11. Rozenfel'd, I. L., Kuznetsov, Yu. I., Kerbeleva, I. Ya., and Persiantseva, V. P., *Zashch. Met.* **10,** 612 (1975).
12. Antropov, L. I. and Kuleshova, N. F., *Zashch. Met.* **3,** 166 (1967).
13. Grigorev, V. P. and Ekilik, G. N., *Izv. Vyssh. Uchebn. Zaved. Khim. Khim. Tekhnol.* **20,** 1171, (1977).
14. Kolotyrkin, Ya. M., *Uspekh. Khim.* **31,** 322 (1962).
15. Remi, G., *Inorganic Chemistry Course* (in Russian) (Inostr. Lit., Moscow, 1963).
16. Maslikova, M. A., and Chemodanov, D. I., *Neorg. Mater.* **7,** 1773 (1971).
17. Podgornova, L. P., Gorodetskii, A. E., Kuznetsov, Yu. I., and Kolesnichenko, A. A., *Zashch. Met.* **4,** 500 (1980).
18. Hawkins, C. J., and Perrin, D. D., *J. Chem. Soc.* **4,** 1351 (1962).
19. Podgornova, L. P., Shirokov, V. F., Kazanskii, L. P., and Kuznetsov, Yu. I., *Izv. Akad. Nauk SSSR, Khim.* **10,** 2221 (1987).
20. Evans, U. R., *Corrosion and Oxidation of Metals* (Arnold, London, 1961).
21. Bregman, G. I., *Corrosion Inhibitors* (MacMillan, New York, 1963).
22. Tomashov, N. D., and Chernova, G. P., *Passivity and Protection of Metals from Corrosion* (in Russian) (Nauka, Moscow, 1977).
23. Rozenfel'd, I. L., *Ingibitory Korrozii* (Khimiya, Moscow, 1977).

24. Frumkin, A. N., Bagotskii, V. S., Iofa, Z. A., and Kabanov, V. N., *Kinetics of Electrode Processes* (in Russian) (Moscow State University, Moscow, 1952).
25. Iofa, Z. A., Batrakov, V. V., and Ba, K. N., *Zashch. Met.* **1,** 55 (1965).
26. Damaskin, B. B., Petrii, O. A., and Batrakov, V. V., *Adsorption of Organic Compounds on Electrodes* (in Russian) (Nauka, Moscow, 1968).
27. Krishtalik, L. I., *Electrode Reactions–Mechanism of the Elementary Act* (in Russian) (Nauka, Moscow, 1982).
28. Reshetnikov, S. M., *Inhibitors of the Acid Corrosion of Metals* (in Russian) (Khimiya, Leningrad, 1986).
29. Kolotyrkin, Ya. M., Popov, Yu. A., and Alekseev, Yu. V., *Elektrokhimiya* **8,** 1725, (1972); **9,** 629, (1973).
30. Rozenfel'd, I. L., Kuznetsov, Yu. I., and Belov, A. V., *Zashch. Met.* **13,** 448 (1977).
31. Damaskin, B. B. and Afanas'ev, B. N., *Elektrokhimiya* **13,** 1099 (1977).
32. Timashev, S. F., *Elektrokhimiya* **15,** 333 (1979).
33. Grigor'ev, V. P. and Ekilik, V. V., *Chemical Structure and Protective Action of Corrosion Inhibitors* (in Russian) (Rostov State University, Rostov-on-Don, 1978).
34. Yamaoka, H., Lorenz, W., and Fisher, H., in: *Extended Abstracts of 2nd International Congress on Metallic Corrosion* (NACE, Houston, 1963, p. 17.)
35. Kolotyrkin, Ya. M., in: *Tr. Mosk. Inst. Khim. Mashinostr.,* No. 67 (1975).
36. Heusler, K. E., and Cartledge, G. H., *J. Electrochem. Soc.,* **108,** 732 (1961).
37. Bech-Nielsen, G., *Electrochim. Acta.* **21,** 627 (1976).
38. Kuznetsov, Yu. I. and Podgornova, L. P., *Zashch. Met.* **19,** 98 (1983); *Elektrokhimiya* **20,** 982 (1984).
39. Zytner, Ya. D. and Rotinyan, A. L., *Elektrokhimiya* **2,** 1371 (1966).
40. Gileadi, E. and Conway, B. E., in: *Modern Aspects of Electrochemistry* (Russian translation) (Mir, Moscow, 1967), p. 392.
41. Rotinyan, A. L., Tikhonov, K. E., and Shoshina, I. A., *Theoretical Electrochemistry* (in Russian) (Khimiya, Leningrad, 1981).
42. Krylov, O. V. and Kiselev, V. F., *Adsorption and Catalysis in Transition Metals and Their Oxides* (in Russian) (Khimiya, Moscow, 1981).
43. Kuznetsov, Yu. I. and Garmanov, M. E., *Elektrokhimiya* **23,** 381 (1987).
44. Barre, P., *Kinetics of Heterogeneous Processes* (in Russian) (Mir, Moscow, 1976) p. 399.
45. Kuznetsov, Yu. I., Bogomolov, D. B., Gorodentskii, A. E., Oleinik, S. V., and Kolesnichenko, A. A., *Poverkh. Fiz. Khim. Mekh.* **198,** 129 (1983).
46. Bockris, J. O'M., Genshaw, M. A., Brusic, U. and Wroblowa, H., *Electrochim. Acta.* **16,** 1859 (1971).
47. Reinhard, G., *Corrosion* (DDR) **5,** 230 (1978).
48. Frankenthal. R. P., *Electrochimica Acta,* **16,** 1845 (1971).
49. Tsuru, T., Zaitsu, T., Masamuru, K. and Haruyuma, S., *J. Jap. Inst. Metals* **42,** 197 (1978).
50. Nagayama, N. and Cohen, M., *J. Electorchem. Soc.,* **109,** 781 (1962).
51. Sato, N. *The Passivity of Metals and Passivating Films.* Proceedings of the 4th International Symposium on Passivity, Warenton Va., 1977, Princeton (1978), p. 29.
52. Akimov, A. G., *Elektrokhimiya* 15, 1888 (1979).
53. Fischer, M., *Werks. Korros.* **29,** 188 (1978).
54. Fehler, F. P. and Mott, N. F., *Oxid. Met* **2,** 59 (1970).
55. Chao, C. Y., Lin, L. F., and Macdonald, D. D., *J. Electrochem. Soc.* **128,** 1187 (1981).
56. Goswani, K. N. and Staehle, R. W., *Electrochim. Acta* **16,** 1895 (1971).
57. Brasher, D. M., *Tribune de Cebedeau,* 1968, (300), 1.
58. Akimov, G. V., in: *Studies on Stainless Steels* (in Russian) USSR Academy of Sciences, Moscow, 1956). p. 166.

59. Novakovskii, V. M., in: *Corrosion and Protection of Metals* (VNITI, Moscow, (1973). p. 5.
60. Skorcheletti, V. V., in: *Theoretical Basis of the Corrosion of Metals* (in Russian) (Khimiya, Leningrad, 1973).
61. Mayne, J. E. O. and Menter, J. W., *J. Chem. Soc.* **1954,** 103 (1954).
62. Kolotyrkin, Ya. M., Kononova, M. D. and Florianovich, G. M., *Zashch. Met.* **2,** 609 (1966).
63. Kolotyrkin, Ya. M., Popov, Yu. A. and Florianovich, G. M., *Elektrokhimiya* **12,** 527 (1976).
64. Sukhotin, A. M. and Osipenko, I. G., in: *Scientific Proceedings of the State Institute of Applied Chemistry: Corrosion Behavior and the Passive State of Metals,* A. M. Sukhotin (ed.) (Izd. GIPKh Leningrad, 1977), p. 3.
65. Varenko, E. S. and Galushko, V. P., *Zashch. Met.* **9,** 460 (1973).
66. Kuznetsov, Yu. I., Rozenfel'd, I. L., and Podgornova, L. P., *Zashch. Met.* **15,** 561 (1978).
67. Tsai, K. C. and Hackerman, N., *J. Electrochem. Soc.* **118,** 28 (1971).
68. Ogura, K. and Kobayashi, M., *Corros. Sci.* **20,** 587 (1980).
69. Kuznetsov, Yu. I., Oleinik, S. V., and Rozenfel'd, I. L., *Elektrokhimiya* **17,** 942 (1981).
70. Kuznetsov, Yu. I., Rozenfel'd, I. L., Kerbeleva, I. Yu., Naidenko, E. V., and Balashova, N. N., *Zashch. Met.* **14,** 253 (1978).
71. Sato, N., Kudo, K. and Nishimura, R., *J. Electrochem. Soc.* **123,** 1419, (1976).
72. Kuznetsov, Yu. I., Oleinik, S. V., Veselyi, S. S., Kazanskii, L. P., and Kerbeleva, I. Ya., *Zashch. Met.* **21,** 553 (1985).
73. Andreeva, N. P. and Kuznetsov, Yu. I., *Zashch. Met.* **23,** 601 (1987); **25,** 214 (1989).
74. Rozenfel'd, I. L., Loskutov, A. I., Kuznetsov, Yu. I., Popova, L. I., Oleinik, S. V., and Alekseev, V. N., *Zashch. Met.* **17,** 699 (1981).
75. Dur'e, Yu. Yu., *Handbook of Analytical Chemistry* (in Russian) (Khimiya, Moscow 1979).
76. Sokolova, L. A., Kossyi, G. G., Ovcharenko, V. I., and Kolotyrkin, Ya., *Zashch. Met.* **12,** 145 (1976).
77. Sukhotin, A. M. and Bereskin, M. Yu., *Zashch. Met.* **18,** 562 (1982).
78. Sastri, V. S. and Packwood, R. H., *Werks. Korros.* **38,** 77 (1987).
79. Voshida, T., Sato, Y., and Sawada, Sh., *J. Chem. Soc. Jpn.* **(11),** 1975, 2031 (1975).
80. Klyuchnikov, N. G., Ushenina, V. F., and Barabash, I. A., *Uchebn. Zap. MGPI,* (1971), (340), p. 278.
81. Jeannin, S., *Proc. 5 SEIC, Ann. Univ. Ferrara, N.S.,* Sez. V, Suppl. No. 7, **Vol. 4,** (1980), p. 1125.
82. Oshawa, Sh., Hiroc, K., and Takeda, M., *J. Fac. Eng., Ibarak Univ.,* **29,** 25 (1981).
83. Chadwick, D. and Hashemi, T., *Surf. Sci.* **89,** 649 (1979).
84. Boffardi, B. P., *Power* **129,** 57, (1985).
85. Ohsawa, M. and Suëtaka, W. *Corros. Sci.* **19,** 709 (1979).
86. Lynch, P. F., Brown, C. W., and Heiderbach, R., *Corrosion* **39,** 357 (1983).
87. Heakal El-Taib, F. and Haruyama, S., *Corros. Sci.* **20,** 887 (1980).
88. Fleischman, M., Hill, I. R., Mengoli, G., and Musiani, M. M., *Electrochim. Acta.* **28,** 1325 (1983); **30,** 1591 (1985).
89. Notoya, T. and Poling, G. W., *Corrosion* **35,** 193 (1979).
90. Notoya, T. and Tatsuo, I., 8th International Congress on Metal Corrosion, Toronto (1984), p. 333.
91. Vysokovskaya, M. S. and Belousov, Yu. A., *Inhibitors of the Corrosion of Metals* (in Russian) (Lenin Moscow State Pedagogical Institute, Moscow, 1985), p. 58.
92. Youda, R., Nishihara, H., and Aramaki, K., *Corros. Sci.* **28,** 87 (1988).
93. Rother, H.-J., Kuron, D., and Storp, S., *Proc. 6 SEIC, Ann. Univ. Ferrara, N.S.* Sez V, Suppl. No 8, **Vol. 1,** (1985), p. 951.

94. Cechini, M. A. G., and Tanaka, D. K., *6th Int. Congr. Met. Corros.* (extended abstract), (1975).
95. Videm, K., *3rd. International Congress on Metal Corrosion,* **Vol. 1,** (Mir, Moscow, 1968), p. 526.
96. Marshakov, I., K., Karavaeva, A. P., Koroleva, T. I., Alekseenko, T. D., and Chernykh, O. P., *Zashch. Met.* **11,** 3 (1975).
97. Pryor, M. I., *Localised Corrosion,* (1971 Williamsburg Conference proceedings) (NACE, Houston, 1974), p. 679.
98. Richardson, J. A. and Wood, G. C., *J. Electrochem. Soc.* **120,** 193 (1973).
99. Davydov, A. D., *Elektrokhimiya* **16,** 1542 (1980).
100. Vanyukova, L. V. and Kabanov, B. N., *Doklad. Akad. Nauk SSSR* **59,** 917 (1948).
101. Kolotyrkin, Ya. M. and Freiman, L. I., *Summaries in Science and Technology* (in Russian) (VINITI, Moscow, Vol. 6 (1979), p. 5.
102. Sato, N. *Electrochim. Acta* **16,** 1683 (1971).
103. Kaesche, H., *Corrosion of Metals* (in Russian) (Metallurgiya, Moscow, 1984).
104. Vermilyea, D., *J. Electrochem. Soc.* **118,** 529 (1971).
105. Schwabe, K., 3rd. International Congress on Metal Corrosion (Mir, Moscow, 1968), p. 54.
106. Strehblow, H.-J. and Titze, B., *Corros. Sci.* **17,** 461 (1977).
107. Kuznetsov, Yu. I. and Popova, L. I., *Zashch. Met.* **19,** 727 (1983).
108. Antropov, L. I., *Zhur. Fiz. Khim.* **37,** 1631 (1953).
109. Katrevich, A. N., Florianovich, G. M., and Kolotyrkin, Ya. M., *Zashch. Met.* **10,** 369 (1974).
110. Kuznetsov, Yu. I., Rozenfel'd, I. L., Podgornova, L. P., and Balashova, N. N., *Elektrokhimiya* **14,** 1869 (1978).
111. Mikhailovskii, Yu. N. and Spitsyn, K. I., *Dokl. Akad. Nauk SSSR.* **266,** 1184 (1982); *Zashch. Met.* **18,** 701 (1982); **23,** 195 (1987).
112. Taub, M., *Mechanism of Inorganic Reactions* (in Russian) (Mir, Moscow, 1975).
113. Hammett, L. P., *Physical Organic Chemistry* (McGraw Hill, New York, 1970).
114. Frumkin, A. N., *Selected Works: Hydrogen Overvoltage* (in Russian) (Nauka, Moscow, 1988).
115. Klopman, G., *Reactivity and Reaction Paths* (in Russian) (Mir, Moscow 1977), p. 63.
116. Hansch, C. and Leo, A., *Correlation Analysis in Chemistry and Biology* (J. Wiley, New York, 1981).
117. Hackerman, N., *Corrosion* **18,** 332 (1962).
118. Donahue, F. M. and Nobe, K., *J. Electrochem. Soc.* **112,** 886 (1965).
119. Kuznetsov, Yu. I. and Podgornova, J. P., Scientific and Technical Summaries, Vol. 15 (in Russian) (Izd. VINITI, Moscow) (1989), p.132.
120. Leffler, J. E. and Grunwald, E., *Rates and Equilibria of Organic Reactions,* J. Wiley, New York, (1963).
121. Pearson, R. G., *J. Amer. Chem. Soc.* **85,** 3533 (1963); *Uspekh. Khim.* **40,** 1259 (1971).
122. Edwards, J. O., *J. Amer. Chem. Soc.* **76,** 1540 (1954).
123. Pal'm, V. A., *Bases of the Quantitative Theory of Organic Reactions* (in Russian) (Khimiya, Leningrad, 1977).
124. Nys, G. G. and Rekker, R. F., *Eur. J. Med. Chem.* **9,** 361 (1979).
125. Kabanov, B. N. and Leikis, D. I., *Dokl. Akad. Nauk SSSR.* **58,** 1685 (1947).
126. Florianovich, G. M., Sokolova, L. A., and Kolotyrkin, Ya. M., *Elektrokhimiya* **3,** 1359 (1967).
127. Ingold, C. K., *Structure and Mechanism in Organic Chemistry* (Cornell University Press, Ithaca, 1969).
128. Lorking, K. F. and Mayne, J. E. O., *J. Appl. Chem.* **11,** 170 (1961).
129. Foroulis, Z. A., *Werks. Korros.* **26,** 350 (1975).

130. Novakovskii, V. M., *Zashch. Met.* **15,** 3 (1979).
131. Rahner, D., Vorkh, K., Forker, W., and Garts, I., *Zashch. Met.* **18,** 527 (1982).
132. Le Anh, H., Brown, B. F., and Foley, R. T., *Corrosion* **36,** 673 (1980).
133. Foley, R. T., *J. Electrochem. Soc.* **122,** 1493 (1975); *Ind. Eng. Chem.* 17, 14, 1978.
134. Nguyen, T. N. and Foley, R. T., *J. Electrochem. Soc.* **127,** 2563 (1980).
135. Samuels, B. W., Sotuden, K., and Foley, R. T., *Corrosion* **37,** 92 (1981).
136. Kolotyrkin, Ya. M., Popov, Yu. A., and Alekseev, Yu. V., *Summaries of Science and Technology,* Vol. 9 (in Russian) (VINITI, Moscow, 1982), p. 88.
137. Rozenfel'd, I. L. and Maksimchuk, V. P., *Dokl. Akad. Nauk SSSR.* **119,** 980 (1958).
138. Kuznetsov, Yu. I., and Luk'yanchikov, O. A., *Dokl. Akad. Nauk SSSR.* **291,** 894 (1986); *Elektrokhimiya* **22,** 1225 (1987).
139. Kuznetsov, Yu. I., Oleinik, S. V., and Andreev, N. N., *Dokl. Akad. Nauk SSSR* **277,** 906 (1984).
140. Pelikan, J., Vosta, J., and Smrz, M., *Werks. Korroz.* **28,** 85 (1977).
141. Szklarska-Smialowska, S., *Pitting Corrosion of Metals,* (NACE, Houston, 1986), p. 431.
142. Robitaille, D. R., *Ind. Water Eng.* **16,** 14 (1979).
143. Sugimoto, K. and Sawada, Yo., *Corrosion* **332,** 347 (1976).
144. Herbsleb, G. and Schwenk, W., *Metalloberflache* **30,** 1 (1976).
145. Batroff, H. P. and Tostmann, K. H., *Proc. 5 SEIC, Ann. Univ. Ferrara N.S.,* Sez V, Suppl.No.7, **Vol. 4,** (1980), p. 1103.
146. Rajagopolan, K. S. and Venkatachari, G., *Proc. 4 SEIC, Ann. Univ. Ferrara N.S.,* Sez V, Suppl. No.6, **Vol. 1,** (1975), p. 60.
147. Grigor'ev, V. P., Gershanova, I. M., Reznikova, L. N., Kravchenko, V. M., and Gontmakher, N. M., *Inhibition and Passivation of Metals* (in Russian) (Izd. Rostov. Gos. Univ., Rostov-on-Don, 1976), p. 34.
148. Kuznetsov, Yu. I, *Summaries of Science and Technology,* Vol. 7 (VINITI, Moscow, 1978), p. 159.
149. Agladze, T. R., *Summaries of Science and Technology,* Vol. 9 (VINITI, Moscow, 1982), p. 3.
150. Shmid, R. and Sapunov, V. N., *Non-Formal Kinetics: Searches for Chemical Reaction Routes* (Mir, Moscow, 1984).
151. Burger, K., *Solvation, Ionic Reactions and Complex Formation in Non-aqueous Media* (in Russian) (Mir, Moscow, 1985).
152. Gonik, A. A., *Corrosion of Oil Industry Equipment and Measures for its Prevention* (Nedra, Moscow 1976).
153. Kolotyrkin, Ya. M. and Kossyi, G. G., *Zashch. Met.* **1,** 272 (1965).
154. Shatma, G., Pandey, G. N., and Sanyal, B., *Corros. Prev. Control* **26,** 5 (1979).
155. Rabinovitz, E., Hermann, V., and Lebin, R.-J., *Corrosion* **38,** 510 (1982).
156. Davydov, A. D., Kamkin, A. N., and Konoplyantseva, N. A., *Elektrokhimiya* **17,** 637, (1982).
157. Ekilik, V. V. and Grigor'ev, V. P., *The Nature of the Solvent and the Protective Action of Corrosion Inhibitors* (in Russian) (RGU, Rostov-on-Don, 1984).
158. Winterton, N., Brooks, W. N., Pugh, A. C., Salter, D. F., and Moreland, P. J., International Congress of Corrosion and Electrochemistry, (Capri, Italy, *Behaviour of Metals in Non-Aqueous Solutions* 1985), p. 8.
159. Fudzii, S. and Sugano, P., *Boseku Gidzyutsy* (in Japanese) **20,** 127 (1971).
160. Gupta, S. S. and Sanyal, B., *Brit. Corros. J.* **14,** 155 (1979).
161. Gerasyutina, L. I., Fedash, V. P., and Tremba, I. N., *Zashch. Met.* **24,** 508 (1988).
162. Persiantseva, V. P., Raiudene, V. E., Sergeikin, A. Ya., and Pospelov, M. V., 3rd National Scientific and Technical Congress (in Russian) (Varna, Bulgaria, 1982).

163. Nechaev, E. A., *Chemisorption of Organic Substances on Oxides and Metals* (in Russian) (Kar'kov Gos. Univ., 1989).
164. Nechaev, E. A. and Urbakh, M. N., *Elektrokhimiya* **16,** 1264 (1980).
165. Kuprin, V. P. and Nachaev, E. A., *Summaries of Science and Technology,* (VINITI, Moscow, 1989).
166. Vermilyea, D. A., Brown, J. F., and Jrochar, D. R., *J. Electrochem. Soc.* **117,** 783 (1970).
167. Kuznetsov, Yu. I., Rozenfel'd, I. L., and Popova, L. I., *Korroz. Zashch. Neft. Gaz. Prom.* **4,** 6 (1978); *Zhur. Prikl. Khim.,* **LIV 2,** 276 (1981).
168. Shinoda, K., Nakagava, T., Tamamusi, V., and Isemura, T., *Colloidal Surface Active Substances* (in Russian) (Mir, Moscow, 1966).
169. Fokin, M. N., and Kotenev, V. A., *Zashch. Met.* **21,** 914 (1985).
170. Berzins, A., Lowson, R. T., and Mizams, K. I., *Austral. J. Chem.* **30,** 1891 (1977).
171. Kuznetsov, Yu. I., Kazanskii, L. P., and Dubrova, M. I., *Zashch. Met.* **18,** 942 (1982).
172. Kuznetsov, Yu. I., Popova, L. I., and Makarychev, Yu. B., *Zh. Prikl. Khim.,* **LIX 5,** 105 (1986).
173. Gatos, H., *Proc. SEIC Ann. Univ. Ferrara, N.S.,* Sez V, Suppl. No.3 (1962), p. 739.
174. *Adsorption from Solution on the Surface of Solid Bodies* (in Russian) (Mir, Moscow 1986), p. 289.
175. Oleinik, S. V., Kuznetsov, Yu. I., Vesselyi, S. S., and Komakhidze, M. G., *Elektrokhimiya* **28,** 856 (1990).
176. Kriss, E. E., Kuznetsov, Yu. I., Kuznetsova, I. G., Kazanski, L. P., Gridor'eva, A. S., Kopakhovitch, N. F., and Fialkov, Yu. A., *Koordinats. Khim.* **11,** 462 (1985).
177. Bockris, J. O'M., Habib, M. A., and Carbajal, J. L., *J. Electrochem. Soc.* **131,** 3032 (1984).
178. Bockris, J. O'M., Scharifker, B. R., and Carbajal, J. L., *Electrochim. Acta.* **32,** 799 (1987).
179. Goledzinowski, M. M., Rolle, D., and Schultze, J. W., *Werks. Korroz.* **36,** 381 (1985).
180. Engell, H. J. and Stolica, N. D., *Z. Phys. Chem.* **20,** 113 (1959).
181. Heusler, K. F., and Fisher, L., *Werks. Korroz.* **27,** 551 (1976).
182. Hoar, T. P. and Jacob, W. R., *Nature* **216,** 1299 (1967).
183. Kuznetsov, Yu. I., *Zashch. Met.* **13,** 739 (1978).
184. Kuznetsov, Yu. I. and Valuev, I. A., *Zashch. Met.* **22,** 89 (1986); **23,** 138 (1987); **23,** 822 (1987).
185. Zhdanov, Yu. A. and Minkin, V. I., *Correlation Analysis in Organic Chemistry* (in Russian) (Rostov State University, Rostov-on-Don, 1966).
186. Kuznetsov, Yu. I., Kerbeleva, I. Ya., Brusnikina, V. M., Rozenfel'd, I. L., *Elektrokhimiya* **15,** 1703 (1979).
187. Kuznetsov, Yu. I., *Zashch. Met.* **20,** 359 (1984).
188. Kuznetsov, Yu. I. and Oleinik, S. V., *Zashch. Met.* **19,** 92 (1983); **20,** 224 (1984).
189. Kuznetsov, Yu. I., Rozenfel'd, I. L., Kuznetsova, I. G., and Brusnikina, V. M., *Elektrokhimiya* **28,** 1592 (1982).
190. Valuev, I. A., Kuznetsov, Yu. I., and Tyr, E. V., *Zashch. Met.* **26,** 564 (1990).
191. Kuznetsov, Yu. I. and Andreev, N. N., *Zashch. Met.* **21,** 484 (1987); **28,** 96 (1992).
192. Kuznetsov, Yu. I., Rozenfel'd, I. L., and Dubrova, M. I., *Dokl. Akad. Nauk, SSSR* **256,** 1418 (1980).
193. Kuznetsov, Yu. I. and Andreev, N. N., *Zashch. Met.* **23,** 495 (1987).
194. Kuznetsov, Yu. I., Valuev, I. A., Popova, L. I., and Bruskina, V. M., *Zashch. Met.* **22,** 144 (1986).
195. Kuznetsov, Yu. I. and Luk'yanchikov, O. A., *Zashch. Met.* **24,** 930 (1988); **27,** 64 (1991).
196. Weisstuch, A. and Lange, K. K., *Mat. Prot. Perform.* **10,** 29, (1971).
197. Kuznetsov, Yu. I., Luk'yanchikov, O. A., and Fialkov, Yu. G., *Zashch. Met.* **21,** 816 (1985).
198. Deberry, D. W., *Corrosion 89,* Paper No. 447 (NACE, Houston, Texas, 1989).

199. Altsybeeva, A. I. and Levin, S. I., *Inhibitors of the Corrosion of Metals* (handbook, in Russian) (Khimiya, Moscow, 1968).
200. Kuznetsov, Yu. I., Rozenfel'd, I. L., and Kerbeleva, I. Ya., in: *Studies of the Electrochemistry and Corrosion of Metals* (TPI, Tula, 1977), p. 26.
201. Kuznetsov, Yu. I., *Corrosion 89,* Paper No. 134, NACE, Houston, Texas, 1989.
202. Kuznetsov, Yu. I., Oleynik, S. V., and Vesely, S. S., *Bul. Elektrochem.* **3,** 591 (1987); *Proc. 7 SEIC, Ann. Univ. Ferrara N.S.,* Sez V, Suppl. No. 9, (1990), p. 175.
203. Klyuchnikov, N. G. and Ushnina, V. F., *Zh. Prikl. Khim.* **44,** 191 (1971).
204. Kuznetsov, Yu. I., *Proc 7 SEIC, Ann. Univ. Ferrara,* N.S. Sez V, Suppl. No.7, Vol. 1, (1990), p. 1.
205. Margulova, T. Kh., Use of Complexones in Thermal Power, [in Russian], Energoatomizdat, Moscow, (1986). 279 pages.
206. Noak, M. *Corrosion 89,* Paper No. 436 (NACE, Houston, Texas, 1989).
207. Dyatlova, N. M., Temkina, V. Ya., and Popov, K. I., *Complexones and Complexonates of Metals* (in Russian) (Khimiya, Moscow, 1988).
208. Kuznetsov, Yu, Trunov, E. A., and Isaev, V. A., *Zashch. Met.* **23,** 86 (1987).
209. Kuznetsov, Yu. I. and Isaev, V. A., *Zh. Prikl. Khim.* **12,** 2645 (1987).
210. Kuznetsov, Yu. I. and Bardasheva, T. I., *Zashch. Met.* **24,** 234 (1988).
211. Kuznetsov, Yu. I., Trunov, E. A., and Starobinskaya, I. V., *Zashch. Met.* **24,** 389 (1988).
212. Kuznetsov, Yu. I. and Trunov, E. A., *Zh. Prikl. Khim.* **57,** 498 (1984).
213. Kuznetsov, Yu. I., Trunov, E. A., Rozenfel'd, I. L., and Belik, R. V., *Korroz. Zashch. Neftegaz. Prom.* **7,** 5 (1980); **2,** 6 (1981).
214. Kabachnik, M. I., Medved', T. Ya., Dyatlova, N. M., and Rudomino, M. V., *Uspek. Khim.* **8,** 1561 (1974).
215. Horner, L., *Chem. Ztg.* **6,** 247 (1976).
216. Kabachnik, M. I., Lastovskii, R. P., Medved', T. Ya., Medyntsev, V. V., Kolrakova, I. D., and Diatlova, M. M., *Dokl. Akad. Nauk SSSR.* **177,** 582 (1967).
217. Kuznetsov, Yu. I., Isaev, V.A., and Trunov, E. A., *Zashch. Met.* **26,** 798 (1990).
218. Kusnetsov, Yu. I. and Raskol'nikov, A. F., *Zashch. Met.* **28,** 707 (1992).
219. Rozenfel'd, I. L., *Corrosion Inhibitors for Natural Media* (in Russian) (Akad. Nauk, SSSR. Moscow, 1953).
220. Grigor'ev, V. P. and Popov, S. Ya., *Zh. Prikl. Khim.* **35,** 1621 (1962).
221. Kuznetsov, Yu. I., Rozenfel'd, I. L., Filimonova, G. V., and Tsygankova, N. K., *Izv. Vyssh. Uchebn. Zaved. Khim. Khim. Tekhnol.* No. 6, 706 (1979).
222. Kuznetsov, Yu. I., Rozenfel'd, I. L., and Agalarova, T. A., *Zashch. Met.* **18,** 562 (1982).
223. Agalarova, T. A., Bobrova, M. K., Gorodetskii, A. E., and Kuznetsov, Yu. I., *Zashch. Met.* **24,** 788 (1988).
224. Reizin, B. L., Strizhevskii, I. V., and Shepelev, O. A., *Corrosion and Protection of Municipal Water Pipes* (in Russian) (Stroizdat, Moscow, 1979).
225. Troup, D. H. and Richardson, J. A., *Werks Korros.* **29,** 312 (1978).
226. Tsutomu, T. and Kadzuo, T., *J. Inst. Eng. Austral.* **50,** 9 (1978).
227. Midzumoto, I., Yamasaki, Ts., Yamamoto, A., and Fudzin, K., *Mitsyubishi Dzyuko Gikho* **11,** 692 (1974).
228. Subbotina, N. P., *Water Regimes and Chemical Control in Thermal Power Stations,* Energiya, Moscow, 1974.
229. Kanioternyki, A. C., Hayashi, S., and Shirota, T., *Goge Esun* 237, 33 (1978).
230. Nikitina, R. A. and Ivanov, E. S., *Korroz. Zashch. Neftegaz. Prom.* 3, 9 (1983).
231. Zecher, D., *Mat. Perf.* **15,** 33 (1976).
232. Japanese Patent, Class 12A82, No. 53-27696 (10.08.78).

233. U.S. Patent 4018702, 252–389R (19.04.77).
234. Japanese Patent, Class 12A82, No. 53-28863, (17.08.78).
235. Robinson, D. S., *Corrosion Inhibitors, Recent Developments* (Noyes Data Corporation, New Jersey, 1979).
236. U.S. Patent 3803048, 252–389A (9.04.74); 3738806, 21–27, (12.06.73).
237. Burlov, V. V., Teslya, V. M., Ermolina, E. Yu., and Zarinova, N. A., *Neft. Neftekhimm.* **9,** 8 (1986).
238. U.S. Patent 3803047, 252-389A, (9.04.74).
239. Robitaile, D. R., *Mat. Perf.* **11,** 40 (1976).
240. Hiromoto, H., *Rust. Prev. Control* **22,** 26 (1978).
241. Thomas, C. and Breske, C., *Mat. Perf.* No. 2, 17 (1977).
242. U.S. Patent 4401587, C23F 11/18, (30.08.83).
243. U.S. Patent 4317744, C09K 3/00 (2.03.82).
244. Kuznetsov, Yu. I., Isaev, V. A., Starobinskaya, I. V., and Bardasheva, T. I., *Zashch. Met.* **26,** 965 (1990).
245. Marshall, A. and Greaves, B. *Corrosion 89,* Paper No.439 (NACE, Houston, Texas).
246. Marshall, A., *Anti-Corros. Meth. Mater.* **31,** 10, (1984).
247. Marshall, A. and German, R., *VGB Kraftwerkstechen* **10,** 969 (1987).
248. U.S. Patent 4664884, C23F 11/12, (12.05.87).
249. Japanese Patents, 62-96683, (6.05.87); 52-146942, (21.12.78).
250. German Democratic Republic Patents 24952 and 24953, 02F 5/14, (2.09.87).
251. U.S. Patent 4497713, C02F 5/14, (5.02.85).
252. Japanese Patent 56-62974 (29.05.81).
253. Japanese Patent Application 58-58285 (6.04.83).
284. Japanese Patent Application, 62-70586 (1.04.87).
255. Sedahmed, G. H., Soliman, M., and Nagy el Kholy, N. S., *J. Appl. Electrochem.* **12,** 479 (1982).
256. Sedahmed, G. H., Abd-El-Naby, B. A., and Abdel-Khalik, A., *Proc. 5 SEIC Ann. Univ. Ferrara,* N.S., Sez V, Suppl. 7, Vol. 1 (1980), p. 15.
257. Kuznetsov, Yu. I. and Lyublinskii, E. Ya., *Inhibitors for Corrosion Protection During Settling, Storage and Transport of Oil* (in Russian) (VNIIOENG, 1980).
258. Sukhotin, A. M., (ed), *Non-combustible Heat Transfer Agents and Hydraulic Fluids* (in Russian) (Khimiya, Leningrad, 1979), p. 360.
259. Duprat, M., Dubosi, F., Moran, F., and Rocher, S., *Corrosion* **37,** 262 (1981).
260. U.S. Patent 364133, C23f 11/16, (16.09.72).
261. Sukhotin, A. M. and Borshchevskii, A. M., *Developments in the Methods of Protecting Metals and Components from Corrosion* (Tez. Doklad. Vses. Nauchn.-Tekhn. Konf., Moscow, 1988), p. 90.
262. Bugai, P. M., Kachanov, V. A., Makarov, V. A., *et al., Vestn. Khar'kov. Politekhn. Inst.* **6,** 74 (1975).
263. U.S. Patent 4536302, E21B 41/02, (20.08.85).
264. U.S. Patent 4668416, C09K 3/18, (26.05.87).
265. Mercer, A. D., *Br. Corros. J.* **14,** 179 (1979).
266. Butler, G. and Mercer, A. D., *Br. Corros. J.* **12,** 171 (1977).
267. Sigurdson, H. and Tusvik, R., *Bull. Korrosionsinstitutet,* (Sweden), **101,** 65 (1986).
268. U.S. Patent 4497364, F28F 19/00, (5.02.85).
269. Orlova, O. S., Shub, G. A., Kharlamnovich, G. D., *et al., Trudy Ural. Nauchno-Issled. Khim. Inst.,* **55,** 41 (1983).
270. Zhuravlev, A. B., Zarubin, P. I., Kantsler, E. V., *et al. Trudy Ural. Nauchno-Issled. Khim. Inst.* **55,** 41, (1983).

271. Davies, D. E. and Prigmore, R. M., *Proc. 6 SEIC, Ann Univ. Ferrara,* N.S., Sez V, Suppl. 8, (1985), p. 15.
272. Rother, H-J. and Kuron, D., *Proc. 5 SEIC, Ann. Univ. Ferrara,* N.S., Sez V, Suppl. 7, (1980), p. 989.
273. Ivanov, B. I., *Cleaning of Metal Surfaces with Non-flammable Compositions* (in Russian) (Mashinostroenie, Moscow 1979).
274. Doufin, G., Zable, J. P., Mithel, F., Pagetti, J., and Trikit, *Metaux* (1977), 53,253,623,
275. A. c. NRB N 20705 523V 1/04 (18.01.78)
276. Ekilik, V. V., Grigor'ev, V. P., Sement'leva, N. I., *Studies in the Field of Corrosion and Protection of Metals* (in Russian) (Kalmytskoe Knizhnoe Izdatel'stvo, Elista, 1971), p. 76.
277. Gur'yanova, T. P. and Lavrova, A. M., *Prib. Sistem Upravl.* **2,** 34 (1979).
278. USSR Patent 1127918, Byul. Izobr., No. 45, (1984)
279. USSR Patent 667589, Byul. Izobr., No. 22, (1979).
280. USSR Patent 662578, Byul. Izobr., No. 18, (1979).
281. USSR Patent 745925, S11D 1/6, (10.07.1980).
282. USSR Patent 536261, Byul. Izobr., No. 43, (1976).
283. Japanese Patent 53-6093, 12A13, (4.03.78).
284. U.S. Patent 4010086, 204-141, (01.03.77).
285. Czechoslovakian Patent 218389, (12.02.85).
286. Japanese Patent 53-13204, 19F2, (9.05.78).
287. Cherednichenko, G. I., in: *Fuels, Lubricant Materials, and Technical Liquids* (in Russian, V. M. Shkol'nikova, ed.) (Khimiya, Moscow, 1989), p. 344.
288. Osten-Sacken, J., Pompe, R., Johanson, L. G., and Shold, R., *Thermochimica Acta.* **114,** 287 (1987).
289. Berliner, E. M., Burdin, M. V., Pertsov, N. V., *et al., Vestnik Mosk. Gos. Univ., Khimiya* **28,** 582 (1987).
290. Watanabe, S., Fucuda, K., *et al., Chem. Ind.,* **7,** 286 (1980); *J. Amer. Chem. Soc.,* **62,** 1607 (1985); *Mater. Chem. Phys.* **15,** 89 (1986).
291. Japanese Patent claim 59-183678, S01M 173/00, (31.03.86).
292. Abdeenko, A. P., Karnaukh, A. F., and Semyanyakova, L. V., *Vopr. Khim. Khim. Tekhnol.* **78,** 50 (1985).
293. Japanese Patent claim 60-3371, C23F 11/14, (2.02.85).
294. U.S. Patent 4626367, C16M 173/02, (2.12.86).
295. U.S. Patent 4724071, C23F 1/00, (9.02.88).
296. Kuznetsov, Yu. I., *Inhibitors of Atmospheric Corrosion of Semi-products of Non-ferrous Metals and Alloys* (in Russian) (Tsvetmetinformatsiya, Moscow, 1977).
297. Rozenfel'd, I. L. and Persiantseva, V. P., *Inhibitors of Atmospheric Corrosion* (in Russian) (Nauka, Moscow, 1985).
298. Donovan, P. D., *Protection of Metals from Corrosion in Storage and Transit* (Wiley, New York, 1986).
299. Bogdanova, T. I. and Shekhter, Yu. N., *Inhibited Oil Products for Corrosion Protection* (in Russian) (Khimiya, Moscow, 1984).
300. Gerasimenko, A. A., Aleksandrov, Ya. I., Andreev, I. N., Batalov, A. K., Beloglazov, S. M., Pogoyavlenskii, A. Ph., Valeev, N. N., Gerasimova, V. V., Gerasimov, V. V., Gonik, A. A., Elinov, N. P., Krupina, E. M., Michaylova, L. A., Moiseev, Yu. V., Ozeryanova, I. P., Oriov, V. A., Ponomoreva, O. V., Steklov, O. I., Tsirlin, M. C., and Chistyakov, V. V., *Protection from Corrosion, Aging, and Biodeterioration of Machines, Plants and Equipment* (in Russian) (Mashinostroenie, Moscow, 1987), p. 784,
301. Ivanow, J., Kuznetsov, Yu. I., and Setkovich, K. *Proc. 7 SEIC, Ann. Univ. Ferrara,* N.S., Ser V, Suppl. 9, (1990), p. 795.

302. U.S. Patent 4131583, C23F 11/14, (26.12.78).
303. Rozenfel'd, I. L. and Rubinshtein, F. I., *Anticorrosion Primers and Inhibited Paint Coatings* (in Russian) (Khimiya, Moscow, 1980).
304. Andrejew, N., Ivanow, I., Kuzniecow, I., Oleinik, S., Stezala, S., and Wlodarchyk, S., *Powloki Ochrone* **4,** 59 (1988).
305. Severnyi, A. E., Efimova, I. A., *et al.,* in: *Anticorrosion Paint Coating and Contemporary Technical Paints* (in Russian) (Moscow, 1983), p. 6; *Khim. Tekhnol. Topl. Masel* **10,** 8 (1985).
306. Japanese Patent claims 56-57863, (20.05.81).; 55-44661, (7.11.81),; 57-203782, (14.12.82); 58-117880, (13.07.83).
307. Japanese Patent claim 58-117880 (13.07.83).

Index

www.ingramcontent.com/pod-product-compliance
Ingram Content Group UK Ltd.
Pitfield, Milton Keynes, MK11 3LW, UK
UKHW041858190726
13854UKWH00002B/964

* 9 7 8 1 4 8 9 9 1 9 5 7 1 *